Proteinases
and their inhibitors

STRUCTURE, FUNCTION AND
APPLIED ASPECTS

Proteinases
and their inhibitors

STRUCTURE, FUNCTION AND
APPLIED ASPECTS

Proceedings of the International Symposium, Portorož, Yugoslavia,
September 29 — October 3, 1980

Editors

V. TURK,
Ljubljana

LJ. VITALE,
Zagreb

MLADINSKA KNJIGA — PERGAMON PRESS
LJUBLJANA — OXFORD

U.K.	Pergamon Press Ltd., Headington Hill Hall, Oxford OX3 OBW, England
U.S.A.	Pergamon Press Inc., Maxwell House, Fairview Park, Elmsford, New York 10523, U.S.A.
CANADA	Pergamon Press Canada Ltd., 75 The East Mall, Toronto, Ontario, Canada
AUSTRALIA	Pergamon Press (Aust.) Pty. Ltd., 19a Boundary Street, Rushcutters Bay, N.S.W. 2011, Australia
FRANCE	Pergamon Press SARL, 24 rue des Ecoles, 75240 Paris, Cedex 05, France
FEDERAL REPUBLIC OF GERMANY	Pergamon Press GmbH, 6242 Kronberg, Taunus, Hammerweg 6, Federal Republic of Germany

First Edition 1981

British Library Cataloguing in Publication Data

Proteinases and their inhibitors
1. Proteinase — Congresses
I. Turk, V. II. Vitale, Lj.
574.19'256   QP609.P75
ISBN 0-08-027377-7

This publication contains selected papers from the Symposium on Proteinases and their Inhibitors. In order to make this volume available as economically and as rapidly as possible the authors' typescripts have been reproduced in their original forms, after editing. This method has its typographical limitations but it is hoped that they in no way distract the reader.

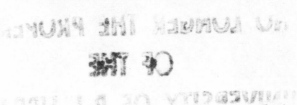

Printed in Yugoslavia by Mladinska knjiga

Contents

PROPERTIES AND STRUCTURE OF ENZYMES

List of Contributors

M. ABRAMIĆ
Department of Organic Chemistry and Biochemistry, Rudjer Bošković Institute,
41000 Zagreb, Yugoslavia

V.K. ANTONOV
Shemyakin Institute of Bioorganic Chemistry, USSR Academy of Sciences,
117988 Moscow, USSR

K. AUNSTRUP
Novo Research Institute, 2880 Bagsvaerd, Denmark

J. BABNIK
Department of Biochemistry, J. Stefan Institute, University E. Kardelj,
61000 Ljubljana, Yugoslavia

F.J. BALLARD
CSIRO Division of Human Nutrition, Kintore Avenue, Adelaide, S.A.5000, Australia

A.J. BARRETT
Biochemistry Department, Strangeways Laboratory, Cambridge CB1 4RN, U.K.

M. BAUDYŠ
Institute of Organic Chemistry and Biochemistry, Czechoslovak Academy of Science,
166 10 Prague, Czechoslovakia

M. BENUCK
Center for Neurochemistry, Rockland Research Institute, Ward's Island, New York
10035, USA

J.W.C. BIRD
Bureau of Biological Research, Rutgers University, Piscataway, N.J. 08854, USA

I. BLOCK
Biochemical Department, Diabetes Forschungsinstitut an der Universität Düsseldorf,
Auf'm Hennekamp 65, D 4000 Düsseldorf, FRG

P. BOHLEY
Physiologisch-Chemisches Institut, Martin-Luther-Universität Halle, DDR 402
Halle(Saale), Hollystrasse 1, GDR

B. BRDAR
Central Institute for Tumors and Allied Diseases, 41000 Zagreb, Yugoslavia

J. BRZIN
Department of Biochemistry, J. Stefan Institute, University E. Kardelj,
61000 Ljubljana, Yugoslavia

J.L. BURGOYNE
CSIRO Division of Human Nutrition, Kintore Avenue, Adelaide, S.A. 5000, Australia

O. CRIVELLARO
Roche Institute of Molecular Biology, Nutley, N.J. 07110, USA

B. DAHLMANN
Biochemical Department, Diabetes Forschungsinstitut an der Universität Düsseldorf,
Auf'm Hennekamp 65, D 4000 Düsseldorf, FRG

C. DAMBMANN
Novo Research Institute, 2880 Bagsvaerd, Denmark

M. DERENČIN
TOK, Factory for Production of Organic Acids, 66250 Ilirska Bistrica, Yugoslavia

M. DROBNIČ-KOŠOROK
Department of Biochemistry, J. Stefan Institute, University E. Kardelj,
61000 Ljubljana, Yugoslavia

A. DUBIN
Institute of Molecular Biology, Jagiellonian University, 31-001 Krakow, Poland

D.J. ETHERINGTON
Muscle Biology Division, Meat Research Institute Langford, Bristol BS18 7DY, U.K.

G. FEINSTEIN
Department of Biochemistry, The George S. Wise Faculty of Life Sciences, Tel Aviv
University, Ramat Aviv, Israel

P.F. FOX
Department of Food Chemistry, University College, Cork, Ireland

R. GEIGER
Abteilung für Klinische Chemie und Klinische Biochemie in der Chirurgischen Klinik der Universität München, Nussbaumstrasse 20, D-8000 München, FRG

T. GIRALDI
Istituto di Farmacologia, Università di Trieste, I-34100 Trieste, Italy

D.E. GOLL
Muscle Biology Group, University of Arizona, Tucson, Az. 85721, USA

F. GUBENŠEK
Department of Biochemistry, J. Stefan Institute, University E. Kardelj,
61000 Ljubljana, Yugoslavia

S. HASHIDA
Department of Enzyme Chemistry, Institute for Enzyme Research, School of Medicine, Tokushima University, Tokushima City, Tokushima, Japan

H. HOLZER
Biochemisches Institut, Universität Freiburg, D-7800 Freiburg i.Br., FRG

B.L. HORECKER
Roche Institute of Molecular Biology, Nutley, N.J. 07110, USA

H.J. KÄRGEL
Zentralinstitut für Molekularbiologie der Akademie der Wissenschaften der DDR, DDR-1115, Berlin-Buch, GDR

N. KATUNUMA
Department of Enzyme Chemistry, Institute for Enzyme Research, School of Medicine, Tokushima University, Tokushima City, Tokushima, Japan

J. KAY
Department of Biochemistry, University College, Cardiff, Wales, U.K.

K.S. KEATON
Bureau of Biological Research, Rutgers University, Piscataway, N.J. 08854, USA

H. KEILOVA
Institute of Organic Chemistry and Biochemistry, Czechoslovak Academy of Science, 166 10 Prague, Czechoslovakia

U. KETTMANN
Physiologisch-Chemisches Institut der Universität Halle(Saale), DDR-402 Halle/S., und Ingenieurhochschule Köthen, DDR

H. KIRSCHKE
Physiologisch-Chemisches Institut der Universität Halle(Saale), DDR-402 Halle/S.

R. KLEINE
Physiologisch-Chemisches Institut der Universität Halle(Saale), DDR-402 Halle/S., und Ingenieurhochschule Köthen, DDR

A. KOJ
Institute of Molecular Biology, Jagiellonian University, 31-001 Krakow, Poland

E. KOMINAMI
Department of Enzyme Chemistry, Institute for Enzyme Research, School of Medicine, Tokushima University, Tokushima City, Tokushima, Japan

M. KOPITAR
Department of Biochemistry, J. Stefan Institute, University E. Kardelj, 61000 Ljubljana, Yugoslavia

V. KOSTKA
Institute of Organic Chemistry and Biochemistry, Czechoslovak Academy of Science, 166 10 Prague, Czechoslovakia

I. KREGAR
Department of Biochemistry, J. Stefan Institute, University E. Kardelj, 61000 Ljubljana, Yugoslavia

L. KUEHN
Biochemical Department, Diabetes Forschungsinstitut an der Universität Düsseldorf,
Auf'm Hennekamp 65, D 4000 Düsseldorf, FRG

T. LAH
Department of Biochemistry, J. Stefan Institute, University E. Kardelj,
61000 Ljubljana, Yugoslavia

P.S. LAZO
Roche Institute of Molecular Biology, Nutley, N.J. 07110, USA

J.A. LEE
Bureau of Biological Research, Rutgers University, Piscataway, N.J. 08854, USA

H. LEVY
Department of Biochemistry, The George S. Wise Faculty of Life Sciences, Tel Aviv
University, Ramat Aviv, Israel

P. LOČNIKAR
Department of Biochemistry, J. Stefan Institute, University E. Kardelj,
61000 Ljubljana, Yugoslavia

M. LONGER
Department of Biochemistry, J. Stefan Institute, University E. Kardelj,
61000 Ljubljana, Yugoslavia

I. MALJEVAC
TOK, Factory for Production of Organic Acids, 66250 Ilirska Bistrica, Yugoslavia

N. MARKS
Center for Neurochemistry, Rockland Research Institute, Ward's Island, N.Y.10035,
USA

M.A. McELLIGOTT
Bureau of Biological Research, Rutgers University, Piscataway, N.J. 08854, USA

K. MORIHARA
Shionogi Research Laboratories, Shionogi and Co., Ltd., Fukushima-ku, Osaka 553,
Japan

A.J. MURRAY
CSIRO Division of Human Nutrition, Kintore Avenue, Adelaide, S.A.5000, Australia

C.E. MURRAY
CSIRO Division of Human Nutrition, Kintore Avenue, Adelaide, S.A.5000, Australia

M. MÜLLER
Institut für Toksikologie, Biochem. Gesellschaft für Strahlen- und Umweltforschung
München, D-8042 Neuherberg, FRG

Y. OTSUKA
Muscle Biology Group, University of Arizona, Tucson, Az. 85721, USA

M. POKORNY
Research and Development Institute, Krka, Pharmaceutical and Chemical Works,
68000 Novo mesto, Yugoslavia

S. PONTREMOLI
Institute of Biological Chemistry, University of Genoa, Genoa, Italy

L.M. POPE
CSIRO Division of Human Nutrition, Kintore Avenue, Adelaide, S.A.5000, Australia

T. POPOVIĆ
Department of Biochemistry, J. Stefan Institute, University E. Kardelj,
61000 Ljubljana, Yugoslavia

A. PUC
LEK, Pharmaceutical and Chemical Works, 61000 Ljubljana, Yugoslavia

V. PUIZDAR
Department of Biochemistry, J. Stefan Institute, University E. Kardelj,
61000 Ljubljana, Yugoslavia

E. REGOECZI
Department of Pathology, McMaster University, Hamilton, Ont., Canada

H. REINAUER
Biochemical Department, Diabetes Forschungsinstitut an der Universität Düsseldorf,
Auf'm Hennekamp 65, D 4000 Düsseldorf, FRG

M. RENKO
Department of Biochemistry, J. Stefan Institute, University E. Kardelj,
61000 Ljubljana, Yugoslavia

S. RIEMANN
Physiologisch-Chemisches Institut der Martin-Luther-Universität Halle, DDR-402
Halle(Saale), Hollystrasse 1, GDR

A. RITONJA
Department of Biochemistry, J. Stefan Institute, University E. Kardelj,
61000 Ljubljana, Yugoslavia

F.J. ROISEN
Dept.of Anatomy, Rutgers Medical School, CMDNJ, Piscataway, N.J.08854, USA

U. ROTHE
Physiologisch-Chemisches Institut der Universität Halle(Saale), DDR-402 Halle/S.,
und Ingenieurhochschule Köthen, DDR

M. RUTSCHMANN
Biochemical Department, Diabetes Forschungsinstitut an der Universität Düsseldorf,
Auf'm Hennekamp 65, D 4000 Düsseldorf, FRG

G. SAVA
Istituto di Farmacologia, Università di Trieste, I-34100 Trieste, Italy

H. SCHELLE
Physiologisch-Chemisches Institut der Universität Halle(Saale), DDR-402 Halle/S.,
und Ingenieurhochschule Köthen, DDR

I. SHAKED
Department of Biochemistry, The George S. Wise Faculty of Life Sciences, Tel Aviv
University, Ramat Aviv, Israel

J. SORIĆ
Central Institute for Tumors and Allied Diseases, 41000 Zagreb, Yugoslavia

V.M. STEPANOV
Institute of Genetics and Selection of Industrial Microorganisms, Moscow 113545,
USSR

F.S. STEVEN
Department of Medical Biochemistry, Stopford Building, University of Manchester,
Manchester M13 9PT, U.K.

A.C. ST. JOHN
Bureau of Biological Research, Rutgers University, Piscataway, N.J. 08854, USA

A. SUHAR
Department of Biochemistry, J. Stefan Institute, University E. Kardelj,
61000 Ljubljana, Yugoslavia

S.C. SUN
Roche Institute of Molecular Biology, Nutley, N.J. 07110, USA

A. SZPACENKO
Muscle Biology Group, University of Arizona, Tucson, Az. 85721, USA

F.M. TOMAS
CSIRO Division of Human Nutrition, Kintore Avenue, Adelaide, S.A. 5000, Australia

T. TOWATARI
Department of Enzyme Chemistry, Institute for Enzyme Research, School of Medicine,
Tokushima University, Tokushima City, Tokushima, Japan

O. TSOLAS
Department of Biological Chemistry, University of Ioannina, Medical School,
Ioannina, Greece

V. TURK
Department of Biochemistry, J. Stefan Institute, University E. Kardelj,
61000 Ljubljana, Yugoslavia

J. TURKOVA
Institute of Organic Chemistry and Biochemistry, Czechoslovak Academy of Science,
166 10 Prague 6, Czechoslovakia

M.J. VALLER
Department of Biochemistry, University College, Cardiff, Wales, U.K.

Lj. VITALE
Department of Organic Chemistry and Biochemistry, Rudjer Bošković Institute,
41000 Zagreb, Yugoslavia

D.M. WARNES
CSIRO Division of Human Nutrition, Kintore Avenue, Adelaide, S.A. 5000, Australia

B. WIEDERANDERS
Physiologisch-Chemisches Institut, Martin-Luther-Universität Halle, DDR 402
Halle(Saale), Hollystrasse 1, GDR

G. YORKE
Department of Anatomy, Rutgers Medical School, CMDNJ, Piscataway, N.J. 08854,
USA

M. ZUBANOVIĆ
Pliva, Pharmaceutical and Chemical Works, 41000 Zagreb, Yugoslavia

Preface

The papers in this book comprise the proceedings of a Symposium under the same title, organized by the Department of Biochemistry, Jožef Stefan Institute, E. Kardelj University, Ljubljana, and the Department of Organic Chemistry and Biochemistry, Rudjer Bošković Institute, Zagreb which was held on September 29 to October 3, 1981, in Portorož, Yugoslavia.

During the past decade the study of proteinases, particularly of tissue origin, and of proteinase inhibitors has advanced rapidly, mainly because of the recognition of their role in a variety of physiological and pathological conditions. Awareness of the importance of these biologically active substances in the regulation of metabolic processes, and of the growing possibilities for their application in industry and everyday life, gave an impetus for the organization of the Symposium.

The papers in the book are divided into four sections: Biological Functions, Properties and Structure of Enzymes, Industrial Application and Proteinase Inhibitors. They contain new data, concepts and visions which we hope will not only prove useful, but will also stimulate further investigations of proteinases and their inhibitors.

We wish to thank authors for their prompt submission of manuscripts, and all participants of the Symposium for creating a cordial atmosphere and frank discussion.

The organizers of the Symposium gratefully acknowledge the sponsorship of Tovarna organskih kislin, Ilirska Bistrica, and the financial support of the Union of Selfmanaged Communities of Interest for Scientific Activities of SFR Yugoslavia. Also, financial support from Krka, Pharmaceutical Works, Novo mesto, Lek, Pharmaceutical Works, Ljubljana, and Beckmann, Geneva, Switzerland, was appreciated.

Vito Turk
Ljubinka Vitale

Biological Functions

MUSCLE PROTEIN BREAKDOWN IN HUMAN CATABOLIC STATES

**F.J. Ballard, C.E. Murray, A.J. Murray, D.M. Warnes,
J.L. Burgoyne, L.M. Pope and F.M. Tomas**

*CSIRO Division of Human Nutrition, Kintore Avenue,
Adelaide, S.A. 5000, Australia*

ABSTRACT

The urinary excretion ratio of 3-methylhistidine (3MH) to creatinine has been used
as a measure of the fractional rate of muscle protein breakdown. Relative to age-
matched controls, this fractional rate was increased about three-fold in Duchenne
dystrophy, while lesser but significant increases were observed in Becker,
Duchenne-like autosomal recessive and limb-girdle dystrophies. Spinal muscular
atrophy, dystrophia myotonica, myotonia congenita, dermatomyositis, central core
disease and motor neurone disease were also associated with high rates of muscle
protein breakdown. Physical trauma resulted in high excretion ratios of 3MH to
creatinine. High muscle protein breakdown rates in hyperthyroid patients and in
premature infants who were losing weight or who were receiving inadequate amounts
of dietary nitrogen fell to normal values upon satisfactory treatment. Our
results indicate the usefulness of the 3MH/creatinine method as a monitor in the
treatment of human catabolic conditions.

KEYWORDS

Muscle protein breakdown; 3-methylhistidine; creatinine; muscular dystrophies;
neuromuscular diseases; premature infants; thyroid disorders; trauma

INTRODUCTION

Although there is considerable indirect evidence that increases in muscle protein
breakdown accompany various catabolic conditions, direct evidence is very limited.
This deficiency occurs because virtually all methods used for quantifying protein
breakdown are extremely complex, require the continual infusion of labelled precursor and are not suitable for repeated measurements. However, the urinary
excretion ratio of 3MH to creatinine provides a satisfactory measure of the
fractional rate of muscle protein breakdown since 3MH is derived exclusively from
contractile proteins and is quantitatively excreted (Young and Munro, 1978).
Creatinine production is a useful index of body muscle mass (Graystone, 1968).

Several assumptions are needed in order to obtain the fractional rate of muscle
protein breakdown from the 3MH to creatinine excretion ratio. Thus (1) both 3MH

and creatinine must be derived exclusively from muscle proteins; (2) neither 3MH nor creatinine should be re-utilized; (3) dietary sources of 3MH and creatinine must be eliminated; (4) 3MH is assumed to be present at a constant amount in muscle protein; (5) creatinine is assumed to be produced at a rate proportional to the amount of contractile tissue; (6) fluctuations in blood or urine flow must affect 3MH and creatinine production equally; (7) the catabolic rate of actin and myosin is representative of total muscle proteins. The reliability of these assumptions has been discussed (Young and Munro, 1978; Tomas, Ballard and Pope, 1979) and there is some consensus that the assumptions are valid for human studies or the errors involved are relatively minor. The major concerns are that signif-icant amounts of 3MH are derived from actin in non-muscle tissues and under some conditions creatinine excretion may not reflect muscle mass. Nevertheless, these problems can only be evaluated after the method has been used extensively in a variety of situations. Accordingly we have calculated rates of muscle protein breakdown from 3MH and creatinine excretion rates in several human conditions associated with acute or chronic muscle wasting.

METHODS

Except as noted below subjects consumed a meat-free diet of their own choice for five days and were counselled to maintain protein and energy intakes at previous levels. Total urine collections were obtained on the fourth and fifth days of the diet (Tomas, Ballard and Pope, 1979). This timing was adopted to exclude dietary sources of 3MH or creatinine. The oral or parenteral diet of subjects receiving intensive care (injury patients and premature infants) was based entirely on clinical criteria but did not contain 3MH or creatinine.

Urine samples were stored at $-20^{o}C$. Creatinine was measured using Technicon Auto-Analyser method N116 and 3MH by amino acid analysis of deproteinised urine (Tomas, Ballard and Pope, 1979). Fractional rates of muscle protein breakdown were calculated from the expression:

$$\% \text{ Muscle protein degraded per day} = 100 \times \frac{\mu mol\ 3MH.ml^{-1}}{\mu mol\ creatinine.ml^{-1}} \times \frac{1000}{3.63 \times 4 \times 113}$$

In this expression 3.63 is the μmol of 3MH per g of muscle protein (Tomas, Ballard and Pope, 1979); 4 is derived from the observation that 1 mg of creatinine per kg per day is derived from 4 g of muscle protein (Graystone, 1968) while 113 is the molecular weight of creatinine. When the equation is applied to premature infants 2.42 replaces 3.63 because of the lower 3MH content of muscle protein at this developmental stage (Ballard and others, 1979).

RESULTS AND DISCUSSION

Muscular Dystrophies and Inherited Myopathies

An age curve for normal subjects is shown in Fig. 1. Data from both males and females of ages between 27 weeks gestation (premature infants) and 50 years are included. A sharp fall in the fractional rate of muscle protein breakdown occurs within the first two years of life followed by a more gradual decline up to age 20. Provided the values are expressed as the excretion ratio of 3MH to creatinine no sex difference is evident. However, 3MH excretion in μmol per kg body weight per day is much lower in females (Tomas, Ballard and Pope, 1979) due to the larger amount of adipose tissue in females. The age curve is required for comparison with individuals with muscle wasting conditions, and all the values reported

subsequently have been expressed relative to age-matched controls.
The fractional rate of muscle protein breakdown for 20 boys with Duchenne
dystrophy is also shown in Fig. 1. It is evident that the rate of muscle protein

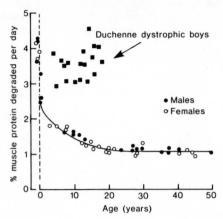

Fig. 1. Muscle protein breakdown in Duchenne dystrophic boys and in controls

breakdown in Duchenne dystrophy is between two and three times the value in
controls of similar ages. However, absolute rates of muscle protein breakdown,
determined from 3MH excretion rates, are lower than in normal individuals because
the amount of contractile tissue per kg of body weight is far lower than in normal
children (Ballard, Tomas and Stern, 1979). The reliability of the calculation of
muscle protein breakdown requires that either 3MH is present in muscle protein at
the same concentration in dystrophic subjects as in normals, or any deviation from
the normal 3MH content is accompanied by a comparable change in creatinine
excretion. At the present time we have insufficient information to evaluate these
possibilities, although the second alternative would be likely if infiltration of
connective tissue into the muscle of dystrophic subjects occurred and was
associated with comparably lower concentrations of creatine phosphate, the
precursor of creatinine. A rigorous conclusion that calculated rates of muscle
protein breakdown represent real values would require direct measurement of muscle
3MH and muscle mass for each myopathic state, clearly an impossible restriction.

The measurements on Duchenne subjects have been extended to other dystrophies and
myopathies (Table 1). We find increased fractional rates of muscle protein
breakdown in Becker dystrophy, Duchenne-like autosomal recessive dystrophy, facioscapulohumeral dystrophy, dystrophia myotonica, myotonia congenita, most cases of
spinal muscular atrophy, peroneal muscular atrophy, central core disease,
myasthenia gravis, motor neurone disease and dermatomyositis. The 95% confidence
limits of a group of normal individuals are ± 18% with catabolic rates in all the
above conditions lying above this range. Two subjects with familial spastic para-
plegia, two with facioscapulohumeral dystrophy and one with dystonia musculorum
deformans had values within the normal range and are not included in Table 1. Two
points should be noted with respect to the cited examples. First, a reasonable
correlation occurs between the fractional rate of muscle protein breakdown and the
disability index of Gardner-Medwin and Walton (1974), implying that the increase
in protein breakdown is related to the degree of muscle wasting. Second, none of
the subjects described in Table 1 had rates of 3MH excretion per kg body weight
that were above normal. Thus the increased fractional rate of muscle protein
breakdown was invariably accompanied by lower calculated amounts of muscle mass.
Although this finding could be interpreted as lower than normal rates of

Although this finding could be interpreted as lower than normal rates of creatinine excretion per kg of muscle, but normal rates of muscle protein breakdown, we do not believe such an interpretation is valid because the amount of muscle mass calculated from a creatinine excretion rate of 1 g per 20 kg muscle (or 4 kg muscle protein) per day (Graystone, 1968) was generally consistent with clinical observations. McKeran and co-workers (1979) have also reported high rates of muscle protein catabolism in subjects with various neuromuscular diseases.

TABLE 1 Muscle Protein Breakdown in Dystrophies and Inherited Myopathies

Diagnosis	Sex,Age	Disability rating	Muscle Protein Breakdown	
			% per day	% of age controls
Duchenne dystrophy (n=20)		3-8	3.59±0.09	243±6
Becker dystrophy	M,19	3	2.28	204
	M,19	3	1.72	153
	M,37	8	3.16	283
Duchenne-like, autosomal recessive dystrophy	F,13	6	2.14	161
Limbgirdle dystrophy	F,51	4	1.60	143
	F,50	1	1.31	117
Dystrophia myotonica	M,39	3	1.54	137
Myotonia congenita	F,22	1	1.36	122
	M,46	0	1.28	123
Spinal muscular atrophy	M,5	1	1.75	95
	M,8	1	1.40	87
	F,28	4	2.43	217
	F,10	8	2.81	189
	F,5	8	2.52	137
	F,15	7	2.10	170
	F,51	4	1.65	148
	M,75	5	1.83	163
Myasthenia gravis	F,49	1	1.34	120
Peroneal muscular atrophy	F,48	7	1.77	160
Central core disease	M,31	2	1.48	132
	F,32	3	1.62	144
Motor neurone disease	F,20		1.56	139
Dermatomyositis	F,6		3.73	216

Thyroid Myopathies

Three hyperthyroid subjects had fractional rates of muscle protein breakdown well above normal (Table 2). Treatment of the first of these subjects to reduce

thyroxine concentrations was accompanied by a progressive fall in the protein
breakdown rate (#2 and #3 in Table 2).

TABLE 2   Muscle Protein Breakdown in Thyroid-related Myopathies

Diagnosis	Sex,Age	Muscle Mass % of body wt	Muscle Protein Breakdown	
			% per day	% of age controls
Hyperthyroid	F,30 #1	28.4	2.19	196
	#2	29.7	1.64	146
	#3	32.4	1.09	97
	F,13	21.8	2.34	176
	F,58	19.2	1.92	171
Hypothyroid	F,79	16.5	0.91	81
	F,48	22.6	0.69	62

These results are notable because, unlike the dystrophies and inherited myopathies, the hyperthyroid subjects had calculated amounts of muscle mass within the
normal range for women. Thus the rates of muscle protein breakdown are not likely
to be artifactually high due to underestimation of the muscle mass. Lower fractional rates of muscle protein breakdown were found in the two hypothyroid
subjects.

The administration of thyroid hormones has been shown by others to increase muscle
protein breakdown in euthyroid subjects (Burman and others, 1979).

Myopathy Associated with Trauma

Williamson and co-workers (1977) measured 3MH excretion in several patients who
had sustained injuries ranging from a simple fracture to multisystem trauma.
The patients were in negative nitrogen balance and had 3MH excretion rates averag-
ing twice the normal value. Although a meat-free diet was not given to these
subjects, the large increase in 3MH excretion certainly suggests a commensurate
rise in muscle protein breakdown. Increased 3MH excretion has also been described
in other trauma or infection conditions including sepsis (Long and others, 1977),
experimentally-induced sandfly fever (Wannemacher and others, 1975) and following
burns (Bilmazes and others, 1978).

We have measured 3MH and creatinine excretion in 5 patients receiving total
parenteral nutrition in an intensive care ward. No muscle products were included
in their diet. Calculated fractional rates of muscle protein breakdown were more
than twice the normal value in all cases (Table 3). Furthermore, in the fifth
case listed in Table 3, the amount of muscle mass calculated from the creatinine
excretion rate seems unreasonably high as compared to a mean of 54.2% of body
weight, in 6 non-obese normal males (Tomas, Ballard and Pope, 1979) and the true
fractional rate of muscle protein breakdown is probably greater than the already
high value calculated for this subject. Creatinine excretion is increased above
normal by fever (Wannemacher and others, 1975; Long and others, 1975) in man and
by glucocorticoids (Tomas, Munro and Young, 1979) in rats.

Since trauma is frequently associated with elevated blood concentrations of
glucocorticoids, the increased creatinine excretion in the fifth subject listed in
Table 3 may be a consequence of such an endocrine response.

TABLE 3   Muscle Protein Breakdown during Severe Trauma

Diagnosis	Sex,Age	Muscle mass % of body wt	Muscle Protein Breakdown	
			% per day	% of age controls
Tetanus	M,69	20.0	2.37	212
Multiple injuries	M,16	43.8	2.86	238
	M,29	51.5	2.43	217
	M,35	52.5	2.31	206
Head injuries	M,15	76.5	3.65	304

Muscle Protein Breakdown in Premature Infants

Although a major goal in the management of very premature infants is the provision of optimum levels of all nutrients, a variety of unforeseen events frequently prevents that goal being attained. Accordingly, the infants will be exposed to periods of low nitrogen and/or low caloric intake.

Starvation of obese adults is accompanied by a fall in muscle protein breakdown as indicated by 3MH excretion (Young and others, 1973). However, these individuals have substantial alternate energy reserves in addition to muscle protein and the response can be viewed as a preservation of muscle protein at the expense of adipose tissue. Similarly children with protein energy malnutrition had low rates of muscle protein breakdown and total protein breakdown which increased above normal during therapy (Golden, Waterlow and Picou, 1977; Nagabhushan and Rao, 1978). The situation is even more complicated because undernourished children who did not show any clinical signs of protein deficiency had high rates of muscle protein breakdown as calculated from 3MH to creatinine ratios. As found with the protein-energy malnutrition group, administration of an adequate diet was accompanied by an increase in muscle protein breakdown (Nagabhushan and Rao, 1978). In these studies with humans muscle protein breakdown increased when more dietary protein was provided, a result consistent with experiments where nitrogen intake was varied in rats (Haverberg and others, 1975; Funabiki and others, 1976; Nishizawa and others, 1977; Millward and others, 1976).

With premature infants, on the other hand, the provision of a diet that did not provide a substantial positive nitrogen balance was associated with an increase in the fractional rate of muscle protein breakdown (Table 4; see also Ballard and others, 1979). This response presumably reflects the minimal energy reserves in

TABLE 4   Extent of Nitrogen Retention and Muscle Protein Breakdown

Nitrogen retention (mmol per kg per day)	Muscle protein breakdown % per day
<0	6.57±0.83 (4)
>0 <10	5.51±0.49 (10)
>10 <15	4.74±0.38 (8)
>15 <20	4.55±0.23 (15)
>20 <25	4.40±0.23 (11)
>25	3.83±0.12 (8)

Values are means±SEM with the number of balance studies given in parentheses.

premature infants weighing about 0.9 kg. The increased muscle protein breakdown and an assumed decreased rate of muscle protein synthesis would have extremely deleterious effects if it persisted for an extended period.

We noted that the amount of calories given to the infants did not correlate with rates of muscle protein breakdown in the study with premature infants (Ballard and others, 1979). In support of the proposed relationship between nitrogen retention and muscle protein breakdown, infants who were gaining weight at the time of the balance study had lower fractional rates of muscle protein breakdown than those whose weight was stable, and substantially lower than observed in infants who were losing weight (Ballard and others, 1979).

All but one of the 36 premature infants in the study retained a substantial proportion of the nitrogen in the diet. The exception, although receiving what would otherwise be considered as an adequate amount of nitrogen (29 mmol per kg per day), was one of the small group in negative nitrogen balance. This infant excreted much more creatinine than expected, more than twice the amount of 3MH per kg body weight than any other infant, and died within 2 days of the urine collections. Our interpretation is that this infant was in a severely catabolic state somewhat analogous to the patient with head injuries listed in Table 3. These two examples are the only ones amongst the more than 200 human subjects we have tested where the 3MH to creatinine ratio clearly resulted in an underestimation of the fractional amount of muscle protein degraded.

Future Prospects for the 3MH to Creatinine Ratio Method

There are three practical disadvantages of the 3MH to creatinine excretion method. First, we consider that a muscle-free diet must be used for at least three days prior to and during the urine collection period. This restriction may not be needed if a standardised diet is used or if muscle-containing foods normally make up only a small part of the total food intake. However, dietary sources of 3MH accounted for about a third of the total 3MH excretion in a study of six males who were consuming the amount of meat normal for an Australian adult (Tomas, Ballard and Pope, 1979). This fraction would be more in women, or any group of subjects who had lower proportions of body muscle.

Second, two sequential 24-hour urine collections are sometimes difficult to obtain with hospital patients or due to subject non-compliance. Also this extended time period is unsatisfactory for situations where rapid changes in protein breakdown rate are expected. Accordingly, we have investigated the possibility of using single urine voidings, a procedure which should be satisfactory provided that creatinine and 3MH are released from muscle and excreted at constant proportions. Individual voidings were collected over a two to three day period from several subjects with different myopathic conditions and from normals. The data from four such experiments are shown in Fig. 2.

For the Becker dystrophy, spinal muscular dystrophy, premature and normal subjects described in Fig. 2 (and for the other studies not illustrated), 3MH concentrations were highly correlated with creatinine concentrations even though absolute concentrations varied over a very wide range. The bracketed figures in Fig. 2 represent the average molar ratio of 3MH to creatinine. Accordingly, we consider that the second practical difficulty of the method can be overcome simply by taking single voidings after at least 3 days of a muscle-free diet. The only disadvantage of this simplified procedure is that muscle mass cannot be calculated and errors due to inordinately high excretion rates of creatinine will not be detected.

The third practical difficulty of the method is the time taken for 3MH measurement

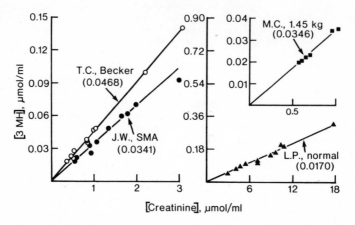

Fig. 2.   3MH and creatinine in sequential urine collections.

by analysis on the basic column of an amino acid analyser. In our hands a total
instrument time of about 12 hours is needed for 4 measurements (duplicate analyses
on each of two 24-hour urine samples). This time would be reduced by half if
measurements were made on a single urine voiding but with some loss of precision.
Other alternatives include high performance liquid chromatography or gas chromatography. These should reduce the total instrument time for four measurements from
12 hours to 2 hours or less but would require pre-purification or derivatisation
of samples prior to analysis. A more satisfactory alternative would be enzymatic
or fluorometric quantitation of urinary 3MH following removal of histidine. Such
methods are certainly possible and their development will surely increase the use
of 3MH measurements in clinical practice.

ACKNOWLEDGEMENTS

We thank the many Adelaide physicians for their assistance in encouraging subjects
to participate in this research and the Muscular Dystrophy Association (USA) for
financial support.

REFERENCES

Ballard, F.J., F.M. Tomas, and L.M. Stern (1979). Increased turnover of muscle
      contractile proteins in Duchenne muscular dystrophy as assessed by 3-methyl-
      histidine and creatinine excretion. Clin. Sci., 56, 347-352.
Ballard, F.J., F.M. Tomas, L.M. Pope, P.G. Henry, B.E. James, and R.A. MacMahon
      (1979). Muscle protein degradation in premature human infants. Clin. Sci.,
      57, 535-544.
Bilmazes, C., C.L. Kien, D.K. Rohrbaugh, R. Uauy, J.F. Bourke, H.N. Munro, and V.R.
      Young (1978). Quantitative contribution by skeletal muscle to elevated rates
      of whole-body protein breakdown in burned children as measured by N^τ-methyl-
      histidine output. Metabolism, 27, 671-676.
Burman, K.D., L. Wartofsky, R.E. Dinterman, P. Kesler, and R.W. Wannemacher, Jr.
      (1979). The effect of T_3 and reverse T_3 administration on muscle protein
      catabolism during fasting as measured by 3-methylhistidine excretion.
      Metabolism, 28, 805-813.
Funabiki, R., Y. Watanabe, N. Nishizawa, and S. Hareyama (1976). Quantitative
      aspect of the myofibrillar protein turnover in transient state on dietary
      protein depletion and repletion revealed by urinary excretion of N^τ-methyl-

histidine. Biochim. Biophys. Acta, 451, 143-150.

Gardner-Medwin, D., and J.N. Walton (1974). The clinical examination of the voluntary muscles. In J.N. Walton (Ed.), Disorders of Voluntary Muscle. Churchill Livingstone, Edinburgh. pp. 517-560.

Golden, M.H.N., J.C. Waterlow,and D. Picou (1977). Protein turnover, synthesis and breakdown before and after recovery from protein-energy malnutrition. Clin. Sci. Mol. Med., 53, 473-477.

Graystone, J.E. (1968). Creatinine excretion during growth. In D.B. Cheek (Ed.), Human Growth, Lea & Febiger, Philadelphia. pp. 182-197.

Haverberg, L.N., L. Deckelbaum, C. Bilmazes, H.N. Munro,and V.R. Young (1975). Myofibrillar protein turnover and urinary Nτ-methylhistidine output. Response to dietary supply of protein and energy. Biochem. J., 152, 503-510.

Long, C.L., W.R. Schiller, W.S. Blakemore, J.M. Geiger, M. O'Dell,and K. Henderson (1977). Muscle protein catabolism in the septic patient as measured by 3-methylhistidine excretion. Am. J. Clin. Nutr. 30, 1349-1355.

McKeran, R.O., D. Halliday, P. Purkiss,and P. Royston (1979). 3-Methylhistidine excretion as an index of myofibrillar protein catabolism in neuromuscular disease. J. Neurol. Neurosurg. Psychiat., 42, 536-541.

Millward, D.J., P.J. Garlick, D.O. Nnanyelugo,and J.C. Waterlow (1976). The relative importance of muscle protein synthesis and breakdown in the regulation of muscle mass. Biochem. J.., 156, 185-188.

Nagabhushan, V.S.,and B.S.N. Rao (1978). Studies on 3-methylhistidine metabolism in children with protein-energy malnutrition. Am. J. Clin. Nutr., 31, 1322-1327.

Nishizawa, N., M. Shimbo, S. Hareyama,and R. Funabiki (1977). Fractional catabolic rates of myosin and actin estimated by urinary excretion of Nτ-methylhistidine: the effect of dietary protein level on catabolic rates under conditions of restricted food intake. Br. J. Nutr., 37, 345-353.

Tomas, F.M., F.J. Ballard,and L.M. Pope (1979). Age-dependent changes in the rate of myofibrillar protein degradation in humans as assessed by 3- methylhistidine and creatinine excretion. Clin. Sci., 56, 341-346.

Tomas, F.M. H.N. Munro,and V.R. Young (1979). Effect of glucocorticoid administration on the in vivo rate of muscle protein breakdown in rats, as measured by urinary excretion of Nτ-methylhistidine. Biochem. J., 178, 139-146.

Wannemacher, R.W., Jr., R.E. Dinterman, R.S. Pekarek, P.J. Bartelloni,and W.R. Beisel (1975). Urinary amino acid excretion during experimentally induced sandfly fever in man. Am. J. Clin. Nutr., 28, 110-118.

Williamson, D.H., R. Farrell, A. Kerr,and R. Smith (1977). Muscle-protein catabolism after injury in man, as measured by urinary excretion of 3-methylhistidine. Clin. Sci. Mol. Med., 52, 527-533.

Young, V.R., and H.N. Munro (1978). Nτ-Methylhistidine (3-methylhistidine) and muscle protein turnover: an overview. Fed. Proc., 37, 2291-2300.

Young, V.R., L.N. Haverberg, C. Bilmazes,and H.N. Munro (1973). Potential use of 3-methylhistidine excretion as an index of progressive reduction in muscle protein catabolism during starvation. Metabolism, 22, 1429-1436.

LYSOSOMAL PROTEINASES IN CULTURED MUSCLE CELLS

A.C.St. John, M.A. McElligott, J.A. Lee, K.S. Keaton, G. Yorke*, F.J. Roisen* and J.W.C. Bird

Bureau of Biological Research, Rutgers University and
*Department of Anatomy, Rutgers Medical School, CMDNJ
Piscataway, NJ 08854, USA

ABSTRACT

The characteristics of lysosomal proteinases were examined in cultured skeletal muscle cells. Rat myoblasts from the L_6 myogenic line were cultured to provide three morphologically distinct populations: pre-fusion, post-fusion and non-fusion, a subclone that did not fuse even at high cell density. The lysosomal cathepsins B, D, H and L exhibited high specific activities in cell homogenates. Activities of cathepsin B and H were highest in post-fusion cells. The Michaelis-Menton constant (K_m) was determined for cathepsins B, D and H and these values were the same in the different morphological populations. The lysosomal proteinases demonstrated classical distribution patterns after cell fractionation by differential centrifugation. Cathepsin B was partially purified from the L_6 cells and shown to be capable of degrading native and denatured myosin. This myosin-degrading activity was completely blocked by leupeptin. All the L_6 cultures exhibited active rates of protein degradation. Protein catabolism was most rapid in pre-fusion myoblasts.

KEYWORDS

Cultured skeletal muscle cells; lysosomal proteinases; cell fractionation; myosin-degrading activity; protein degradation

INTRODUCTION

Although the precise role of the lysosomal apparatus and its complement of proteinases in muscle protein turnover remains to be elucidated, a role for lysosomes in the degradation of intracellular proteins is highly probable in muscle, especially since their function in the degradation of extracellular macromolecules in this tissue is minimal. The cathepsins B, D, H and L can degrade native or denatured purified actin and myosin (Schwartz and Bird, 1977; Sohar, et al., 1979; Bird, et al., 1980). The role of specific lysosomal or non-lysosomal proteases in

13

the degradative pathway for myofibrillar proteins has proved difficult to determine because of the cellular heterogeneity of muscle tissue. For example,
recently Woodbury, et al. (1978) showed that the alkaline serine proteinase in
muscle homogenates originated from mast cells. These results have been confirmed
by McKee, et al. (1979) and McElligott and Bird (1980). To circumvent the problems caused by the mixed cell populations in muscle, the present study examined
the proteolytic enzyme activities and rates of protein turnover in an established
myogenic line, which has been shown to have a well developed lysosomal apparatus
(Bird, et al., 1981).

MATERIALS AND METHODS

Muscle Cell Line L_6

The rat myogenic line L_6 originally cloned by Yaffee (1968), and generously
supplied by Schubert (Salk Institute, San Diego, CA), was maintained as stock in
plastic flasks in 90% Dulbecco's Modified Eagle Medium (DMEM) containing L-
glutamine and 4.5 gm/L glucose (GIBCO), 10% fetal calf serum (Irvine Scientific,
Irvine, CA) and 10 mg % gentamicin. L_6 cells were released from the stock flasks
with 0.05% trypsin containing 1 mM EDTA, resuspended in media and plated on
collagen-coated plastic petri dishes. Three different cell populations were
grown, consisting of pre-fusion, post-fusion and confluent non-fused cells. The
pre-fusion cells were plated at a sufficiently low density to prevent fusion
during the culture period. The post-fusion cells were maintained for 14 days in
vitro, at which time they were confluent, actively fusing (30-80% fused cells)
and frequently exhibited spontaneous contractions. The third group consisted of
a subclone of L_6 cells which under culture conditions identical to the post-fusion
group did not fuse even when maintained at extremely high densities.

Harvesting of Cells and Homogenization Procedures

The cells were washed in the petri dishes with CMF-HBSS (37° C), followed by a
rinse with the homogenization solution (0.25 M sucrose, 0.002 M Na_2 EDTA and
0.02 M KCl, pH 6.8, 37° C). The cells from three 60 mm petri dishes were released by gentle scraping with a rubber policeman. All the following procedures
were carried out at 4° C. The cells were centrifuged at 800 g x 5 min. The
supernatant containing damaged cells was discarded. The cell pellet was suspended in 10 mM phosphate buffer, 1 mM EDTA, pH 6.0 and homogenized in an ice-
cold Duall tissue homogenizer. Triton X-100 was added to the homogenate at a
final concentration of 0.2%. The homogenate was centrifuged at 32,000 g x 15
min, and the supernatant solution removed and saved. The pellet was extracted
two additional times. All the supernatant solutions were pooled and brought to
a final volume of 10 ml for the enzyme assays.

Differential Centrifugation Studies

Confluent cultures of L_6 non-fused cells were used for differential centrifugation
studies. Cells were grown for 4-5 days in 75 cm^2 flasks without collagen coating
and collected by incubating in homogenization solution at 37° C for 10 minutes.
The cells were gently removed from six flasks with a Pasteur pipette. After
centrifugation of the cell suspension at 800 g x 5 min, the cell pellet was
suspended in 1.5 ml of homogenizing solution without 0.02 M KCl and homogenized
at low speed in a VirTis 45 tissue homogenizer. Disruption of cells was monitored
with phase contrast microscopy and continued until only free nuclei were

observed. The cell homogenate was diluted to 3 mls, centrifuged at 800 g x 10 min, and the resulting nuclear pellet (N) was suspended in 0.5 ml of homogenization solution and centrifuged. Supernatant fractions from both centrifugations were pooled and centrifuged at 11,400 g x 5 min to obtain the heavy mitochondrial pellet (H); 22,500 g x 15 min to obtain the light mitochondrial pellet (L); and, 100,000 g x 45 min to obtain the microsomal pellet (P) and the final supernatant fraction (S) containing unsedimentable enzymes. Each subcellular fraction was assayed for various enzyme activities.

Enzymology

Enzyme activities were measured within the limits of linearity at 37°C. Cathepsin B (E.C. 3.4.22.1) was measured by the method of Barrett (1972). Substrates for the reaction were α-N-benzoyl-DL-arginine-2-naphthylamide (BANA) and CBZ-arg-arg-4-methoxy-2-naphthylamide (Arg-Arg), pH 6.0. Cathepsin D (E.C. 3.4.23.5) was determined by the method of Anson (1938) as modified by Bird, et al. (1968) for muscle, using denatured hemoglobin as substrate, pH 3.8. Cathepsin H (E.C. 3.4.22.-) was measured by the method of Kirschke, et al. (1977a), using L-leucyl-2-naphthylamide (Leu-NA) as substrate at pH 6.0. N-acetyl-β-D-glucosaminidase (NAβG) was determined by the procedure of Bosmann (1969), using p-nitrophenyl-N-acetyl-β-D-glucosaminide as substrate, pH 4.3. Cytochrome c oxidase activity was measured by the method of Applemans, et al. (1955). Protein concentrations were approximated by the method of Lowry, et al. (1951) with bovine serum albumin as standard.

Degradation of Myosin

For myosin degradation studies, cathepsin B was partially purified from muscle cell homogenates and separated from cathepsins D and H by passage through an ACA-54 column followed by elution from a 2B CM Sepharose column with a gradient of 20 mM-50 mM acetate buffer, pH 5.0, containing 0.3 M NaCl. Myosin purified from rabbit muscle (Sigma Chemical Co.) (0.23 mg/ml) was incubated with cathepsin B (14.4 mU/ml) in 50 mM Na acetate buffer, pH 5.2, containing 1.5 mM dithiothreitol and 1.5 mM Na$_2$ EDTA. Disc electrophoresis in polyacrylamide slab gels containing sodium dodecyl sulfate (SDS) was performed in the buffer system of Lammeli and Favre (1973). Gels were electrophoretically fixed, stained and destained as previously described (Schwartz and Bird, 1977). Denatured myosin (0.84 mg/ml) was prepared according to the procedure of Schwartz and Bird (1977).

Protein Degradation Studies

The degradation of proteins was measured by a modification of the technique used by Poole and Wibo (1973) to measure protein catabolism in rat fibroblasts. L$_6$ myoblasts were cultured in 35 mm petri dishes in 1.5 ml DMEM containing 10% dialyzed heat-inactivated fetal calf serum and 1 μCi/ml [^{3}H]phenylalanine for an 18 hour period. The cultures were washed two times with 2 ml of HBSS and incubated in Eagle's MEM (Earle's salts) containing 4 mM phenylalanine and 5% fetal calf serum. At zero time and various times later, the medium in 4 replicate dishes was brought to a final concentration of 10% trichloroacetic acid (TCA). The amount of radioactivity in the TCA soluble and TCA precipitable fractions was determined as described previously (St. John, et al., 1980). The standard errors of the determinations were generally within ±5% of the mean. The TCA soluble radioactivity released at each time was measured and expressed as a percentage of the total radioactivity in the TCA precipitable fraction at zero time (percent protein degradation).

A.C. St. John et al.

TABLE 1   Specific Activities of Lysosomal Enzymes in L_6 Cells

ENZYME	PRE-FUSION (n = 5)	POST-FUSION (n = 5)	NON-FUSION (n = 4)
Cathepsin B (A-A)	3.28 ±0.14	5.86 ±0.46*	4.45 ±0.23*
Cathepsin B (BANA)	0.56 ±0.05	1.04 ±0.09*	0.67 ±0.07
Cathepsin H (Leu-NA)	0.20 ±0.03	0.43 ±0.04*	0.23 ±0.02
Cathepsin D	79.2 ±9.3	79.8 ±9.18	65.5 ±15.0
NAβG	1.09	2.67	1.66

Values are the mean ± SEM. Significance: *$p \leq 0.01$ vs. pre-fusion values, Student's t test.

Units of enzyme activity are: Cathepsin B (Arg-Arg), μmole 4-methoxy-2-NA/mg/hr; Cathepsin B (BANA), μmole 2-NA/mg/hr; Cathepsin H (Leu-NA), μmole 2-NA/mg/hr; Cathepsin D, μg tyrosine/mg/hr; NAβG, μmole N-nitrophenyl/mg/hr.

RESULTS AND DISCUSSION

Our recent morphological studies of striated muscle cell cultures (Bird, et al., 1981) have demonstrated that both primary chick embryonic muscle cultures and the rat myogenic line, L_6, have an extensive and early development of the lysosomal apparatus. In the current study, cells from the rat myogenic line (L_6) were plated at different densities and maintained through several divisions to provide

TABLE 2   Enzyme Kinetics (K_m) in Cultured Muscle Cells (L_6)

SUBSTRATE	PRE-FUSION (n = 3)	POST-FUSION (n = 6)	NON-FUSION (n = 3)
Cathepsin B (A-A)	0.28 ±0.08	0.23 ±0.03	0.29 ±0.02
Cathepsin B (BANA)	3.30 ±0.40	3.90 ±0.60	3.20 ±0.40
Cathepsin H (Leu-NA)	0.26 ±0.05	0.19 ±0.02	0.19 ±0.16

Values are the mean ± SEM.

Fig. 1.   Rat L$_6$ myogenic line was cultured to provide 3 morphologically distinct
homogeneous cell populations, 14 days in vitro. Nomarski optics,
Bar = 20µ.

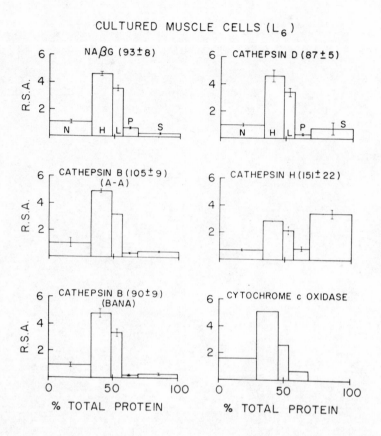

Fig. 2.   Fractionation of L₆ cells. Cells were fractionated as described in text.
          Relative specific activities (R.S.A.) were calculated by the method of
          de Duve et al. (1955) and expressed as a percentage of total protein.

three different morphological conditions. Pre-fusion cultures were plated at a
sufficiently low density to prevent fusion during the culture period. In this
group, frequent mitotic figures were seen (Fig. 1a, arrows). Post-fusion L₆
myoblasts were plated at a high cell density to promote cell-to-cell contact
and fusion. These cultures exhibited the early stages of myoblast fusion and
mitotic acitvity as well as the more highly differentiated contractile myotubes
characterized by their parallel rows of clustered nuclei, prominent myofibrils
and frequent cross-striations (Fig. 1b). Non-fusion cultures were obtained from
a subclone of the L6 myogenic line which had lost its ability to undergo fusion
even at very high densities. These cells formed a multilayered sheet when
plated at high density and continued their high level of mitotic activity
(Fig. 1c).

We have reported recently the total activities of proteolytic enzymes in cultured
muscle cells (Bird, et al., 1981). A comparison of the specific activities of

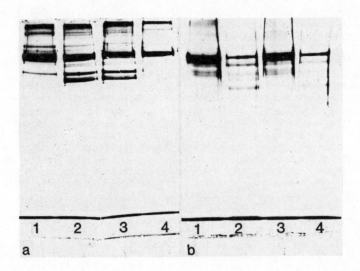

Fig. 3.   Degradation of myosin by cathepsin B.
          a.   Native myosin incubated for 4 hours as described in text; 1. myosin
               alone, 2. myosin and cathepsin B, 3. myosin, cathepsin B and 5 μM
               pepstatin, 4. myosin, cathepsin B and 10 μM leupeptin.
          b.   Denatured myosin incubated for 4 hours as described in text;
               1. myosin alone, 2. myosin and cathepsin B, 3. myosin, cathepsin B
               and 10 μM leupeptin, 4. myosin, cathepsin B and 5 μM pepstatin.

several lysosomal enzymes (cathepsins B, D, H and NAβG) in pre-fusion, post-
fusion and non-fusion cultured L_6 cells is shown in Table 1. The total
activities in these myoblast cultures were several orders of magnitude greater
than those reported for rat muscle homogenates (Bird, et al., 1981). Pre-fusion
myoblasts had the lowest activities of cathepsins B and H. The activities of
these cysteine proteinases were significantly higher ($p \leq 0.01$) in post-fusion
cells. The Michaelis-Menton constants (K_m) were calculated for cathepsin B
(against both Arg-Arg and BANA as substrates) and for cathepsin H (against Leu-NA)
for all three morphological states (Table 2). These K_m's might not represent
absolute values since they were calculated from the activities of crude high
speed supernatants. The observation that the K_m's were the same in different
cell types strongly suggested that the same enzymes were active in all three
populations.

The subcellular distribution of these cathepsin activities was investigated in
differential centrifugation studies on non-fusion L_6 myoblasts. The enzyme
markers, cathepsins B, D and NAβG, were sedimented in the heavy and light mitochondrial fractions (Fig. 2), which is similar to the classical distribution
pattern for lysosomal enzymes (de Duve, 1955). The distribution of cathepsin H
suggested a lysosomal localization although a very high percentage of activity
was present in the unsedimentable fraction. While this finding might suggest a
non-lysosomal location of cathepsin H in vivo, it is also possible that there is a
population of lysosomes in myoblasts that are rich in cathepsin H and more
sensitive to homogenization stress. The recoveries of cathepsin B, D and NAβG
were 87-105% of the total activities in cell homogenates. In contrast, the 150%

recovery of cathepsin H activity after fractionation suggested that an endogenous inhibitor to this enzyme might be present in these muscle cells. Additional studies are examining this possibility.

Previous studies by Schwartz and Bird (1977) demonstrated that purified cathepsins B and D from muscle had myosin-degrading activity. To prove that this activity was a characteristic of cathepsin B that originated solely from muscle cells, we examined the myosin-degrading ability of cathepsin B partially purified from L_6 non-fusion myoblasts. Purified cathepsin B degraded native myosin at pH 5.2 to smaller peptide fragments (Fig. 3a). The degradation of denatured myosin was more extensive but produced the same degradation products (Fig. 3b) The proteolysis of myosin was inhibited by 10 μM leupeptin (a cathepsin B inhibitor) but not by 5 μM pepstatin (a cathepsin D inhibitor). The pattern of myosin degradation peptides produced by cathepsin B from cultured myoblasts was similar to that produced by cathepsin B purified from whole muscle (Schwartz and Bird, 1977).

Our studies have shown that dramatic morphological changes as well as changes in the specific activities of lysosomal enzymes were associated with the differentia-tion of the pre-fusion, actively dividing myoblasts to the contractile multi-nucleated myotubes. The extensive development of the lysosomal apparatus in these cultures suggested that active protein turnover occurred during this developmental process. To examine this possibility the rate of intracellular protein degradation in cultured L_6 myoblasts was examined during several stages of development (Fig. 4). The L_6 cultures were prelabeled with [^{3}H]phenylalanine over an 18 hour period, washed to remove unincorporated radioactivity and then incubated in medium con-taining 4 mM unlabeled phenylalanine for up to 24 hours. This concentration of phenylalanine should not influence the overall rate of protein catabolism. Previous studies by Fulks, et al. (1975) using skeletal muscle, Chua, et al. (1979) using heart muscle and Vandenburgh and Kaufman (1980) using cultured chick embryonic muscle demonstrated that high levels of phenylalanine do not alter the rates of protein catabolism in muscle.

Pre-fusion L_6 myoblasts demonstrated the most rapid rate of protein degradation (Fig. 4, Table 3). After 24 hours approximately 60% of the labeled proteins had been degraded. This rate was similar to that found with other dividing cells in culture, e.g. rat fibroblasts (Poole and Wibo, 1973) and Balb/c 3T3 fibroblasts (St. John, et al., 1980). The degradation of proteins in the post-fusion cultures and in the non-fusion myoblasts was significantly lower than in pre-fusion myo-blasts (p < 0.001).

The differences in rates of protein degradation among these three morphologically distinct cultures were not influenced by changes in isotope reutilization by the cells. When protein synthesis was blocked by cycloheximide (10 μg/ml), the re-lease of radioactivity into an acid-soluble form did not increase. In fact, the addition of cycloheximide reduced the rate of protein degradation 10-50%. This result is consistent with the observations by a number of groups (Hershko and Tompkins, 1971; Epstein, et al., 1975; Goldberg and St. John, 1976; Vandenburgh and Kaufman, 1980), that protein degradation can be lowered in cells treated with inhibitors of protein synthesis.

There appears to be an inverse relationship between the specific activities of lysosomal cathepsins and the rates of protein catabolism in different morpholo-gical stages of the L_6 myoblasts. These findings suggest that: (1) the lysosomal cathepsins may not be the rate-limiting enzymes involved in protein turnover; (2) different levels of intracellular inhibitors may be present in the various stages of development and therefore the total enzyme activity in cell

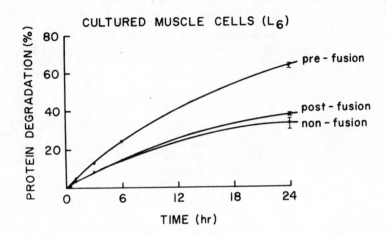

Fig. 4.   Rates of protein degradation in cultured L_6 cells. Values were obtained
from 6-8 replicate plates.

extracts may not be a true measure of total enzyme levels in the cultured cells;
and/or (3) the differences in protein degradation reflect the different stabilities
of the proteins labeled during the 18 hour pre-incubation period. Thus the
rapidly dividing pre-fusion myoblasts may have synthesized proteins with shorter
half-lives than those synthesized by post-fusion or non-fusion cells. As myoblasts
reach a high cell density and fuse, they begin to synthesize high levels of the
components of the contractile apparatus: actin, myosin, myokinase. These proteins
are relatively stable in adult muscle. Therefore even though extensive cellular
rearrangements are occurring in post-fusion myotubes, the proteins synthesized in
the 18 hour period prior to measurement of protein degradative rates may be
relatively stable cell components.

TABLE 3 Percent Protein Degradation in Cultured Muscle Cells (L6 Clone)

	1 HOUR	3 HOUR	6 HOUR	24 HOUR
Pre-fusion (n = 8)	5.0 ±0.2	12.8 ±0.4	20.5 ±0.3	60.4 ±0.5
Post-fusion (n = 6)	3.2 ±0.1	8.0 ±0.2	14.0 ±0.3	38.0 ±0.4
Non-fusion (n = 7)	3.0 ±0.2	8.3 ±0.7	13.6 ±1.4	33.3 ±3.0

Values are the mean ± SEM of replicate assays from two independent
experiments.

Our studies have demonstrated that in the L_6 myogenic line (1) cell homogenates exhibit high specific activities of lysosomal enzymes; (2) lysosomal proteinases demonstrate classical distribution patterns after cellular fractionation by differential centrifugation; (3) myofibrillar proteins are degraded by lysosomal proteinases isolated from cultured muscle cells; and, (4) the most rapid rates of protein degradation are observed in pre-fusion myoblasts. Therefore, cultured myoblasts present a unique model, unencumbered by contamination from other cell types, for the study of specific enzymes, inhibitors, control mechanisms, and morphological changes associated with the turnover of muscle proteins.

ACKNOWLEDGEMENTS

The authors thank Ms. Laura Wood for technical assistance and Ms. Jane Sherwood and Ms. Linda Ciak for aiding us in the preparation of the manuscript. The work in this study was supported in part by generous grants from the Muscular Dystrophy Association and the Charles and Johanna Busch Endowment.

REFERENCES

Anson, M. L. (1938). The estimation of pepsin, trypsin, papain, and cathepsin with hemoglobin. J. Gen. Physiol., 22, 79-89.

Applemans, F., R. Wattiaux, and C. de Duve (1955). Tissue fractionation studies. V. The association of acid phosphatase with a special class of cytoplasmic granules in rat liver. Biochem. J., 59, 438-445.

Barrett, A. J. (1972). A new assay for cathepsin B and other thiol proteinases. Anal. Biochem., 47, 280-293.

Bird, J. W. C., T. Berg, and J. H. Leathem (1968). Cathepsin activity of liver and muscle fractions of adrenalectomized rats. Proc. Soc. Exptl. Biol. Med., 127, 182-188.

Bird, J. W. C., J. Carter, R. E. Triemer, R. M. Brooks, and A. M. Spanier (1980). Proteinases in cardiac and skeletal muscle. Fed. Proc., 39, 20-25.

Bird, J. W. C., F. J. Roisen, G. Yorke, J. A. Lee, M. A. McElligott, D. F. Triemer, and A. C. St. John (1981). Lysosomes and proteolytic enzyme activities in cultured striated muscle cells. J. Cytochem. Histochem., (in press).

Bosmann, H. B. (1969). Glycoprotein degradation: Glycosidases in fibroblasts transformed by oncogenic viruses. Exptl. Cell Res., 54, 217.

Chua, B., D. L. Siehl, and H. E. Morgan (1979). Effect of leucine and metabolites of branched chain amino acids on protein turnover in heart. J. Biol. Chem., 254, 8353-8362.

de Duve, C., B. C. Pressman, R. Gianetto, R. Wattiaux, and E. A. Applemans (1955). Tissue fractionation studies. VI. Intracellular distribution patterns of enzymes in rat liver tissue. Biochem. J., 60, 604-617.

Epstein, D., S. Elias-Bishko, and A. Hershko (1975). Requirement for protein synthesis in the regulation of protein breakdown in cultured hepatoma cells. Biochem., 14, 5199-5204.

Fulks, R. M., J. B. Li, and A. L. Goldberg (1975). Effects of insulin, glucose and amino acids on protein turnover in rat diaphragm. J. Biol. Chem., 250, 290-298.

Goldberg, A. L., and A. C. St. John (1976). Intracellular protein degradation in mammalian and bacterial cells. Ann. Rev. Biochem., 45, 747-803.

Hershko, A., and G. M. Tompkins (1971). Studies on the degradation of tyrosine amino transferase in hepatoma cells in culture: Influence of the composition of the medium and adenosine triphosphate dependence. J. Biol. Chem., 246, 710-714.

Kirschke, H., J. Langner, B. Wiederanders, S. Ansorge, P. Bohley, H. Hanson
   (1977a). Cathepsin H: An endoaminopeptidase from rat liver lysosomes. Acta
   Biol. Med. Germ., 36, 185-199.
Kirschke, H., J. Langner, B. Wiederanders, S. Ansorge, P. Bohley (1977b).
   Cathepsin L: A new proteinase from rat-liver lysosomes. Eur. J. Biochem.,
   74, 293-301.
Lammeli, U. K., and M. Favre (1973). Maturation of bacteriophage T4. I. DNA
   packaging events. J. Mol. Biol., 80, 575-599.
Lowry, O. H., N. J. Rosenbrough, A. L. Farr, R. J. Randall (1951). Protein
   measurement with the folin phenol reagent. J. Biol. Chem., 193, 265-275.
McElligott, M. A., and J. W. C. Bird (1980). Effect of streptozotocin-induced
   diabetes on proteolytic activity in rat muscle. Fed. Proc., 39, 635.
McKee, E. E., M. G. Clark, C. J. Beinlich, J. A. Lins, and H. E. Morgan (1979).
   Neutral-alkaline proteases and protein degradation in rat heart. J. Mol.
   Cell Cardiol., 11, 1033-1051.
Poole, B., and M. Wibo (1973). Protein degradation in cultured cells: The
   effect of fresh medium, fluoride, and iodoacetate on the digestion of
   cellular protein of rat fibroblasts. J. Biol. Chem., 248, 6221-6226.
St. John, A. C., M. Merion, R. E. Triemer, R. J. Kuchler, J. W. C. Bird,
   J. H. Carter, and R. D. Poretz (1980). Internalization and intracellular
   fate of Wistaria floribunda agglutinin in Balb/c 3T3 fibroblasts. Exp. Cell.
   Res., 128, 143-150.
Schwartz, W. N., and J. W. C. Bird (1977). Degradation of myofibrillar proteins
   B and D. Biochem. J., 167, 811-820.
Sohar, I., O. Takaos, F. Guba, H. Kirschke, and P. Bohley (1979). Degradation of
   myofibrillar proteins by cathepsins from rat liver lysosomes. Szeged-
   Halle Symposium. Martin Luther University Press, Halle.
Vandenburgh, H., and S. Kaufman (1980). Protein degradation in embryonic
   skeletal muscle: Effect of medium, cell type, inhibitors, and passive
   stretch. J. Biol. Chem., 255, 5826-5833.
Woodbury, R. G., M. Everitt, Y. Sanada, N. Katunuma, D. Langunoff, and H. Neurath
   (1978). A major serine protease in rat skeletal muscle: Evidence for mast
   cell origin. Proc. Natl. Acad. Sci., 75, 5311-5313.
Yaffe, D. (1968). Retention of differentiation potentialities during prolonged
   cultivation of myogenic cells. Proc. Nat. Acad. Sci., 61, 477-483.

AGE-DEPENDENT CHANGES IN INTRACELLULAR PROTEIN TURNOVER

B. Wiederanders, P. Bohley and H. Kirschke

Physiologisch-chemisches Institut, Martin-Luther-Universität Halle, DDR 402 Halle, Hollystr. 1

ABSTRACT

The following methods were used to compare the intracellular protein break-down in the livers of rats of different age: Estimation of the in-vivo half-lives of cytosol proteins by pulse labelling with $H^{14}CO_3^-$; determinations of the specific proteolytic activities at pH 3.0 and 6.0, resp.; characterization of the cytosol proteins by SDS-PAGE; calculation of the bulk hydrophobicity of the cytosol proteins by hydrophobic chromatography; in vitro digestion of the cytosol proteins by cell-own lysosomal proteinases. The apparent half-lives of the cytosol proteins seem to be longer in old animals. The SDS-pattern are similar in different age groups, but the bulk hydrophobicity of the cytosol proteins of old rats is remarkably less compared with those of young animals. This change is accompanied by reduced digestibility of the cytosol proteins of old animals in vitro. These findings favour the hypothesis, that the hydrophobicity of the substrate proteins play an important role for the rate at which proteins are degraded within the cell.

KEYWORDS

Rat liver cytosol; half-life; proteinases and ageing; superficial hydrophobicity and digestion rate of proteins.

INTRODUCTION

The intracellular protein degradation is one of the mechanisms by which a cell can adapt to changing internal and external conditions. The maintenance of the steady state in the cell requires the precise regulation of protein synthesis as well as that of protein degradation. Although our knowledge about this metabolic way became broader in the last years (Waterlow and co-workers, 1978; Millward, 1979; Holzer and Tschesche, 1979) there are still some problems to be solved, e. g. how to explain the big differences in the rate, by which various enzymes are removed from the cell. Suitable model systems are required to study a question like this. We chose as one of the possible models the comparison of protein breakdown in both adult and

senescent rats. If there are changes in the rate of protein degradation between both groups of animals it should be interesting to search for the reasons responsible. Additionally, the dispute of experimental gerontologists
on a unique hypothesis of ageing will achieve some new facts.

HALF-LIFE-DETERMINATIONS

The first question to be answered was whether or not is the breakdown of
intracellular proteins in old animals as fast as in young animals. The answer
was not quite clear. The study of protein turnover was undertaken in experimental gerontology mostly as a mean for the purpose. We learned from the
literature that either results are expressed as change in the

$^3H/^{14}C$ ratio of proteins (if both isotopes are given before as labelled amino

acid in a definite time interval) (Comolli and co-workers, 1972) or the
authors did not take notice of the extent of reutilization of labelled amino
acids. The latter can simulate longer half-lives, as discussed extensively
by Waterlow and co-workers (1978). The most reliable data came from two
laboratories. Millward (1979) reported on developmental changes of half-
lives of skeletal muscle proteins. Whereas in 23 days old rats the half-life
is only 3.1 days, they determined the half-lives in 330 and 700 days old rats
to be 16.9 and 23.9 days, resp. Uzi Reiss and co-workers (1977) observed an
increase in the half-life of ornithin decarboxylase from 10 to 20 minutes and
of aldolase from 25 to 37 hours comparing young and old mice.

We determined the degradation rate of liver cytosol proteins in rats of 21
months of age. 160 MBq $H^{14}CO_3^-$ per 100 g body weight was injected intraperitoneally. The animals were killed 90 min and 10 days later, resp. The cytosol
was prepared after perfusion of the liver. The amino acids have been removed
by gel filtration, and the resulting cytosol was hydrolysed in 6 N HCl.

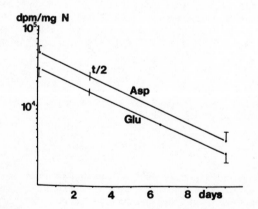

Fig. 1.   Decay of radioactivity in glutamate and
           aspartate, resp. of rat liver cytosol
           proteins after injection of radioactive
           bicarbonate. n = 6 per group.

The amino acids were separated on Dowex I X 8 to get glutamate and aspartate which carry about 50 % of the total radioactivity in the proteins after la- belling with bicarbonate (Swick and Ip, 1974). Figure 1 shows the decay line from which we calculated the half-life. This kind of calculation seems to be very simple. But Garlick, Waterlow and Swick (1976) compared three indepen- dent methods of calculating the degradation rate of protein mixtures like cytosol. Measuring the decay of radioactivity at each of the following days after the injection of the isotope is more complicated. But it gave similar k-values as a straight line drawn on a semilogarithmic plot from the maximum of incorporation to that time point when 90 % of the label are lost from the protein. Using this method we found a half-life of 2.8 days. This is more than 50 % longer than 1.8 days reported by Swick (1958) or the 1.6 days re- ported by Garlick and co-workers (1976) using the same isotope but working with young adult rats. As conclusion the answer to the first question is: the rate of protein breakdown in senescent animals is presumably declining.

THE ENZYMES

The literature dealing with data on changes of proteolytic enzyme activities accompanying the aging process has to be reviewed very critically. Nearly all of the authors determined the hemoglobin hydrolysis around pH 3.0 and called the activity they found Cathepsin D. Actually, this activity often has been found to increase with increasing age of the animals in different organs (see e. g. Lundholm and Schersten, 1975; Platt and Gross-Fengels, 1979). Therefore the conclusion was drawn, the intracellular proteolysis in old animals is accelerated. But in the liver this enzyme seems to play its essential role in the degradation of extracellular proteins. Bohley and co- workers (1979) called this intercellular proteolysis. This role of Cathepsin D was suggested by Huisman and co-workers(1974) and by our study

TABLE 1 Distribution of Cathepsin D in Different Cell Types of Rat Liver[+] and in Homogenates Prepared from Animals of Different Age[++].

Cell type	Cathepsin D (μg/mg protein)	
Parenchymal cells	1.4	n = 1
Non-parenchymal cells	10.8 ± 1.6	n = 8
Endothelial cells	4.6 ± 0.5	n = 6
Endothelial cells	4.1 ± 0.8	n = 4
Endothelial cells	3.1 ± 1.4	n = 2
Kupffer cells	33.0 ± 3.0	n = 4
Kupffer cells	17.4 ± 4.0	n = 3
Kupffer cells	6.8 ± 1.4	n = 5
Homogenate (5 months old rats)	0.9 ± 0.2	n = 4
Homogenate (21 months old rats)	1.2 ± 0.2	n = 7

[+]The cells were prepared in Rijswijk, n means the number of determinations.

[++]The homogenates were prepared in Halle, n means the number of animals.

(Wiederanders and co-workers, 1976), and it has been made probable by the
recent finding of Knook (1977) that it is mainly localized in the phagocytic
cells of the liver. We determined the amount of Cathepsin D by means of
rocket electrophoresis in different rat liver cells in cooperation with
D. L. Knook and A. Brouwer from the TNO Institute for Experimental Gerontology in Rijswijk (The Netherlands) (Ansorge and co-workers, 1979). The
results are shown in Table 1. They are in good accordance with the finding
of Knook (1977). The table shows further a higher concentration of Cathepsin
D in livers of 21 months old rats compared with young adult animals. It can
be discussed as a compensatory mechanism for the drastically elevated synthesis of albumin in very old rats (v. Bezooijen and co-workers, 1977), This
analogy as well as the localization of Cathepsin D in different liver cell
types support the suggestion of its role made above.

Lysosomal enzymes taking part in the degradation of cell own proteins
should be active at pH values around 6.0, since the intralysosomal pH is in
this region (Reijngoud and co-workers, 1976). Cathepsin L (Kirschke and co-
workers, 1977) shows a very high specific activity towards cytosol proteins
at this pH, too. It plays probably the main role in the lysosomal way of
intracellular protein breakdown in cooperation with Cathepsins B and H,
resp. As long we had neither a specific test system nor antibodies against
the different cathepsins we determined the proteolytic activity at pH 6.0
towards azocasein in liver cell fractions in animals of different age. The
results are shown in Table 2. There is a remarkable loss of azocasein hydrolysis per mg protein in the old group in all cell fractions except cytosol
and microsomes, resp. Although these results seem to be very clear, they
have to be interpreted carefully. We cannot say much about endogenous specific inhibitors of these enzymes. The proteolytic activity at pH 3.0 is also
lower in the liver cell fractions of old animals, but the <u>amount</u> of Cathepsin
D determined immunologically was higher in old animals, as described above.

Table 2 Proteolytic Activity (mU^+/mg protein) in Liver
Cell Fractions Prepared from Animals of 4 Months and
18 Months of Age, resp. pH 6.0.

Cell Fraction	Proteolytic Activity	
	Young	Old
Homogenate	11.2	10.5
Nuclei	11.5	6.6
Mitochondria	8.1	7.5
Lysosomes	328.0	234.0
Microsomes	0.9	1.7
Cytosol	2.7	1.6

$^+$1 U means hydrolysis of 87 µg azocasein per min.

Cathepsins B, H and L are cysteine proteinases. So, the concentration of
free -SH groups in the cell might be of significance for their activity.
Therefore we determined the concentration of free and protein bound -SH
group in liver homogenates of animals 5 and 18 months old, and their dependence on the food supply. Table 3 shows the results. The differences were
not significant, the standard deviation was above 20 % in the old group, and
the number of animals in each group was 5. Nevertheless the old animals

probably seem to depend much more on a sufficient food supply for keeping
their reduced thiol concentration constant. So, we cannot state a remarkable
change in the concentration of free or protein bound -SH groups in dependence
on the animals age. This is in accordance with recent findings of Nohl and
co-workers (1979) and Stohs and co-workers (1980).

Table 3 Concentration of Free and Protein Bound
-SH Groups (nMol/mg Protein) in Liver Homogenates
of 5 Months and 18 Months old Male Rats.

	Old	Young
Free -SH		
Rats fed continously	73,7	59.1
24 h fasting rats	43.8	51.2
Protein bound -SH		
Rats fed continously	144.1	121.8
24 h fasting rats	149.2	158.1

As a conclusion our results as well as the data from the literature do not
permit the explanation of the reduced intracellular protein breakdown in old
animals in terms of reduced activity of proteolytic enzymes.

THE SUBSTRATES

The specificity of proteolytic enzymes is not restricted to only one substrate. Therefore the big differences in the half-lives of some proteins must
be explained by their intrinsic properties which render them more or less
stabil to the action of proteinases. As Bohley suggested (1968) the superficial hydrophobicity of a protein is a determinant of the rate by which
this protein can be removed from the cell. This suggestion has now been pro-
ven by some independent laboratories (for review see Bohley and co-workers,
1979). Other properties of proteins, which are positively correlated with the
rate of their degradation are their size, their subunit size, their negative
charge, their content of misincorporated amino acids and their content of
amino acid analogues incorporated arteficially (for review see Goldberg and St.
John, 1976).

Starting with the early observation of Zeelon and co-workers (1973) there is
a steadily increasing number of reports dealing with the occurrence of modified enzyme molecules in old animals.Rothstein (1977) reviewed the knowledge
on these modifications. In summary, they consist in the occurrence of
a) enzymatically inactive molecules which are still reactive immunologically;
b) heat labile enzyme molecules; c) enzyme forms with different kinetic properties. The electrophoretic mobilities of enzymes isolated from old subjects
mostly cannot be distinguished from those of the young counterpart. Sharma
and Rothstein (1978) were even able to show the identical amino acid composition of both forms of enolase, isolated from young and old nematodes. So,
there are obviously conformational changes of the enzymes giving rise to the
variations in the properties mentioned above. Uzi Reiss and co-workers (1977)
and Sharma and Rothstein (1978) reported on differences of UV spectra of
enzymes isolated from young and old subjects, respectively.

Such changes in the conformations of proteins can influence the rate of their degradation. Therefore we incubated liver cytosol proteins of 4 months and 30 months old rats with a lysosomal enzyme mixture at pH 6.0. Figure 2 shows a representative example. Cytosol proteins of old rats are degraded about 20 % less effectively than those of young animals. We observed the faster digestion of young cytosol proteins over a wide range of substrate concentrations (2 - 40 mg/ml), but only at pH 6.0. Pepsin, Trypsin, Pronase and Papain did not show a similar faster digestion of "young" cytosol (Wiederanders and co-workers, 1977).

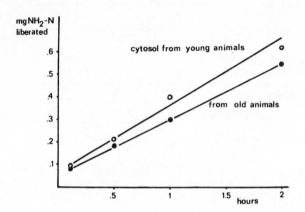

Fig. 2. Digestion of cytosol proteins by a lysosomal enzyme mixture at pH 6.0. The liberation of 0.5 μg amino-N represents 10 % hydrolysis of the substrate present.

The nature of the conformational changes is yet to be determined. As suggested, we were not able to show differences in the protein pattern of both, old and young cytosol, by means of SDS-PAGE, as shown in Fig. 3.1. The sensitive crossed immunelectrophoresis in antibody containing agarose gels yielded qualitative differences (Wiederanders and Römer, 1977). Antibodies against "young" cytosol proteins gave a reduced reaction with "old" cytosol proteins. The number of precipitation lines as well as their peak height was diminished, indicating differences in the protein pattern of both mixtures which were not visible on SDS gels. But these immunological differences don't explain the different digestibility of these cytosol proteins. Therefore we determined the amount of proteins bound to Phenyl-Sepharose using "old" and "young" cytosol proteins with reference to the hydrophobic chromatography of cytosol proteins reported by Bohley and co-workers (1979). We found indeed remarkable differences in these experiments. 38 % of the "young" proteins were bound to the hydrophobic matrix, whereas only 22 % of the "old" proteins are retarded by the gel (Wiederanders, 1978). These differences are illustrated in Fig. 3.2. On SDS gels much more proteins of the high molecular weight region are lacking in the "young" cytosol sample after mixing with Phenyl-Sepharose, i. e. more proteins are bound to the hydrophobic matrix than from the "old" sample. More insight in the extent and the kind of conformational changes taking place during the ageing process can be achieved perhaps by physico-chemical methods using purified proteins.

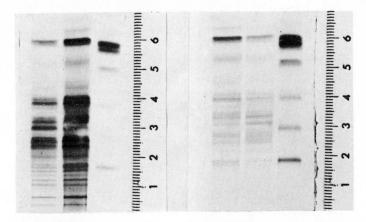

Fig. 3.1.
SDS-PAGE of liver cytosol
proteins of 5 months old
(middle) and 21 months
old rats (right).

Fig. 3.2.
The same as in Fig. 3.1.
after shaking with Phenyl-
Sepharose. The amounts of
protein mixed with the
Phenyl-Sepharose were
identical in both samples.

At the left hand side of both figures molecular weight
markers are shown. From the top: bovine serum albumin,
ovalbumin, chymotrypsinogen, soyabean trypsin inhibitor, myoglobin, cytochrom c.

The occurence of proteins with varying properties in old subjects is probably
due to conformational changes influencing their own degradation rate in the
direction of a longer half-life.

REFERENCES

Ansorge, S., B. Wiederanders, A. Brouwer and D. L. Knook (1979). Immunelek-
   trophoretische Enzymbestimmungen in verschiedenen Zelltypen der Rattenleber. 11. Jahrestagung der Biochemischen Gesellschaft der DDR, Abstract
   No. E 47.
v. Bezooijen, C. F. A., T. Grell and D. L. Knook (1977). The effect of age
on protein synthesis by isolated liver parenchymal cells. Mech. age.
develop., 6, 293 - 304.
Bohley, P., H. Kirschke, J. Langner, M. Miehe, S. Riemann, Z. Salama,
   E. Schön, B. Wiederanders and S. Ansorge (1979). Intracellular Protein
   turnover. In H. Holzer and H. Tschesche (Eds.) Biological Function of
   proteinases. 30th Colloquium der Gesellschaft für Biologische Chemie.
   Springer, Heidelberg, pp. 17 - 34.
Comolli, R., M. E. Ferioli and S. Azzola (1972). Protein turnover of the
   lysosomal and mitochondrial fractions of rat liver during aging. Exp.
   Gerontol., 7, 369 - 376.
Garlick, P. J., J. C. Waterlow and R. W. Swick (1976). Measurement of protein
   turnover in rat liver. Analysis of the complex curve for decay of label in
   a mixture of proteins. Biochem. J., 156, 657 - 663.

Goldberg, A. L. and A. C. St. John (1976). Intracellular protein degradation
    in mammalian and bacterial cells: part 2. Ann. Rev. Biochem., 45, 747 - 803.
Holzer, H. and H. Tschesche (Eds.) (1979). Biological function of proteinases.
    30th Colloquium der Gesellschaft für Biologische Chemie. Springer, Heidel-
    berg.
Huisman, W., L. Lanting, H. J. Doddema, J. M. W. Bouma and M. Gruber (1974).
    Role of individual cathepsins in lysosomal protein digestion as tested by
    specific inhibitors. Biochim. Biopyhs. Acta, 370, 297 - 307.
Kirschke, H., J. Langner, B. Wiederanders, S. Ansorge and P. Bohley (1977).
    Cathepsin L. A new proteinase from rat liver lysosomes. Europ. J. Biochem.,
    74, 293 - 301.
Knook, D. L. (1977). The role of lysosomal enzymes in protein degradation in
    different types of rat liver cells. Acta biol. med. germ., 36, 1747 - 1752.
Lundholm, K. and T. Schersten (1975). Leucine incorporation into proteins and
    cathepsin-D activity in human skeletal muscle. The influence of the age of
    the subject. Exp. Gerontol., 10, 155 - 159.
Millward, D. J. (1979). Protein degradation in muscle and liver. In M. Florkin,
    L. L. M. van Deenen and A. Neuberger (Eds.), Comprehensive Biochemistry,
    Vol. 19 B, Pt. I, Elsevier Scient. Publ., Amsterdam, pp. 153 - 232.
Nohl, H., D. Hegner and K. H. Summer (1979). Responses of mitochondrial super-
    oxide dismutase, catalase and glutathione peroxidase activities to aging.
    Mech. Age. develop, 11, 145 - 151.
Platt, D. and F. Gross-Fengels (1979). Influence of age and spironolactone on
    lysosomal enzyme activities, DNA and protein content of rat liver after
    partial hepatectomy. Gerontology, 25, 87 - 93.
Reiss, U., L. Lavie, S. Jacobus, J. Dresnick, H. Gershon and D. Gershon
    (1977). Studies on altered enzyme molecules from livers of aging animals.
    In D. Platt (Ed.) Liver and Ageing. 4th International Giessener Symposium
    on Experimental Gerontology. Schattauer, Stuttgart, pp. 199 - 207.
Reijngoud, D.-J., P. S. Oud, J. Kas and J. M. Tager (1976). Relationship
    between medium pH and that of lysosomal matrix as studied by two indepen-
    dent methods. Biochim. Biophys. Acta, 448, 290 - 302.
Rothstein, M. (1977). Recent developments in age related alterations of
    enzymes. Mech. age. develop., 6, 241 - 257.
Sharma, H. K. and M. Rothstein (1978). Age-related changes in the properties
    of enolase from Turbatrix aceti. Biochemistry, 17, 2869 - 2876.
Stohs, S. J., J. M. Hassing, W. A. Al-Turk and A. N. Masoud (1980).
    Glutathione levels in hepatic and extra-hepatic tissues of mice as a
    function of age. Age, 3, 11 - 14.
Swick, R. W. (1958). Measurement of protein turnover in rat liver. J. Biol.
    Chem., 231, 751 - 764.
Swick, R. W. and M. M. Ip (1974). Measurement of protein turnover in rat
    liver with (14 C)-carbonate. J. Biol. Chem., 249, 6836 - 6841.
Waterlow, J. C., P. J. Garlick and D. J. Millward (Eds.) (1978). Protein
    turnover in mammalian tissues. North-Holland, Amsterdam.
Wiederanders, B., S. Ansorge, P. Bohley, U. Broghammer, H. Kirschke and
    J. Langner (1976). Intrazellulärer Proteinabbau VI. Isolierung, Eigen-
    schaften und biologische Bedeutung von Kathepsin D aus der Rattenleber.
    Acta biol. med. germ., 35, 269 - 283.
Wiederanders, B., S. Ansorge, P. Bohley, H. Kirschke, J. Langner and H. Hanson
    (1977). The age dependence of intracellular proteolysis in the rat liver.
    In V. Turk and N. Marks (Eds.) Intracellular Protein Catabolism II. Plenum
    Press, New York. pp. 144 - 147.
Wiederanders, B. and I. Römer (1977). Ageing changes in intracellular protein
    breakdown. Acta biol. med. germ., 36, 1837 - 1841.
Wiederanders, B. (1978). Characterization of rat liver cytosol proteins of
    old and young animals with respect to their digestibility by proteases.
    12th FEBS Meeting Dresden, Abstracts No. 1928.

ON THE METABOLISM OF OPIATE PEPTIDES BY BRAIN PROTEOLYTIC ENZYMES

A. Suhar, N. Marks,* V. Turk and M. Benuck*

*Department of Biochemistry, J.Stefan Institute, University E.Kardelj,
6100 Ljubljana, YU*
*Center for Neurochemistry, Rockland Research Institute, Ward's Island,
New York 10035, USA*

ABSTRACT

The properties of two classes of proteolytic enzymes present in brain are described one of which is a cysteine (thiol) proteinase and the other a dipeptidyl carboxypeptidase. Both types of enzymes are extractable with buffers containing detergent and can be shown to degrade opiate peptides or their precursor forms. Cysteine proteinases were purified 500 fold from calf brain and exhibit heterogeneity when submitted to CM-cellex column chromatography. Degradation of β-lipotropin was accompanied by release of an 8 kilodalton unit, indicating cleavage at the LPH 76-80 region. The enzyme hydrolyzed basic protein of myelin, histones, glucagon and protamine. The best synthetic substrate was $Z-(Arg\_)_3-4M-\beta Nap$. It was inhibited competitively by leupeptin and non-competitively by a tripeptidyl aldehyde Boc-D-Phe-Pro-Arg. Rabbit brain contained two dipeptidyl carboxypeptidases one of which resembled angiotensin converting enzyme (ACE), while the other degraded enkephalin but not angiotensin-1. The two enzymes were separated by immunoaffinity chromatography and differentiated by the use of specific ACE inhibitors. The enzyme degrading enkephalin but not angiotensin-1 was also present in kidney at a higher concentration than ACE.

KEYWORDS

Cysteine (thiol) proteinase; dipeptidyl carboxypeptidase; angiotensin converting enzyme (ACE); lipotropin breakdown; enkephalin breakdown; inhibition of cysteine proteinases; inhibition of dipeptidyl carboxypeptidases; opiate peptide metabolism.

INTRODUCTION

Neuropeptides are agents that have potent pharmacological and often behavioral effects and have attracted interest in terms of their metabolism (Marks, 1978). Such peptides are formed as a result of the action of brain proteolytic enzymes that convert precursor polypeptides to the active form followed by enzymatic inactivation. This is illustrated in recent studies on the formation of opiate peptides from a polypeptide precursor known as 31K found in animal pituitaries and which contain sequences associated with corticotropins, melantotropins and endorphins (Mains and Eipper, 1980; Herbert et al., 1980). Conversion of the original gene product includes removal of the first 26 amino acids presequence by a membrane bound protease (signal peptidase) followed by cleavage at sites adjacent to pairs of basic amino acids by an as yet uncharacterized 'tryptic-like' enzyme. Such an enzyme is considered of

A. Suhar et al.

importance since many polypeptide precursors contain one or more pairs of basic residues adjacent to the biologically active fragments (see Steiner et al, 1980). Prohormones such as insulin can be activated by trypsin or cathepsin B followed by the action of a carboxypeptidase B-like enzyme. Studies on cathepsin B, a cysteine proteinase, lead to the expectation that it might also play a role in conversion of neuropeptide precursors such as β-lipotropin. To examine such aspects we have attempted to purify cysteine proteinases from brain and pituitary and studied their specificity using known native and synthetic substrates (Suhar and Marks, 1979). In the present study we report on the heterogeneity of brain enzymes and kinetics of inhibition using specific inhibitors such as leupeptin and a tripeptidyl aldehyde analog.

In addition to cathepsins we have examined also the effect of membrane bound dipeptidyl carboxypeptidases on breakdown of enkephalins. Previous studies established that brain contains angiotensin converting enzyme (ACE) that can be separated from other enzymes by immunoaffinity chromatography (Benuck and Marks, 1979, 1980). We show in the present study that brain contains a second dipeptidyl carboxypeptidase that does not recognize ACE antibody and which degraded enkephalins by removal of the C-terminal dipeptide. As a consequence this second dipeptidyl carboxypeptidase may be relevant to regulation of enkephalin levels in peptidergic pathways that utilize this peptide as a transmitter substance.

CYSTEINE PROTEINASES OF BRAIN

Previously we showed that cysteine proteinases are present in brain tissues and can be measured conveniently with the substrate Bz-Arg-βNap (Suhar and Marks, 1979; Marks et al., 1980). Cysteine proteinases extracted from lysosomes differ considerably in substrate specificity (Kirschke et al. 1980) and thus may play distinct roles in the formation and/or inactivation of neuropeptides. To examine the question of heterogeneity in more detail we extracted fresh rather than frozen calf brain and purified the enzymes by a modified procedure similar to that described previously. The steps included extraction with 10 volumes of 0.1 M sodium acetate buffer pH 4.5 that contained 1 mM EDTA, followed by incubation of the homogenate for 2 h at 37°C (autolytic step), ammonium sulfate fractionaton 45-65% w/v, dialysis and passage through a column of CM-cellex (BioRad grade) extensively washed with cysteine hydrochloride. Elution with a NaCl gradient in 50 mM sodium acetate buffer pH 5.0 gave evidence of greater heterogeneity than previously described (Fig. 1).

The elution profile showed that the largest amount of activity resided in the first peak but owing to the relatively high content of protein the specific activity was lower than in Peak II. After gel filtration of enzyme contained in peak II, the overall purification was 560 fold and the yield was 2.6%. An exact balance sheet for lysosomal (cysteine) proteinases in terms of recovery is obscured by the presence of inhibitory materials, and the activation that follows the autolytic step. It is not known if this involves solubilization of enzyme or removal of an inhibitory substance. Gel-electrophoresis of peak II enzyme revealed that it was essentially homogenous. The molecular weights of the enzymes found in peaks I-III were in the range of 26-28 kilodaltons and in line with that observed in our previous study, and for cysteine proteinases (cathepsin B) of other tissues (Barratt, 1977). Enzyme in peak I was extremely labile when stored at 0°C. This may account for its absence in the previous study when frozen tissue was used.

PROPERTIES OF PURIFIED BRAIN CYSTEINE (THIOL) PROTEINASES

Cysteine proteinases can be differentiated on the basis of their substrate specificity, pH optima and effects of inhibitors (Kirschke et al. 1980). Activity present in peaks I-III (Fig. 1) were active optimally at pH 6.0-6.6 and required the presence of cysteine (maximal at 2 mM in the presence of 1 mM EDTA). In terms of relative rates of hydrolysis for different protein substrates, peak I showed maximum activity

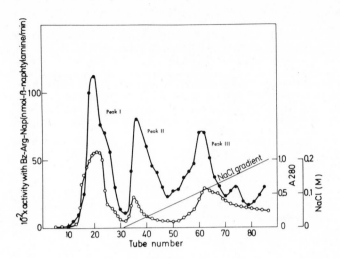

Fig. 1. Elution profile of calf brain extracts assayed with Bz-Arg-βNap.
The assay mixture of 1.25 ml contained 50 µl of enzyme from each fraction
50 µl of substrate (dissolved in DMSO) and was incubated at pH 6.5 for
10-30 min at 37°C. The buffer used was potassium phosphate (0.1 M,
0.75 ml used) and contained 1 mM EDTA and 2 mM cysteine. The release
of β-naphthylamine was measured colorimetrically at 520 nm after coupling
with fast Garnet GBC. ●——●, enzymatic activity, o——o, protein; ——
gradient. The column size was 1.8 x 30 cm. For other details see Table I.

TABLE I   Purification of Cysteine Proteinases

Fraction	Spec. activity (units/mg)	Yield (%)
1. Homogenate	0.37	89
2. Autolysis extract	2.90	100
3. $(NH_4)_2SO_4$ (45-65%)	11.40	36
4. CM-cellex		
peak I	13.4	12.5
peak II	26.6	6.1
peak III	5.8	3.3
5. Sephadex G-75		
I	160.0	7.8
II	208.0	2.6
III	80.0	1.4

Enzymes were purified from 100 g of calf brain. Peak I represents
activity eluted with 50 mM acetate buffer pH 5.0, containing 1 mM
EDTA, and peaks II and III were eluted with 0-0.2 M sodium chloride
gradient in the acetate buffer. The data represent the means of two
separate experiments. One unit of activity is defined as the release
of 1 nmol beta-naphthylamine in 1 min at 37°C. For other details see
Fig. 1.

A. Suhar et al.

with denatured hemoglobin followed by lysine rich histones, total histones, protamine, glucagon and only low activity with myelin basic protein. In contrast, peaks II and III showed higher activity with myelin basic protein as the substrate. The best substrate for peak II was protamine, followed by lysine rich and total histones: the best substrate for peak III also was protamine followed by myelin basic protein, glucagon and low for total histones (Table 2). Breakdown of myelin basic protein by cysteine proteinases may be a factor in neuropathies associated with demyelination. Immunoreactive forms of glucagon exist in hypothalamus and other areas of brain and thus degradation by cysteine proteinases may be involved in their turnover (Tager et al. 1980). The mechanisms involved in cleavage of glucagon were not examined in detail but appear to differ from that reported by Aronson et al. (1978) (removal of C-terminal dipeptides at low pH).

TABLE 2   Substrate Specificity of Brain Cysteine Proteinase

Substrate	Relative activity of CM-cellex fractions		
	I	II	III
Lys-rich histones	100	58	26
Total histones	58	47	9
Hemoglobin	171	0	0
Protamine	44	69	33
Glucagon	47	40	12
Basic protein (bovine myelin)	6	23	25

The reaction mixture containing 1.5-6.5 µg purified enzyme and 300 nmol of substrate was incubated at pH 6.5 for 3 h at 37°C, heated at 100°C for 5 min, and the soluble ninhydrin positive products measured. The results are expressed relative to hydrolysis of lysine-rich histones by fraction I (1.06 µM alpha-amino groups liberated min^{-1} mg enzyme $protein^{-1}$).

To explore differences between the cysteine proteinases the enzymes in peaks I-III were incubated with Z-(Arg)$_3$-βNap and rates of hydrolysis compared to Bz-Arg-βNap. Results showed that the best substrate in terms of the efficiency ratio (K_{cat}/K_m) for all three peaks was Z-(Arg)$_3$-4-methoxy-βNap with values 5-20 fold higher than for the conventional substrate (Table 3).

TABLE 3   Kinetic Constants of Purified Cysteine Proteinases with Bz-Arg-βNap and Z-(Arg)$_3$-MβNap

Substrate	CM-cellex fraction	Kcat (min^{-1})	Kcat/Km ($mM^{-1}min^{-1}$)
Bz-Arg-βNap			
	I	15.7	5.0
	II	22.6	10.4
	III	7.0	6.6
Z-(Arg)$_3$-4M βNap			
	I	66.7	94.0
	II	100.0	100.0
	III	98.0	35.0

Kinetic constants for synthetic substrates, hydrolyzed by purified brain thiol proteinases present in CM-cellex peak I, peak II and peak III, respectively. Km values were obtained by Lineweaver-Burk plots and the Kcat calculated assuming a molecular weight of 27,000.

TABLE 4 Effect of Inhibitors and Other Agents on Purified
Cysteine Proteinase.

A.	mM	% activity
ZnSO$_4$	0.1	88
HgCl$_2$	0.1	0
Iodoacetate	1.0	0
PMSF	1.0	76
Pepstatin	0.1	100
Bestatin	0.23	100
Puromycin	0.7	100
Bacitracin	0.32	100
Leupeptin	0.001	0
Antipain	0.006	0
Chymostatin	0.006	18
Boc-D-Phe-Pro-Arg-H	0.007	0
B.		
Cysteine	1.0	30
EDTA	1.0	82
ZnSO$_4$	0.1	10
HgCl$_2$	0.1	37
Iodoacetate	0.1	0

The reaction vol. 1.4 ml contains 0.7 µg enzyme protein, 130 µmol
0.1 M phosphate buffer, pH 6.2 plus the additions as noted below
and incubated with BANA (1 mg in 50 µl DMSO). Results are expressed
relative to enzyme activity in presence of 2.5 mM cysteine and 1 mM
EDTA (A) as 100% (119 nmol β-naphthylamine released per ml per
min per mg protein), or (B) activity in the presence of 1 mM cysteine
(represented 30% of control value).

Effects of inhibitors and other added materials. Studies on effects of added agents
were conducted with enzymes in peak II since this had the highest specific activity,
gave one major band on disc-gel electrophoresis, and was devoid of cathepsin D
contamination (no activity when incubated with hemoglobin at pH 3.2). Enzyme was
inhibited by 0.1 mM Zn^{2+} and Hg^{2+}. Inhibition by iodoacetate confirms the essentiality
of the -SH group for activity (Table 4). Inhibition by leupeptin demonstrates
sensitivty of this cysteine proteinase to peptidyl aldehydes as further confirmed by
actions of antipain and chymostatin. No inhibition was observed for inhibitors of
aminopeptidases (bestatin, bacitracin, puromycin) or for carboxyl proteinases
(pepstain).

Kinetics of inhibition by tripeptidyl aldehydes. Detailed studies with Ac-(Leu)$_2$-
arginal revealed that inhibition was competitive as shown by the Dixon plot (i/v)
(Fig. 2). The Ki value calculated from this data and also from the alternative
procedure of Knight (i/1-a=Ki (1+s/K$_m$ where a = vi/vo) was between 0.95 and 1.5
x 10^{-8} M. In contrast the synthetic tripeptidyl aldehyde Boc-D-Phe-Pro-Arginal was
non-competitive with a Ki value of 8 x 10^{-7} M (Fig. 3). The mechanism of inhibition
as confirmed when the data was analyzed by the procedure of Cornish-Bowden (i vs
s/v): leupeptin gave lines that were paralled but those for the tripeptidyl analog were
divergent. Bajusz et al. (1975) used the synthetic tripeptidyl analog to inhibit
processes involved in blood clotting (conversion fibrinogen-thrombin). Aldehydes
have been used also to differentiate cysteinyl proteinases (Kirschke et al. 1980).

Cleavage of β-lipotropin by brain cysteine proteinase. Incubation of β-lipotropin
(βLPH) with brain cysteine proteinase present in peak II led to the appearance of a

A. Suhar et al.

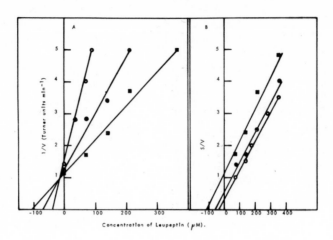

Fig. 2.   Inhibition of brain cathepsin B by leupeptin.
A, Dixon plot; B, Cornish-Bowden plot. Bz-Arg-β-Nap
concentrations are: (o) 0.81 mM; (●) 1.62 mM and (■)
3.24 mM.   V = turner units/min^{-1}, s = mg Bz-Arg-βNap in 1.4 ml.

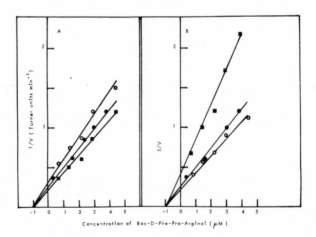

Fig. 3.   Inhibition of brain cathepsin B by Boc-D-Phe-Pro-Arg.
A, Dixon plot; B, Cornish-Bowden plot. Bz-Arg-βNap concentrations
are: (o) 0.81 mM; (●) 1.62 mM and (■) 3.24 mM. (See Fig. 2 for
other details).

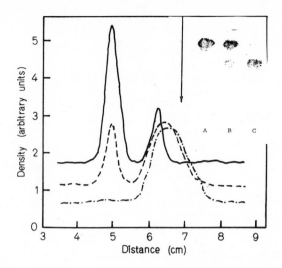

Fig. 4. Densitometric profile of a Coomassie stained slab gel obtained after incubation of porcine β-lipotropin with purified cysteine proteinase of calf brain. Times of incubation were 3 h (---), 20 h (-·-·-·-). The first peak represents β-lipotropin (11.5 kilodaltons) and the second degradation product (8 kilodaltons). Molecular weight markers included iodinated β-endorphin, soya bean trypsin inhibitor, BSA, RNA, polymerase. Gels were prepared with 20% acrylamide and 0.1% SDS according to the procedure described (Suhar and Marks, 1979). The inset shown bands found after radioautography using ^{125}I-β-lipotropin as the substrate; the first band represents the unincubated control, the second found after 3 h and the third after 20 h of incubation.

new product of about 8000 daltons (Fig. 4). This product is larger than that expected for β-endorphin (LPH 61-91, 3500 daltons) so that it can be concluded that the brain enzyme is not involved in the activation of β-LPH to form the smaller active (opiate) component. Production of β-endorphin would require cleavage at the Arg....Tyr bond by a 'tryptic-like' enzyme. There was some expectation based on studies on proinsulin that cathepsin B could act at a similar site but it is evident from present data that insulin is an unusual case and that this site may be vulnerable in lieu of its conformation. It is of interest that MacGregor et al. (1979) found that cathepsin B purified from parathyroid gland failed to activate proparathyroid hormone (ProPTH) by cleavage at the Arg....Ala site with release of a hexapeptide pre-sequence. This site like that of for β-lipotropin (Lys-Arg....Tyr) is adjacent to a pair of basic residues (Lys-Arg....Ala). PTH and β-lipotropin unlike proinsulin do not contain cysteine and have a less rigid conformation since disulfide bridges are absent. The presumed site of cleavage of β-lipotropin by cathepsin B (LPH 76-80, see Fig. 5) appears to be in a region that is susceptible to the action of other proteases that include brain or pituitary cathepsin D (Benuck et al., 1978; Marks, 1978) and ones of plant origin (Austen and Smyth , 1978). As a comment on the method used for study of lipotropin breakdown, it might be noted that slab-gel electrophoresis of iodinated compounds requires exceptionally small amounts of peptide, and is not interfered with by the presence of enzyme impurities. Peptides were iodinated with ^{125}-I by the procedure of Benuck et al. (1978) and separated on slab-gels containing 20% w/v acrylamide in the presence of 0.1% w/v SDS (Boehme et al., 1978).

In addition to β-lipotropin, the brain cysteine proteinase hydrolyzed β-endorphin (LPH 61-91). Cleavage at the Leu[77]-Phe[78] would result in the generation of γ-endorphin (LPH 61-77) a peptide that exhibits opiate properties in vitro, and shown also to act in vivo (Olsen et al., 1979). The des Tyr form of γ-endorphin is reported to have neuroleptic-like actions in man (DeWied et al., 1980). Formation of des Tyr-γ-endorphin can occur as the result of the action of membrane bound and soluble aminopeptidases present in brain fractions (Marks, 1978; Burbach et al., 1979). Another route for the formation of γ-endorphin is breakdown of β-endorphin by brain or pituitary cathepsin D (Benuck et al., 1978). Purified preparations of cathepsin D were shown to cleave Leu[77]-Phe[78] with release of LPH 61-77 in the case of β-endorphin, and LPH 1-77 in the case of β-lipotropin. These examples serve to illustrate a role for cathepsins in conversion of β-lipotropin or its fragments to form peptides that have potent behavioral effects in man.

Glu[1]-Leu-Ala-Gly-Ala-Pro-Pro-Glu-Pro-Ala-Arg-Asp-Pro-Glu-Ala-Pro-Ala-
Glu-Gly-Ala-Ala-Ala-Arg-Ala-Glu-Leu-Glu-Tyr-Gly-Leu-Val-Ala-Glu-Ala-
Gln-Ala-Ala-Glu-Lys-Lys-Asp-Glu-Gly-Pro-Tyr-Lys-Met-Glu-His-Phe-Arg-
Trp-Gly-Ser-Pro-Pro-Lys-Asp-Lys-Arg-Tyr-Gly-Gly-Phe-Met-Thr-Ser-
Glu-Lys-Ser-Gln-Thr-Pro-Leu-Val-Thr-Leu-Phe-Lys-Asn-Ala-Ile-Val-Lys-
Asn-Ala-His-Lys-Lys-Gly-Gln[91]

Fig. 5. Structure of porcine β-LPH. Primary structure of porcine β-LPH. Note that residues underlined represent the Met-enkephalin sequence LPH 61-65. Cleavage by cathepsin D occurred at residues 77-78, and cathepsin B at a bond in the 76-80 region. Formation of β-endorphin (LPH 61-91) will require cleavage at the Arg[60]-Tyr[61] bond by a 'tryptic-like' enzyme.

MEMBRANE BOUND DIPEPTIDYL CARBOXYPEPTIDASES OF BRAIN

The metabolism of enkephalins has attracted interest because of their potent pharmacological and neurotransmitter-like properties. Among enzymes present in brain and other tissues known to degrade this pentapeptide are aminopeptidases, dipeptidyl aminopeptidase (cathepsin C), and dipeptidyl carboxypeptidases but it remains to be established which of thse is the most crucial in situ. Dipeptidyl carboxypeptidases are good candidates since they are present on membranes, the sites for receptor-ligand interactions. In the present study, dipeptidyl carboxypeptidases were purified from rabbit brain by an immunoaffinity procedure utilizing an antibody to purified rabbit lung ACE. Enzyme was extracted by means of a buffer containing Triton X-100 and submitted to DEAE-cellulose chromatography prior to the immunoaffinity step. The separation on the DEAE-cellulose column indicated the presence of three peaks containing enzyme that degraded enkephalin; two of these peaks were found to hydrolyze hippuryl-His-Leu a known substrate of ACE. To separate ACE from other enzymes, the peak containing the largest content of enkephalin-degrading enzyme was passed through a column of Sepharose 4B coupled to the antibody of ACE as described elsewhere (Benuck and Marks, 1979). This resulted in a separation of brain ACE (bound to the column) from other enzymes present in the effluent found to degrade enkephalin. The latter when incubated with enkephalin led to the release of Tyr-Gly-Gly and Phe-Met and indicated the presence of a second dipeptidyl carboxypeptidase. In contrast to enzyme immobilized on the column that present in the effluent was unaffected by ACE inhibitors SQ 20881 and its prolyl analog SQ 14225 (Table 5).

Data in the table show immobilized enzyme (termed dipeptidyl carboxypeptidase A) was identical in tis properties to ACE of other tissues. Thus it degraded angiotensin-I with release of His-Leu and hydrolyzed the synthetic substrate Hippuryl-His-Leu. Its activity was inhibited by SQ 14225 and SQ 20881, and metal chelating reagents such as EDTA and o-phenanthroline. In addition to enkephalin, it cleaved bradykinin

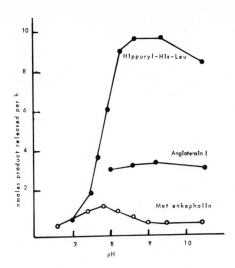

Fig. 6. The pH profile of brain ACE (dipeptidyl carboxypeptidase A).
Enzyme purified from rabbit brain by immunoaffinity chromatography
and tested in an immobilized form as described in the text, and by
Benuck and Marks (1979) with ●, angiotensin-I or hippuryl = His-Leu,
and o, Met-enkephalin. The buffer used was 50 mM HEPES.

TABLE 5 Dipeptidyl Carboxypeptidase of Rabbit Brain and Kidney:
Substrate Specificity and Effects of Inhibitors.

Substrate or addition	Product detected or conc (mM)	Activity (nmol/min/g tissue) Brain		Kidney	
		A	B	A	B
Leu-enkephalin	Tyr-Gly-Gly	2.4	0.8	94	280
Hippuryl-His-Leu	His-Leu	52	tr	330	19
Angiotensin-I	His-Leu	3.2	tr	20	1
Bradykinin	Phe-Arg	1.8	0.8	88	350
EDTA	1.0	−60	−40*	−100	−48
O-phenanthroline	0.1	−60	−50	−57	−37
SQ 14225	0.001	−100	−10	−95	0
SQ 20881	0.005	−90	0	−67	0

A refers to angiotensin converting enzyme (A) and B to a second dipeptidyl
carboxypeptidase cleaving enkephalins. Products were detected by reverse
HPLC or in the case of Hippuryl-His-Leu by a fluorimetric procedure
(Benuck and Marks, 1979, 1980). *Per cent change related to activity with
addition tr = trace.

with release of Phe-Arg. Since the enzyme in the effluent acted like a dipeptidyl
carboxypeptidase but was distinct in its properties from A it is referred to as type
B enzyme. The pH optima for type A enzyme was distinct from that of B since it was
active in the 8-9 range as compared to 7.3-7.6 for the latter (Fig. 6).

ACE occurs in high concentration in peripheral organs such as kidney and lung that
are involved in the renin-angiotensin system. Examination of rabbit kidney by the
same procedure showed that type B enzyme is present in 3 fold higher concentration

than type A enzyme (Table 5). In brain the ratio for these two enzymes was 0.3. The significance of type B enzyme in turnover of neuropeptides is unknown although there is some evidence to suggest that it may act as a more specific enkephalinase than ACE (Malfroy et al., 1978). It is possible that this enzyme plays a general role in conversion of a larger number of precursors by the sequential removal of C-terminal dipeptides (Breckner and Caprioli, 1980). The high concentration of type B in kidney may be related to a role in the regulation of peripheral actions of enkephalins and other opiate peptides. There are reports that enkephalin can affect homeostatic mechanisms related to blood pressure, water balance and transport of Na.

The role of type B enzyme in brain and in peripheral organs such as kidney is unknown except that it is capable of degrading enkephalins among other peptides. In the case of the kidney secretion of this enzyme may be related to inactivation of circulatory forms of enkephalins. Chromaffin granules of adrenal medulla are a rich source of enkephalin or enkephalin precursors and released along with other components such as catecholamines (Lewis et al., 1980). Studies on enkephalin breakdown paved the way for the discovery of a novel dipeptidyl carboxypeptidase and this example serves to provide incentive to investigate breakdown of other biologically active peptides and to ascertain the role of intracellular proteinases or peptidases in regulation in terms of conversion from inactive forms, or inactivation.

ACKNOWLEDGEMENTS

This investigation was supported in part by a grant from U.S.P.H.S. # NS 12578 (NM) and from the J. Stefan Institute, Ljubljana (AS) and Health Research Council of New York # HRC-9-013 (MB).

REFERENCES

Aronson, N.N., and A.J. Barrett (1978). Biochem.J., 17, 759-765.
Austen, B.M., and D.G. Smyth (1977). Biochem.Biophys.Res.Comm., 77, 86-94.
Bajusz, S., E. Barabas, E. Szell, and D. Bagdy (1975). Peptides and Chemistry, Structure, Biology. 603-608. Ann Arbor, Mich. USA.
Barrett, A.J. (1977). In A.J. Barrett (Ed.), Proteinases in Mammalian Cells and Tissues. Elsevier North-Holland Biochemical Press. pp. 181-208.
Benuck, M., and N. Marks (1979). Biochem.Biophys.Res.Comm., 88, 215-221.
Benuck, M., and N. Marks (1980). Biochem.Biophys.Res.Comm., 95, 822-828.
Beckner, C.F., and R.M. Caprioli (1980). Biochem.Biophys.Res.Comm., 93, 1290-1296.
Benuck, M., A. Grynbaum, T.B. Cooper, and N. Marks (1978). Neurosci.Lett., 10, 8-9.
Boehme, H., R. Kosecki, and N. Marks (1978). Brain Res.Bull., 3, 6-15.
Burbach, J.P.H., J.G. Loeber, J. Verhoet, V.M. Wiegant, E.R. de Kloet, and D. de Wied (1980). Nature, 283, 96-97.
de Wied, D., J.M. von Ree, and H.M. Greven (1980). Life Sci., 26, 1575-1579.
Herbert, E., M.Budarf, M.Phillipps, R. Rosa, P. Policastro, and E. Oates (1980). Ann.N.Y.Acad.Sci., 343, 79-94.
Kirschke, H., J. Langner, S. Riemann, B. Wiederanders, S. Ansorge, and P.Bohley (1980). Protein Degradation in Health and Disease. CIBA Fdn. Symp. 75, 15-25. Excerpta Medica.
Lewis, R.V., A.V. Stern, S. Kiman, S.Stein, and S. Udenfriend (1980). Proc.Nat. Acad.Sci., 77, 5018-5020.
Mac Gregor, R.R., J.W. Hamilton, N.G. Kent, E. Shofstall, and D.V. Cohn (1979). J.Biol.Chem., 254, 4428-4433.

Mains, R.E., and B.A. Eipper (1980). Ann.N.Y.Acad.Sci., 343, 95–110.

Malfroy, B., J.P. Swerts, A.Gugon, B.P. Rogues, and J.C. Schwartz (1978). Nature, 276, 523–528.

Marks, N. (1978). Frontiers in Neuroendocrinology, 5, 329–377.

Marks, N., A. Suhar, and M. Benuck (1980). Adv.Biochem.Psychopharm., 22, 205–217.

Olsen, G.A., R.D. Olsen, A.J. Kastin, and D.H. Coy (1979). Neuroscience and Biobehavioral Reviews, 3, 285–299.

Steiner, D.F., P.S. Quinn, S.J. Chan, J. Marsh, and H.S. Tayer (1980). Ann.N.Y. Acad.Sci., 343, 1–16.

Suhar, A., and N. Marks (1979). Eur.J.Biochem., 101, 23–30.

Tager, H., M. Hohenboken, J. Markese, and R.J. Dinerstein (1977). Proc.Natl. Acad.Sci.(USA), 77, 6229–6233.

MALIGNANCY AND TUMOR PROTEINASES: EFFECTS OF PROTEINASE INHIBITORS

T. Giraldi and G. Sava

Istituto di Farmacologia, Universita di Trieste, I—34100 Trieste, Italy

ABSTRACT

Tumor neutral proteinases have been suggested to be responsible for in vivo malignancy (invasiveness and metastasizing capacity) of solid malignant tumors. The in vivo administration of proteinase inhibitors has been reported to reduce chemical carcinogenesis and the growth of transplantable tumors in animals. These data are summarized and discussed. Furthermore, the results obtained in the laboratory of the authors studying the effects of neutral proteinase inhibitors on metastasis formation in mice, are presented. Substances capable to inhibit in vitro neutral proteinases, including inhibitors of natural and synthetic origin, as well as non steroidal antiinflammatory agents and antineoplastic drugs, selectively reduce lung metastasis formation in mice bearing Lewis lung carcinoma. It appears that the use of these agents is therapeutically useful, since the most active compound examined is curative in a large proportion of mice bearing Lewis lung carcinoma and B16 melanoma treated pre- and peroperatively with this agent and subsequent surgical tumor removal.

KEYWORDS

Tumor malignancy; tumor invasion; tumor metastasis; tumor proteinases; proteinase inhibitors; tumor therapy.

INTRODUCTION

The capacity of solid malignant tumors to locally infiltrate healthy sorrounding tissues and to produce systemic metastases is currently limiting the possibilities to cure solid neoplasms in man using surgery and radiotherapy. Local infiltration may be so widespread that radical tumor eradication is impossible. On the other hand, relapses are encountered because of the appearance of systemic metastases also after succesful eradication of the primary neoplastic lesion. In the case of the most common solid tumors metastasizing in man, little or no advantage is obtained by the use of adjuvant chemotherapy with antineoplastic drugs, because of poor responsiveness to these agents.

Proteolytic enzymes have been suggested to partecipate to, or to be responsible for tumor malignancy; proteinase inhibitors have been accordingly examined for their effects on tumor cells and tumor bearing animals. The aim of this paper is therefore that of summarizing the relevant findings obtained studying animals and human tumors, as well as that of reporting the studies performed in the laboratory of the authors on proteinase inhibitors and their antimetastatic effects in mice.

PROTEINASES AND TUMOR MALIGNANCY

Animal and human solid malignant tumors have been examined for their content and secretion of proteinases, in comparison with the corresponding normal tissues and benign tumors. From these studies, summarized in Table 1, it appears that an increased proteolytic activity has been observed in numerous different human and animal neoplasms. In addition to these findings, a neutral proteinase different from collagenase has been recently isolated from animal metastasizing tumors. This

TABLE 1   Activity of Proteinases Increased in Solid Malignant Tumors

Enzyme	Source	Reference
Collagenase	Extracts of human bladder cancer	Wirl and Frick, 1979
	Various human skin tumors including melanoma	Bauer and others, 1977; Abramson and others, 1975; Hashimoto and others, 1973; Yamanishi, Dabbous and Hashimoto, 1972; Yamanishi and others, 1973; Tane and others, 1978; Heuson and others, 1975
	Rabbit VX-2 carcinoma	Dabbous and others, 1977a; Dabbous, Roberts and Brinkley, 1977b; Harris, Faulkner and Wood, 1972; Mc Croskery, Richards and Harris, 1975
	Cell cultures of mouse B16 melanoma	Sauk and Witkop, 1978
	Tissue cultures of rat prostate carcinoma	Huang, Wu and Abramson, 1979
	Tissue cultures of rat Tawa sarcoma	Sakaki and others, 1976
	Extracts of methylcholantrene induced fibrosarcomas in mice	Labrosse and Liener, 1978
Collagenase in active and latent form	Extracts of rabbit V2 carcinoma	Steven and Itzhaki, 1977
Cathepsin B	Tissue culture of human mammary carcinoma	Poole and others, 1978; Rechlies and others, 1980

	Tumor cell surface of several transplanted rodent tumors	Sylvén, Snellman and Straüli, 1974
Cathepsin B, acid proteolytic activity	Interstitial fluid of several transplanted mose tumors	Sylvén and Bois, 1960; Sylvén, 1967, 1968a, 1973
Elastase	Extracts of human mammary carcinoma	Hornebeck and others, 1977
Plasminogen activator plus direct fibrinolysin	Rat ovarian and breast carcinoma cell cultures	Wu and others, 1975; Schultz, Wu and Yunis, 1975.
Plasminogen activator	V2, V7 carcinoma rabbit cell homogenates	Kodama and Tanaka, 1978
	Extracts of human melanoma	Fraki, Niemien and Hopsu-Havu, 1979
	Ascitic fluid, human ovarian carcinoma	Svanberg and Åstead, 1975
	Cell cultures of human neuroblastoma	Wachsman and Biedler, 1974
	Cell cultures of human brain tumors	Tucker and others, 1978
	Extracts of human lung tumors	Markus and others, 1980
	Cell cultures of various human tumors	Nagy, Ban and Brdar, 1977

proteinase is present in active and latent form activable by trypsin treatment, and selectively degrades collagen type IV of basement membrane (Liotta and others, 1980; Garbisa and others, 1980).

The high level of cathepsin B, collagenase and neutral proteinase degrading collagen type IV appears particularly important to account for malignant tumor behaviour. Indeed, the action of these enzymes degrades the two major components of extracellular matrices, namely collagen (Harris and Cartwright, 1977; Liotta and others, 1980; Garbisa and others, 1980) and proteoglycans (Morrison, Barrett and Dingle, 1973; Etherington and Evans, 1977): infiltration of tumor cells into normal adjacent tissue may thus ensue. The action of tumor proteinases may also be responsible for metastasis formation. Collagenase and cathepsin B have been reported to cause tumor cell detachement (Klebe, 1974; Sylvén 1968b). Furthermore, the action of these enzymes and in particular that of the neutral proteinase degrading collagen type IV of basement membrane, may cause vascular invasion by the tumor (Salsbury, 1975). The combined occurrence of tumor cell detachement from primary tumor mass and blood vessel invasion may thus lead to tumor cell entrance into the blood stream (or lymphatic system in the case of lymphatic drainage of detached tumor cells) and to metastasis formation.

The role of the other proteinases listed in Table 1 appears less clear. Plasminogen activator and fibrinolysin may alter the fibrinolytic system of the host. Host fibrin-fibrinolytic system is involved in the process of metastasis formation, as reviewed by Peterson (1977), but this topic falls outside the aims of the present paper and will be not considered in detail.

Lastly, it is noteworthy that activation of latent proteinases (i.e. procollagenase) and consequent amplification of their action may be caused in vivo by the activity of several other proteinases, as already observed in vitro (Vaes, 1980).

PROTEINASE INHIBITORS AND TUMOR MALIGANCY

The suggested role of the increased proteolytic activity of tumor cells for causing tumor malignant behaviour is further supported by the following findings.

Cartilage is a tissue which is naturally resistant to invasion by blood vessels (Eisenstein and others, 1973) and tumors (Kuettner, Pauli and Soble, 1978). A factor considered responsible for resistance to invasion has been extracted from car-

TABLE 2   Inhibitory Effects of Proteinase Inhibitors on
Tumor Development in Animals

Inhibitor	Process inhibited	Reference
Aprotinin	Growth of fibrosarcoma (hamster) and mammary carcinoma (mouse)	Latner, Longstaff and Turner, 1974
Aprotinin	Growth of i.p. Walker 256 carcinosarcoma (rat)	Thompson and others, 1977
Aprotinin	Lewis lung carcinoma and hepatoma 22 (mouse)	Lage, Diaz and Gonzales, 1978
Aprotinin, EACA	Growth of Murphy-Sturm lymphosarcoma (rat)	Back and Steger, 1976
α-2-macroglobulin	Growth of Dana-435 (rat)	Von Ardenne and Chaplain, 1973
Cystein	Growth of malignant thymoma (mouse)	Campbell, Read and Radden, 1974; Matthews, Sardovia and Milo, 1977
EACA, Soybean trypsin inhibitor	Growth of Ehrlich asites carcinoma (mouse)	Back and Leblanc, 1977; Verloes and others, 1978
TLCK, TPCK, TAME, leupeptin, N,N-dimethyl-amino-p-(p'-guanidino-benzoyloxy)benzyl-carbonyoxyglycolate	in vivo carcinogenic effects of 7,12-dimethyl-benz(a)antracene	Troll, Klassen and Janoff, 1970; Hozumi and others, 1972; Pietras, 1978; Yamamura and others, 1978

tilage. It is a cationic protein having a molecular weight of 11,000 dalton (Kuettner and others, 1976), which inhibits in vitro trypsin, chymotrypsin, plasmin (Rifkin and Crowe, 1977) and collagenase from normal skin and human tumors (Kuettner and others, 1977). A similar inhibitor has been isolated from aorta; when administered in vivo to mice bearing transplantable mammary carcinoma and fibrosarcoma, it significantly inhibited tumor growth (Eisenstein and others, 1978).

Other proteinase inhibitors of physiological, natural and synthetic origin have been also tested for their effects on tumor development in animals. Their reported effects, consisting of the inhibition of chemical carcinogenesis and of the growth of various transplantable neoplasms after in vivo administration, are summarized in Table 2.

On the basis of the considerations presented in the previous paragraphs, proteinase inhibitors should be expected to cause inhibition of metastasis formation. This aspect of the pharmacological properties of proteinase inhibitors is being

TABLE 3   Proteinase Inhibitors and Selective Antimetastatic Effects in Mice Bearing Lewis lung carcinoma

Agent	Enzymes inhibited	Activity	References
Aurintricarboxylic acid	Broad spectrum non-specific inhibitor (Bina-Stein and Tritton, 1976)	High	Giraldi and others, 1980b
Chloroquine	Cathepsin B and collagenase (Poole and others, 1977; Cowey and Whitehouse, 1966)	Moderate	Giraldi and others, 1980b
Indomethacin	Collagenase (Wojtecka-Lukasik and Dancewicz, 1974; Wirl, 1977)	Moderate	Giraldi and others, 1980b
Phenylbutazone	Collagenase (Wojtecka-Lukasik and Dancewicz, 1974; Wirl, 1977)	High	Giraldi and others, 1980b
Aprotinin (Trasylol^R)	Broad spectrum inhibitor (Trautschold and others, 1967) active on trypsin, chymotrypsin, kallikrein, plasmin, elastase and cathepsin G (Starkey, 1977)	Moderate	Giraldi, Nisi and Sava, 1977a
N-diazoacetyl-glycinamide	Elastase, chymotrypsin-like neutral proteinase (histonase) (Kopitar and others, 1977a)	Very high	Giraldi, Nisi and Sava, 1977c; Giraldi and others, 1979
p-carboxamidophenyl-3,3-dimethyl-triazene, p-(3,3-	Dimethyltriazenes inhibit elastase, chymotrypsin-like neutral proteinase (histonase)	Very high	Giraldi and others, 1978, 1980a; Sava and others, 1979

dimethyl-1-triaze-no)benzoic acid po-tassium salt (DM-COOK)	(Kopitar and others, 1977b), papain and cathepsin B (unpublished results)		
Leucocyte intra-cellular inhibitor of neutral protei-nases (LNPI)	Elastase, chymotrypsin-like neutral proteinase (histonase) (Kopitar and Lebez, 1975; Kopitar and others, 1977a)	Moderate	Giraldi, Kopitar and Sava, 1977b
Spleen intracel-lular inhibitor of neutral proteinases (SNPI-1)	Trypsin, chymotrypsin-like neutral proteinase (histonase) cathepsin B and H (Brzin and others, 1977; Kopitar and others, 1978)	Moderate	Giraldi and others, 1980b
Spleen intracel-lular inhibitor of neutral proteinases (SNPI-2)	Trypsin, chymotrypsin-like neutral proteinase (his-tonase), α-chymotrypsin (Brzin and others, 1977; Kopitar and others, 1978)	Moderate	Giraldi and others, 1980b
(+)1,2-di(3,5-dioxopiperazin-1-yl)propane (ICRF 159, Razoxane)	Inhibits collagen peptidase activity (assayed using p-phenylazo-benzyoxycarbonyl-L-prolyl-L-leucyl-L-prolyl-D-arginine as substrate) of human tumor homogenates (Bog-gust and Mc Gauley, 1978)	Very high	Bakovsky, 1976; Gi-raldi and others, 1980d
BCNU, CCNU	Elastase and chymotrypsin (by production of iso-cyanates) (Babson and others, 1977	Very high (also on primary tumor)	See Fig. 1.
Chlorozotocin	Does not produce appreciably isocyanates (Kann, 1978)	Moderate (also on primary tumor)	See Fig. 1.

Moderate: weight of spontaneous lung metastases significantly reduced to about 30–50%.
High: weight of spontaneous metastases significantly reduced to 20%; some animals free of metasatases.
Very high: weight of spontaneous metastases significantly reduced to less than 5%; some animals free of metastases.
The growth of subcutaneous tumor is not reduced significantly.

investigated since some years in the laboratory of the authors. The results obtained on pulmonary metastasis formation in mice bearing Lewis lung carcinoma are summarized in Table 3. The substances examined include proteinase inhibitors of physiological and natural origin, as well as non steroidal anti inflammatory agents and antitumor drugs. All of these agents have in common the property to inhibit

in vitro neutral proteinases at in vivo attainable concentrations. When administered in vivo to mice bearing subcutaneous tumors, these agents depress to various
degrees the formation of spontaneous lung metastases, without significantly affecting the growth of subcutaneous primary tumor. For dimethyltriazenes and N-diazo-
acetylglycinamide, which appear to be the most potent agents in the series examined so far, the mechanism of the antimetastatic action has been examined in detail.
From these investigations, it clearly results that the depression of metastasis
formation is exclusively caused by inhibition of tumor cell entrance into the blood
stream (Giraldi and others, 1980c, 1980d; Sava and others, 1979). These findings
are consistent with the expected effects of inhibition of tumor proteinases, as
previously commented.

In agreement with these observations is the fact that ICRF 159 (Razoxane), a selectively antimetastatic drug studied by Hellmann and others (Bakowsky, 1976),
inhibits in vitro tumor collagen peptidase (see Table 3). Nitrosourea derivatives,
which are potent antineoplastic drugs (Connors, 1979), do not exert selective antimetastatic effects, since they depress also primary tumor growth and the formation
of artificial lung metastases (Fig. 1). It is noteworthy, however, that these
drugs decompose generating a metabolite (isocyanate) which is a proteinase inhibitor; chlorozotocin, which is the least active on tumors in the series of three
analogues tested, at the same time does not appreciably decompose to isocyanate
(see Table 3).

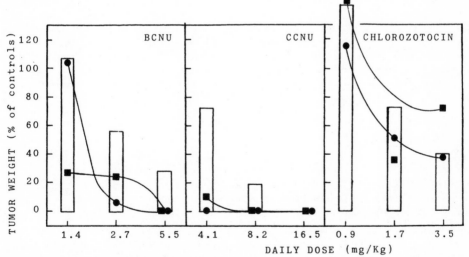

Animals implanted subcutaneously with the tumor were treated daily for the following 14 days. Primary tumor growth (▱) was evaluated on day 15 and the lungs were
examined for spontaneous metastases (●) at sacrifice on day 21. For artificial
metastases (■), the animals were treated daily for 8 days following i.v. injection of tumor cells: sacrifice and metastasis determination were performed on
day 15.

Fig. 1.   Effects of three nitrosoureas on subcutaneous tumor
growth and on the formation of pulmonary metastases.

Since dimethyltriazenes result to prevent metastasis formation in mice bearing Lewis lung carcinoma, they have been examined as prophylactic adjuvants to surgical tumor removal in mice bearing the same tumor and B16 melanoma. As illustrated

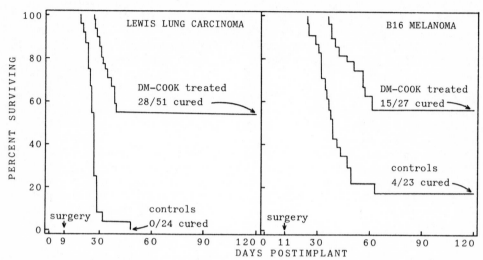

The treatment was performed i.p. daily with 50mg/Kg/day of DM-COOK on days 7-9 and 4-11 for Lewis lung carcinoma and B16 melanoma respectively. Animals having local tumor regrowth at the surgical site have been excluded from the plot: all of the animals died because of metastases as evidenced by necroscopic examination.

Fig. 2. Survival of Lewis lung carcinoma and B16 melanoma bearing mice treated with p-(3,3-dimethyl-1-triazeno)benzoic acid potassium salt (DM-COOK), and surgical removal of primary tumor.

in Fig. 2, surgery alone is not curative in the case of Lewis lung carcinoma, and cures only 17% in the case of B16 melanoma. The pre- and peroperative treatment with DM-COOK causes a high percentage of long term survivors, i.e. cures, for both tumors (55 and 56% respectively).

CONCLUSIONS

Evidence suggesting that an increased production of proteinases by tumor cells may be responsible for their malignant in vivo behaviour is being accumulating during the last years. These data are in general supported by the concomitant observations that in vivo adminitration of neutral proteinase inhibitors depress the malignant growth and behaviour of tumors in laboratory animals.

The transformation of in vitro cell cultures by oncogenic viruses or chemical carcinogens is similarly causing an increased proteolytic activity of the cells; the properties characteristic of the transformed cell phenotype are reverted by proteinase inhibitors (Roblin, Chou and Black, 1975; Barret and others, 1977). Although in vitro cell cultures have not been considered in the present paper, these results are in general supporting the presented data.

Experimental evidence has been obatined in our laboratory that the in vivo administration of agents capable of inhibiting in vitro neutral proteinases reduces, and in the case of the most active substances virtually abolishes, the formation of systemic metastases in mice. Moreover, one of these agents proved to be curative in a large proportion of animals bearing two solid metastasizing tumors, when used as pre- and intraoperative prophylactic adjuvant to surgical removal of primary tumor. Evidence on actual inhibition of tumor proteinases after in vivo administration of therapeutically active proteinase inhibitors, and identification of the enzyme(s) inhibited is presently lacking. Further investigation on these aspects is needed and is partially in progress in our laboratory. The relevant results may contribute to increase the knowledge on the biochemical mechanism of tumor malignancy. The presented results also appear to encourage further studies on the use of proteinase inhibitors in experimental systems as adjuvants to other forms of tumor therapy, which may eventually lead to their future use in man.

ACKNOWLEDGEMENT

This work was supported by Italian National Research Council - Special Project "Control of Neoplastic Growth", Contract n°80.01562.96.

REFERENCES

Abramson, M., R. W. Schilling, C. C. Huang, and R. G. Salome (1975). Ann. Otol., 84, 158-163.

von Ardenne M., and R. A. Chaplain (1973). Experientia, 29, 1271-1272.

Babson, J. R., D. J. Reed, and M. A. Sinkey (1977). Biochemistry, 16, 1584-1589.

Back, N., and R. Steger (1976). Europ. J. Pharmacol., 38, 313-319.

Back, N., and P. Leblanc (1977). Europ. J. Cancer, 13, 947-950.

Bakowski, M. T. (1976). Cancer Treat. Rev., 3, 95-107.

Barrett, J. C., B. D. Crawford, D. L. Grady, L. D. Hester, P. A. Jones, W. F. Benedict, and P. O. P. Ts'o (1977). Cancer Res., 37, 3815-3823.

Bauer, E. A., J. M. Gordon, M. E. Reddick, and A. Z. Eisen (1977). J. Invest. Dermatol., 69, 363-367.

Bina-Stein, M., and T. R. Tritton (1976). Mol. Pharmacol., 12, 191-193.

Boggust, W. A., and H. Gauley (1978). Br. J. Cancer, 38, 329-334.

Brzin, J., M. Kopitar, and V. Turk (1977). Acta Biol. Med. Ger., 36, 1883-1886.

Campbell, N. R., P. C. Reade, and B. G. Radden (1974). Nature, 251, 158-159.

Connors, T. A. (1979). Cancer Chemotherapy. H. M. Pinedo Ed., Excerpta Medica, Amsterdam - Oxford.

Cowey, F. K., and M. W. Whitehouse (1966). Biochem. Pharmacol., 15, 1071-1084.

Dabbous, M. K., C. Sobhy, A. N. Roberts, and Sr. B. Brinkley (1977a). Mol. Cell. Biochemistry, 16, 37-42.

Dabbous, M. K., A. N. Roberts, and Sr. B. Brinkley (1977b). Cancer Res., 37, 3537-3544.

Eisenstein, R., N. Sorgente, L. W. Soble, A. Miller, and K. E. Kuettner (1973). Amer. J. Phatol., 73, 765-772.

Eisenstein, R., B. Schumacher, C. Meineke, B. Matijevitch, and K. Kuettner, (1978). Amer. J. Pathol., 91, 1-8.

Etherington, D. J., and P. J. Evans (1977). Acta Biol. Med. Ger., 36, 1555-

Fraki, J. E., S. Nieminen, and V. K. Hopsu-Havu (1979). J. Cutaneous Pathol., 6, 195-200.

Garbisa, S., K. Kinska, K. Tryggvason, C. Foltz, and L. A. Liotta (1980). Cancer Letters, 9, 359-366.

Giraldi, T., C. Nisi, and G. Sava (1977a). Europ. J. Cancer, 13, 1321-1323.

Giraldi, T., M. Kopitar and G. Sava (1977b). Cancer Res., 37, 3834-3835.

Giraldi, T., C. Nisi, and G. Sava (1977c). J. Natl. Cancer Inst., 58, 1129-1130.

Giraldi, T., P. J. Houghton, D. M. Taylor, and C. Nisi (1978). Cancer Terat. Rep., 62, 721-725.

Giraldi, T., A. M. Guarino, C. Nisi, and L. Baldini (1979). Europ. J. Cancer, 15, 603-607.

Giraldi, T., A. M. Guarino, C. Nisi, and G. Sava (1980a). Pharmacol. Res. Commun., 12, 1-11.

Giraldi, T., G. Sava, M. Kopitar, J. Brzin, and V. Turk (1980b). Europ. J. Cancer, 16, 449-454.

Giraldi, T., G. Sava, R. Cuman, C. Nisi, and L. Lassiani (1980d). Manuscripst submitted to Cancer Res.

Giraldi, T., G. Sava, and C. Nisi (1980c). Europ. J. Cancer, 16, 87-92.

Harris, E. D., C. S. Faulkner, and S. Wood, Jr (1972). Biochem. Biophys. Res. Commun., 48, 1247-1253.

Harris, E. D., E. C. Cartwright (1977). Proteinases in Mammalian Cells and Tissues. A. J. Barrett, Ed., North-Holland Publishing Co., Amsterdam.

Hashimoto, K., Y. Yamanishi, E. Maeyens, M. K. Dabbous, and T. Kanzaki (1973). Cancer Res., 33, 2790-2801.

Heuson, J. C., J. L. Pasteels, N. Legros, J. Heuson-Stiennon, and G. Leclercq (1975). Cancer Res., 35, 2039-2048.

Hornebeck, W., J. C. Derouette, D. Brechemier, J. J. Adnet, and L. Robert (1977). Biomedicine, 26, 48-52.

Hozumi, M., M. Ogawa, T. Sugimura, T. Takeuchi, and H. Umezawa (1972). Cancer Res., 32, 1725-1728.

Huang, C. C., C. H. Wu, and M. Abramson (1979). Biochim. Biophys. Acta, 570, 149-156.

Kann, H. E. Jr., (1978). Cancer Res., 38, 2363-2366.

Klebe, R. J., (1974). Nature, 250, 248.

Kodama, Y., and K. Tanaka (1978). Acta Pathol. Jap., 28, 279-286.

Kopitar, M., and D. Lebez (1975). Europ. J. Biochem., 56, 571-581.

Kopitar, M., J. Babnik, I. Kregar, and A. Suhar (1977a). Movement, Metabolism and Bactericidal Mechanism of Phagocytes. F. Rossi, P. Patriarca, and D. Romeo, Eds., Piccin, Padova.

Kopitar, M., A. Suhar, T. Giraldi, and V. Turk (1977b). Acta Biol. Med. Ger., 36, 1863-1871.

Kopitar, M., J. Brzin, T. Zvonar, P. Locnikar, I. Kregar, and V. Turk (1978). FEBS Letters, 91, 355-359.

Kuettner, K. E., J. Hiti, R. Eisenstein, and E. Harper (1976). Biochem. Biophys. Res. Commun., 72, 40-46.

Kuettner, K. E., L. Soble, R. L. Croxen, B. Marczynska, J. Hiti, and E. Harper (1977). Science, 196, 653-654.

Kuettner, K. E., B. U. Pauli, and L. Soble (1978). Cancer Res., 38, 277-287.

Labrosse, K. R., and I. E. Lienier (1978). Mol. Cell. Biochemistry, 19, 181-189.

Lage, A., J. W. Diaz, and I. Gonzales (1978). Neoplasma, 25, 257-259.

Latner, A. L., E. Longstaff, and G. A. Turner (1974). Br. J. Cancer, 30, 60-67.

Liotta, L. A., K. Tryggvason, S. Garbisa, I. Hart, C. M. Foltz, and S. Shafie (1980). Nature, 284, 67-68.

Markus, G., H. Takita, S. M. Camiolo, J. G. Corasanti, J. L. Evers, and G. H. Hobika (1980). Cancer Res., 40, 841-848.

Matthews, R. H., M. R. Sardovia, and G. E. Milo (1977). Cancer Biochem. Biophys., 2, 65-69.

McCroskery, P. A., J. F. Richards, and E. D. Harris (1975). Biochem. J., 152, 131-142.

Morrison, R. I. G., A. J. Barrett, and J. T. Dingle (1973). Biochim. Biophys. Acta, 302, 411-

Nagy, B., J. Ban, and B. Brdar (1977). Int. J. Cancer, 19, 614-620.

Peterson, H. I. (1977). Cancer Treat. Rev., 4, 231-217.

Pietras, R. J. (1978). Cancer Res., 38, 1019-1030.

Poole, B., S. Ohkuma, and M. J. Warburton (1977). Acta Biol. Med. Ger., 36, 1777-1788.

Poole, A. R., K. J. Tiltman, A. D. Recklies, and T. A. M. Stoker (1978). Nature, 273, 545-547.

Recklies, A. D., K. J. Tiltman, T. A. M. Stoker, and A. R. Poole (1980). Cancer Res., 40, 550-556.

Rifkin, D. B., and R. M. Crowe (1977). Hoppe-Seyler's Z. Physiol. Chem., 358, 1525-1531.

Roblin, R., I. N. Chou, and P. H. Black (1975). Advances in Cancer Research, 22, 203-260.

Sakaki, T., A. Fujita, N. Wakumoto, and S. Murase (1976). Gann, 67, 67-73.

Salsbury, A. J. (1975). Cancer Treat. Rev., 2, 55-72.

Sauk, J. J., and C. J. Witkop, Jr. (1978). Biochim. Biophys. Res. Commun., 83, 144-150.

Sava, G., T. Giraldi, L. Lassiani, and C. Nisi (1979). Cancer Treat. Rep., 63, 93-98.

Schultz, D. R., M. C. Wu, and A. A. Yunis (1975). Exper. Cell Res., 96, 47-57.

Starkey, P. M. (1977). Proteinases in Mammalian Cells and Tissues. A. J. Barrett, Ed., North-Holland Publishing Co., Amsterdam.

Steven, F. S., and S. Itzhaki (1977). Biochim. Biophys. Acta, 496, 241-246.

Svanberg, L., and B. Astedt (1975). Cancer, 35, 1382-1387.

Sylvén, B., and I. Bois (1960). Cancer Res., 20, 831-836.

Sylvén, B. (1967). Some Factors Relating to the Invasiveness and Destructiveness of Solid Malignant Tumors. UICC Monograph Series, Springer Verlag, Berlin.

Sylvén, B. (1968a). Europ. J. Cancer, 4, 463-474.

Sylvén, B. (1968b). Europ. J. Cancer, 4, 559-

Sylvén, B. (1973). Chemotherapy of Cancer Dissemination and Metastasis. S. Garattini, and G. Franchi, Eds., Raven Press, New York.

Sylvén, B., O. Snellman, and P. Strauli (1974). Virchows Arch. B Cell. Path., 17, 97-112.

Tane, N., K. Hashimoto, T. Kanzaki, and H. Ohyama (1978). J. Biochem., 84, 1171-1176.

Thompson, A. W., R. G. P. Pugh-Humphreys, C. H. W. Horne, and D. J. Tweedie (1977). Br. J. Cancer, 35, 454-460.

Trautschold, I., E. Werle, and G. Zickgraf-Rudel (1967). Biochem. Pharmacol., 16, 59-72.

Troll, W., A. Klassen, and A. Janoff (1970). Science, 169, 1211-1213.

Tucker, W. S., W. M. Kirsch, A. Martinez-Hernandez, and L. M. Fink (1978). Cancer Res., 38, 297- 302.

Vaes, G. (1980). Collagenase in Normals and Pathological Connective Tissues. D. E. Wolley, and J. M. Evanson, Eds.,John Wiley & Sons Ltd.

Verloes, R., G. Atassi, P. Dumont, and L. Kanarek (1978). Europ. J. Cancer, 14, 23, 31.

Wachsman, J. T., and J. L. Biedler (1974). Exper. Cell Res., 86, 264-268.

Wirl, G. (1977). Intracellular Protein Catabolism II. V. Turk, and N. Marks, Eds., Plenum Press, New York.

Wirl, G., and J. Frick (1979). Urolog. Res., 7, 103-108.

Wojtecka-Lukasik, E., and A. M. Dancewicz (1974). Biochem. Pharmacol., 23, 2077-2081.

Wu, M. C., D. R. Schultz, G. K. Arimura, M. A. Gross, and A. A. Yunis (1975). Exper. Cell Res., 96, 37-46.

Yamamura, M., N. Nakamura, Y. Fukui, C. Takamura, M. Yamamoto, Y. Minato, Y. Tamura, and S. Fujii (1978). Gann, 69, 749-752.

Yamanishi, Y., M. K. Dabbous, and K. Hashimoto (1972). Cancer Res., 32, 2551-2560.

Yamanishi, Y., E. Maeyens, M. K. Dabbous, H. Ohyama, and K. Hashimoto (1973). Cancer Res., 33, 2507-2512.

PROTEOLYSIS AND CATABOLITE INACTIVATION IN YEAST

M. Müller and H. Holzer

Inst. Toxikol. Biochem. Gesellschaft für Strahlen– und Umweltforschung München, D–8042 Neuherberg, and Biochem. Inst. Univ. Freiburg, D–7800 Freiburg i.Br., Germany

ABSTRACT

Catabolite inactivation is the glucose-induced, rapid inactivation of a group of enzymes in glucose-starved yeast: galactose uptake system, maltose permease, cytoplasmic malate dehydrogenase, fructose-1,6-bisphosphatase, phosphoenolpyruvate carboxykinase, uridine nucleosidase, and vacuolar aminopeptidase. The involvement of proteolysis in catabolite inactivation is suggested by: (1) the requirement of de novo-protein synthesis for reactivation, (2) the inhibition of the inactivation by chloroquine, which is known to block proteolytic processes in mammalians, (3) the parallel loss of enzyme activity and cross-reacting material. Covalent interconversion (fructose-1,6-bisphosphatase) or protein-protein interaction (galactose-, maltose permease) might be possible trigger reactions for a final proteolytic degradation.

KEYWORDS

Yeast; catabolite inactivation; catabolite modification; proteolysis; regulation; immunoprecipitation; covalent interconversion; chloroquine; metabolism of glucose; futile cycle; reactivation.

CATABOLITE INACTIVATION OF YEAST ENZYMES

In yeast, which has been starved for carbon or grown on a non-sugar carbon source, a marked decrease in the activity of several enzymes is induced by the addition of glucose or metabolically related sugars, like mannose, fructose, sucrose, maltose. These enzymes are listed in Table 1.

The loss of catalytic activity is observed even under conditions which prevent cell growth. Thus, loss of activity cannot be merely due to repression of synthesis and dilution of the preexisting enzyme, but is rather the result of rapid inactivation. This inactivation has been termed "catabolite inactivation" (Holzer, 1976) analogously to the term catabolite repression, which had been introduced for the repression of enzyme synthesis by glucose or its catabolites (Magasanik, 1961).

Most of the enzymes which are subject to catabolite inactivation, are involved in the formation of glucose either by gluconeogenesis or by the uptake of hexoses.

57

TABLE 1   Enzymes which are Subject to Catabolite Inactivation in Yeast

1. Galactozymase Galactose uptake system	Spiegelmann and Reiner, 1947 Matern and Holzer, 1977
2. α-Glucoside(maltose) permease	Robertson and Halvorson, 1957
3. Cytoplasmic malate dehydrogenase	Witt, Kronau, and Holzer, 1966a
4. Fructose-1,6-bisphosphatase	Gancedo, 1971
5. δ-Aminolevulinic acid synthase	Labbe and co-workers, 1972
6. Uroporphyrinogen synthase	Labbe and co-workers, 1972
7. Phosphoenolpyruvate carboxykinase	Haarasilta and Oura, 1975; Gancedo and Schwerzmann, 1976
8. Uridine nucleosidase	Magni and co-workers, 1977
9. Vacuolar aminopeptidase	Frey and Röhm, 1979
10. α-Glucosidase (non-induced)	Frey and Röhm, 1979

Consequently their activity becomes unnecessary when these sugars are supplied from the culture medium, or even harmful when the persistance of their activity would lead to an ATP-splitting futile cycle. The latter may explain the requirement of the rapid catabolite inactivation in addition to catabolite repression which is slow.

INVOLVEMENT OF PROTEOLYSIS IN CATABOLITE INACTIVATION

Inactivation of an enzyme may either be caused by a modification which leaves the enzyme peptide backbone intact or by a partial or complete degradation of the enzyme. A combined mechanism is also conceivable, in which a modified molecule is subsequently cleaved by proteinases. The following section summarizes results of experiments, which point to a participation of proteolysis in catabolite inactivation.

Requirement of de novo-Synthesis for Reactivation

Figure 1 shows the reappearance of the activity of the galactose uptake system after glucose-induced inactivation (Matern, Holzer, 1977). Reactivation of the enzyme after transfer to a glucose-free, galactose-containing medium was completely prevented by cycloheximide suggesting the necessity of de novo-protein synthesis for a reestablishment of the enzyme activity. Similar results with other enzymes which are inactivated by glucose have been obtained either by use of cycloheximide (Gancedo, 1971; Gancedo, Schwerzmann, 1976; Görts, 1969) or of amino acid-auxotrophic mutants (Duntze, Neumann, Holzer, 1968; Gancedo, Schwerzmann, 1976).

In vitro-Sensitivity of Glucose-Inactivated Enzymes towards Proteinases

Jušić, Hinze, and Holzer (1976) incubated the purified mitochondrial and cytoplasmic isoenzymes of yeast malate dehydrogenase with preparations of yeast proteinases A and B. Only the cytoplasmic form, which is subject to catabolite inactivation (Witt, Kronau, Holzer, 1966b), was attacked by the proteinases, whereas the mitochondrial isoenzyme, which is not influenced by glucose, remained stable.

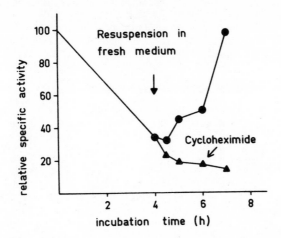

Fig. 1. Reactivation of galactose uptake system after glucose-induced inactivation. After 4 h-incubation with glucose the washed cells were transferred to a galactose-containing optimal medium without (●) or with the addition of cycloheximide (▲). (For details see Matern, Holzer, 1977)

Molano and Gancedo (1974) reported the partial purification of a proteinase from yeast, which inactivated fructose-1,6-bisphosphatase. Several other yeast enzymes, which are not regulated by catabolite inactivation, were not affected by the inactivating factor. Thus regulation by catabolite inactivation and sensitivity towards proteolysis seem to be related.

Inhibition of Catabolite Inactivation by Chloroquine

Chloroquine - a weak base - is known to be an inhibitor of intracellular protein degradation, based on its accumulation in lysosomes with a subsequent raising of the lysosomal pH (Poole, Ohkuma, Warburton, 1977; Wibo, Poole, 1974). Its influence on catabolite inactivation is presented in Table 2 (Lenz, 1980). Catabolite inactivation of all the three enzymes was in fact markedly reduced. In addition, Lenz (1980) demonstrated that chloroquine inhibited the activity of yeast proteinases in vitro and that it blocked protein-turnover in vivo, whereas it did not

TABLE 2   Influence of Chloroquine on Catabolite Inactivation

	Concentration of Chloroquine (mM)					
	0	5	10	20	50	100
Phosphoenolpyruvate carboxykinase	81	78	2	13	1	1
Malate dehydrogenase	37	42	16	17	1	1
Fructose-1,6-bisphosphatase	96	96	92	69	47	35

Values represent % activity lost after 2 h of incubation with glucose

interfere with the glucose uptake or the growth of the cells. One may conclude
from these results, that the chloroquine-induced reduction of catabolite inactivation originates from an inhibition of proteinases. A more conclusive proof, however, would be the demonstration of a selective accumulation of chloroquine in the
yeast vacuole.

Quantitation of Enzyme by Immunoprecipitation during Catabolite Inactivation

Quantitative determination of cross-reacting enzyme by use of specific antibodies
permits distinction between inactivation by modification or by degradation,
provided that the modified enzyme still maintains its immunoreactivity. Phosphoenolpyruvate carboxykinase has been purified from yeast and highly specific antibodies obtained by affinity purification of the antiserum on Sepharose-bound
enzyme (Müller, M., Müller, H., Holzer, 1980). Direct immunoprecipitation of
radioactively labelled phosphoenolpyruvate carboxykinase in crude extracts from
yeast cells harvested at different times during catabolite inactivation is depicted in Fig. 2.

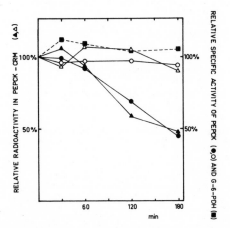

Fig. 2.   Immunoprecipitation of phosphoenolpyruvate carboxykinase
          during catabolite inactivation. At zero time L-$[4,5-^3H]$leucine-
          labelled cells were transferred to a glucose-containing (closed
          symbols) or to a glucose-free medium (open symbols). Non-radio-
          active leucine had been added to both incubations. At the
          times indicated enzymatic activity and immunoprecipitable radio-
          activity in equal amounts of crude extract-protein were deter-
          mined. Quantitative precipitation was verified by the complete
          elimination of enzymatic activity. (▲,△) = radioactivity reco-
          vered from the phosphoenolpyruvate carboxykinase-band (PEPCK)
          of immunoprecipitates separated on SDS-polyacrylamide gels.
          (●,O) = enzymatic activity of phosphoenolpyruvate carboxykinase
          (PEPCK). (■) = enzymatic activity of glucose 6-P dehydrogenase
          (G-6-PDH).

From the loss of immunoprecipitable radioactivity the apparent half-lives of the
enzyme were calculated to be 2.5 h in the presence and 34.5 h in the absence of
glucose. Thus glucose induced a 14-fold increase of the degradation of the enzyme.
The low degradation of the enzyme in the absence of glucose clearly demonstrates
that catabolite inactivation is not the result of repression of enzyme synthesis

together with a rapid rate of degradation. All the cross-reacting material recovered during the course of the catabolite inactivation exhibited the same electrophoretic mobility which indicates that no cleavage product or otherwise modified
antigen had been accumulated. Similar results have been obtained for the catabolite
inactivation of cytoplasmic malate dehydrogenase (Neeff and co-workers, 1978),
vacuolar aminopeptidase (Frey, Röhm, 1979), and fructose-1,6-bisphosphatase
(Funayama, Gancedo, J.M., Gancedo, C., 1980). Although the parallel loss of catalytic activity and immunoreactivity provides strong evidence for a degradative
process during catabolite inactivation, an ultimate proof would be the detection
of cleavage products. Therefore one cannot completely preclude the possibility of
an extensive modification of the enzyme rendering it unrecognizable to the antibody.

PUTATIVE TRIGGER MECHANISMS FOR CATABOLITE INACTIVATION

Protein-Protein Interaction

Matern and Holzer (1977) demonstrated, that addition of glucose to galactose-
grown cells caused an increase in the apparent K_m of the galactose uptake system
from 3.6 to 11 mM galactose, whereas v_{max} remained constant, as shown in Fig. 3.

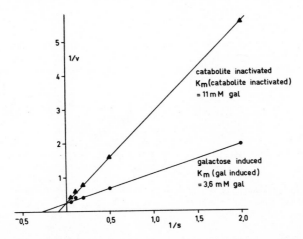

Fig. 3.   Lineweaver-Burk plots of the dependence of the activity of the
          galactose uptake system on galactose concentration after
          treatment (4 h) of galactose-grown yeast cells with (▲) or
          without (●) glucose. v = µMol galactose x min^{-1} x g wet weight^{-1};
          s = mM galactose (gal).

Similar observations for the maltose uptake system were previously reported by
Görts (1969). Thus in these cases glucose induces a modification - "catabolite
modification" - of the enzymes leading to a decrease in the affinity for their
substrates. The dependence of reactivation, i.e. recovery of the high affinity,
on de novo-protein synthesis (Fig. 1) suggests a participation of proteolysis even
in catabolite modification. Therefore a hypothetical modifying protein may be discussed in analogy to α-lactalbumin (Hill and co-workers, 1972). A possible
sequence of events is depicted in Fig. 4. The proteolytic degradation of the modifying protein would be triggered by its dissociation from the permease.

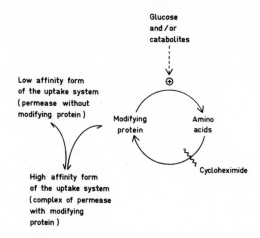

Fig. 4. Hypothetical mechanism for the glucose-effected
         "catabolite modification" of the galactose and
         maltose uptake system, respectively.

Covalent Interconversion

Figure 5 shows the glucose-induced inactivation of fructose-1,6-bisphosphatase
in the yeast strain Saccharomyces cerevisiae M_1 (Lenz, Holzer, 1980). The enzyme
activity very rapidly dropped to less than 40% of the initial value within 3 min
followed by a slow additional decrease within half an hour. In contrast, inactivation of phosphoenolpyruvate carboxykinase and malate dehydrogenase proceeded to a
much lesser extent and at a constant rate under the same experimental conditions
(Lenz, 1980). When transferred to a glucose-free medium 3 min after the beginning
of the inactivation, activity of fructose-1,6-bisphosphatase was completely recovered even in the presence of cycloheximide (Fig. 5). However, after 30 min of
glucose-induced inactivation no considerable reactivation could be observed.
Therefore one has to distinguish between a "short term" inactivation of fructose-
1,6-bisphosphatase, which is reversible without de novo-protein synthesis and a
"long term" inactivation, which presumably leads to degradation of the enzyme as
mentioned above for other enzymes which are subject to catabolite inactivation.
Reactivation after "short term" inactivation was shown to be an ATP-consuming
process (Lenz, Holzer, 1980). This finding might point to an interconversion by
enzyme-catalyzed covalent modification. In fact, the activities of fructose-1,6-
bisphosphatase in crude extracts prepared before or after 3 min of incubation with
glucose showed a different response to the cations Mg^{2+} and Mn^{2+} (Table 3). This
result is consistent with the previous findings, that the Mg^{2+}-dependent activity
of glutamine synthetase in Escherichia coli is rapidly lost by enzyme-catalyzed
adenylation after addition of NH_4^+ to cells grown on other nitrogen sources, in
contrast to the Mn^{2+}-dependent γ-glutamyltransferase activity of the enzyme
(Mecke, Holzer, 1966).

Very recently Tortora and co-workers (1980) followed the fate of fructose-1,6-bis-
phosphatase during "short term" and "long term" inactivation using specific
antibodies. During the first 5 to 10 min after addition of glucose the cross-reacting material did not decrease in contrast to the marked drop of catalytic activity.
This lag-phase was followed by a loss of about 90% of the original amount of
immunoreactive enzyme during the subsequent 50 min. These findings strongly

support the idea, that the "short term" inactivation is caused by covalent interconversion, whereas the "long term" inactivation is the result of proteolytic
degradation.

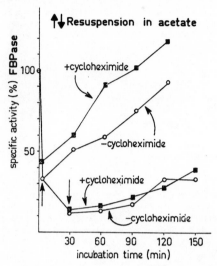

Fig. 5.   Reactivation of fructose-1,6-bisphosphatase activity after
          inactivation by glucose in phosphate buffer containing 2%
          acetate. For experimental details see Lenz and Holzer (1980).

TABLE 3   Effect of Mg^{2+} and Mn^{2+} on the Activity of Fructose-
1,6-bisphosphatase

	Enzyme activity in crude extract (Units/ml)	
	Without incubation with glucose	3 min after incubation with glucose
10 mM Mg^{2+}	0.46	0.24
2 mM Mn^{2+}	0.27	0.24
$\dfrac{\text{Activity with } Mg^{2+}}{\text{Activity with } Mn^{2+}}$	1.7	1.0

It now has to be established, whether the "short term" inactivation only full-
fills the requirement of a rapid, reversible regulation mechanism - in order to
avoid futile cycles, for example - or/and represents the trigger reaction of an
obligatory proteolytic degradation.

ACKNOWLEDGEMENTS

We thank Dr. G. P. Royer for a critical reading of the manuscript. The excellent
assistance of Mrs. D. Montfort during the preparation of the manuscript is gratefully acknowledged. The authors' own research described here has been supported by
the Deutsche Forschungsgemeinschaft, Sonderforschungsbereich 46, and the Verband
der Chemischen Industrie e.V. (Fonds der Chemischen Industrie).

REFERENCES

Duntze, W., D. Neumann, and H. Holzer (1968). Glucose induced inactivation of
    malate dehydrogenase in intact yeast cells. Eur. J. Biochem., 3, 326-331.
Frey, J., and K.-H. Röhm (1979). The glucose-induced inactivation of aminopepti-
    dase I in Saccharomyces cerevisiae. FEBS Lett., 100, 261-264.
Funayama, S., J. M. Gancedo, and C. Gancedo (1980). Turnover of yeast fructose-
    bisphosphatase in different metabolic conditions. Eur. J. Biochem., 109, 61-66.
Gancedo, C. (1971). Inactivation of fructose-1,6-diphosphatase by glucose in
    yeast. J. Bacteriol., 107, 401-405.
Gancedo, C., and K. Schwerzmann (1976). Inactivation by glucose of phosphoenol-
    pyruvate carboxykinase from Saccharomyces cerevisiae. Arch. Microbiol., 109,
    221-225.
Görts, C. P. M. (1969). Effect of glucose on the activity and the kinetics of the
    maltose-uptake system and of α-glucosidase in Saccharomyces cerevisiae.
    Biochim. Biophys. Acta, 184, 299-305.
Haarasilta, S., and E. Oura (1975). On the activity and regulation of anaplerotic
    and gluconeogenetic enzymes during the growth process of baker's yeast. The
    biphasic growth. Eur. J. Biochem. 52, 1-7.
Hill, R. L., R. Barker, K. W. Olsen, J. H. Shaper, and I. P. Trayer (1972).
    Lactose synthetase: Structure and function. In O. Wieland, E. Helmreich, and
    H. Holzer (Eds.), Metabolic Interconversion of Enzymes. Springer-Verlag, Berlin.
    Heidelberg.New York. pp. 331-346.
Holzer, H. (1976). Catabolite inactivation in yeast. Trends Biochem. Sci. (TIBS),
    1, 178-181.
Jušič, M., H. Hinze, and H. Holzer (1976). Inactivation of yeast enzymes by pro-
    teinase A and B and carboxypeptidase Y from yeast. Hoppe-Seyler's Z. Physiol.
    Chem., 357, 735-740.
Labbe, P., G. Dechateaubodeau, and R. Labbe-Bois (1972). Synthèse du protohème par
    la levure Saccharomyces cerevisiae. II. Influence excercée par le glucose sur
    l'adaptation respiratoire. Biochimie, 54, 513-528.
Lenz, A.-G. (1980). Studien zum Mechanismus der Katabolit-Inaktivierung in Hefe.
    Doctoral dissertation, University of Freiburg, Germany.
Lenz, A.-G., and H. Holzer (1980). Rapid reversible inactivation of fructose-1,6-
    bisphosphatase in Saccharomyces cerevisiae by glucose. FEBS Lett., 109, 271-274.
Magasanik, B. (1961). Catabolite repression. Cold Spring Harbor Symp. Quant. Biol.,
    26, 249-256.
Magni, G., I. Santarelli, P. Natalini, S. Ruggieri, and A. Vita (1977). Catabolite
    inactivation of bakers'-yeast uridine nucleosidase. Isolation and partial puri-
    fication of a specific proteolytic inactivase. Eur. J. Biochem. 75, 77-82.
Matern, H., and H. Holzer (1977). Catabolite inactivation of the galactose uptake
    system in yeast. J. Biol. Chem. 252, 6399-6402.
Mecke, D., and H. Holzer (1966). Repression und Inaktivierung von Glutaminsynthe-
    tase in Escherichia coli durch NH_4^+. Biochim. Biophys. Acta, 122, 341-351.
Molano, J., and C. Gancedo (1974). Specific inactivation of fructose 1,6-bisphos-
    phatase from Saccharomyces cerevisiae by a yeast protease. Eur. J. Biochem.,
    44, 213-217.
Müller, M., H. Müller, and H. Holzer (1980/81). Immunochemical studies on catabo-
    lite inactivation of phosphoenolpyruvate carboxykinase in Saccharomyces cerevi-
    siae. J. Biol. Chem., in press.

Neeff, J., E. Hägele, J. Nauhaus, U. Heer, and D. Mecke (1978). Application of an immunoassay to the study of yeast malate dehydrogenase inactivation. Biochem. Biophys. Res. Commun., 80, 276-282.

Poole, B., S. Ohkuma, and M. J. Warburton (1977). The accumulation of weakly basic substances in lysosomes and the inhibition of intracellular protein degradation. Acta biol. med. germ., 36, 1777-1788.

Robertson, J. J., and H. O. Halvorson (1957). The components of maltozymase in yeast, and their behavior during deadaptation. J. Bacteriol. 73, 186-198.

Spiegelman, S., and J. M. Reiner (1947). The formation and stabilization of an adaptive enzyme in the absence of its substrate. J. Gen. Physiol., 31, 175-193.

Tortora, P., H. Holzer, M. J. Mazón, and C. Gancedo.   Unpublished results.

Wibo, M., and B. Poole (1974). Protein degradation in cultured cells. II. The uptake of chloroquine by rat fibroblasts and the inhibition of cellular protein degradation and cathepsin B_1. J. Cell. Biol., 63, 430-440.

Witt, I., R. Kronau, and H. Holzer (1966a). Repression von Alkoholdehydrogenase, Malatdehydrogenase, Isocitratlyase und Malatsynthase in Hefe durch Glucose. Biochim. Biophys. Acta, 118, 522-537.

Witt, I., R. Kronau, and H. Holzer (1966b). Isoenzyme der Malatdehydrogenase und ihre Regulation in Saccharomyces cerevisiae. Biochim. Biophys. Acta, 128, 63-73.

CONVERTING ENZYME OF FRUCTOSE—1,6—BISPHOSPHATASE FROM RAT LIVER

O. Tsolas[1,2], O. Crivellaro[2], P.S. Lazo[2], S.C. Sun[2], S. Pontremoli[3], and B.L. Horecker[2]

[1]Department of Biological Chemistry, University of Ioannina, Medical School, Ioannina, Greece
[2]Roche Institute of Molecular Biology, Nutley, New Jersey 07110, U.S.A.
[3]Institute of Biological Chemistry, University of Genoa, Genoa, Italy

ABSTRACT

Fructose-1,6-bisphosphatase, one of the key enzymes in gluconeogenesis, is activated and converted to an alkaline form by incubation with a protease, converting enzyme, obtained from the heavy particle fraction of rat liver. The converting enzyme is located in the lysosomes and is different from cathepsins A, B, C, and D. The cleavage pattern is complex, resulting in fragments of various molecular weights. Cleavage at $Asn^{64}-Val^{65}$ appears to be associated with the change in catalytic properties. In addition to being activated at alkaline pH, the converted enzyme is less susceptible to AMP inhibition. This is in agreement with the expected changes in the properties of the enzyme when it functions under gluconeogenic conditions.

KEYWORDS

Converting enzyme; fructose-1,6-bisphosphatase; gluconeogenesis; limited proteolysis; protease.

PROPERTIES OF CONVERTING ENZYME

Fructose-1,6-bisphosphatase ($Fru-P_2ase$) is one of the four key enzymes in gluconeogenesis, the other three being pyruvate carboxylase, phosphoenolpyruvate carboxykinase and glucose-6-phosphatase. Therefore, changes in the activity of this enzyme would affect gluconeogenesis. The enzyme is highly susceptible to AMP inhibition (Taketa and Pogell, 1963). In the case of rat liver $Fru-P_2ase$, the concentration of AMP giving 50% inhibition is 21 μM (Tejwani and co-workers, 1976). Since the concentration of AMP in rat liver is 260 μM (Hems and co-workers, 1966) and remains steady under various metabolic conditions, a mecha-

nism that would decrease the sensitivity of the enzyme to AMP inhibition would favor gluconeogenesis.

Rat liver lysosomes contain a protease, converting enzyme (CE), which
increases the activity of rabbit liver $Fru-P_2ase$ assayed at pH 9.2
six-fold. The activity at pH 7.5, the pH optimum of the native enzyme,
decreases only slightly (Crivellaro and co-workers, 1978; Horecker
and co-workers, 1980). The enzyme modified by limited proteolysis is
less susceptible to AMP inhibition. At 40 μM AMP the native enzyme is
55% inhibited, whereas in the converted enzyme this inhibition decreases to 25%.

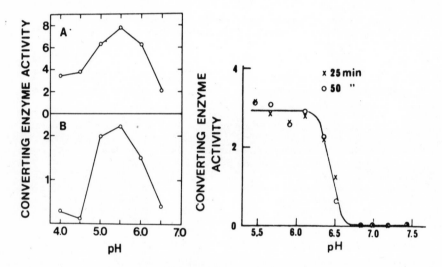

Fig. 1. Left: Effect of pH on the activity of CE
from the soluble fraction (A) and from the
membranes (B). The activity of CE was assayed by incubating the enzyme with
$Fru-P_2ase$ at pH 5.5, $30^{\circ}C$, for 30 min. The
unit of CE activity was defined as the
amount required to increase the specific
activity of $Fru-P_2ase$ at pH 9.2 by 1 μmole
per min per mg protein in the 30-min period.
Right: Effect of pH on the stability of CE
(from Crivellaro and co-workers, 1978).

In rat liver lysosomes the activity of the converting enzyme is cryp-
tic. It is released by freezing and thawing preparations of the heavy
particle fraction containing lysosomes and mitochondria (Crivellaro
and co-workers, 1978). Depending on the number of freeze/thaw cycles,

the activity can be found either in the soluble fraction or in the
membrane precipitate.   After one cycle there is an equal distribution
of activity between the two fractions.   When the cycles are repeated
ten times, 80% of the activity becomes membrane-bound.   This activity
is released by washing with 0.1 M NaCl.   Membrane eluates obtained
after ten cycles of freezing and thawing contain about 1% of the lysosomal activity of cathepsins A, B, and C and 10-14% of cathepsin D.
This indicates that, except for some activity of cathepsin D, all the
other cathepsins are found in the soluble fraction, whereas an appreciable amount of CE is membrane-bound.

The optimum pH of CE activity is 5.5, and it is identical in the soluble and in the membrane fraction (Fig 1, left).   The enzyme from the
soluble fraction is more active at acid pH compared with that from
the membrane fraction.   The converting enzyme is rapidly inactivated
above pH 6.5 (Fig. 1, right).   This property distinguishes it from
the activity of cathepsin D associated with the membranes.   Fractions
containing both CE and cathepsin D incubated at pH 6.5 loose their CE
activity whereas cathepsin D remains stable (Crivellaro and co-workers,
1978).

Converting enzyme is inhibited by iodoacetate, p-tosylamido-2-phenyl-
ethyl chloromethyl ketone (TPCK), and N-a-p-tosyl-L-lysyl chloromethyl
ketone (TLCK).   It is less susceptible to inhibition by phenylmethane-
sulfonyl fluoride (PMSF), and it is not inhibited by leupeptin. Rabbit liver CE has been reported to have cathepsin B activity (Nakashima and Ogino, 1974; Pontremoli and co-workers, submitted for publication).   However, the lack of inhibition or rat liver CE by leupeptin,
a potent inhibitor of cathepsin B, up to concentrations of 2 μg per
ml, indicates that this protease is not identical with cathepsin B.
In conclusion, the evidence indicates that rat liver CE is different
from the classical cathepsins A, B, C, and D.

CLEAVAGE PATTERN OF FRUCTOSE-1,6-BISPHOSPHATASE

The cleavage pattern by CE is complex.   Rabbit liver Fru-P_2ase is a
tetramer of MW 140,000 with four identical subunits of MW 36,000 each
(Traniello and co-workers, 1972). Converting enzyme can cleave Fru-P_2-
ase at Asn^{64}-Val^{65} to give a modified subunit of MW 29,000 and a peptide of MW 7,000 derived from the amino terminal of the native subunit.

This is the major reaction pathway. Alternatively, it can give a modified subunit of MW 26,000 and a peptide of MW 8,000 which is derived from the carboxy terminal of the native subunit. The smaller modified subunit is further cleaved to the MW 7,000 peptide obtained in the first cleavage pathway, and a fragment of Mw 13,000-19,000 (Horecker and co-workers, 1980). The modified Fru-P$_2$ase remains a tetramer at neutral pH and dissociates only under denaturing conditions (Lazo and co-workers, 1978).

It appears that the cleavage bringing about the change in the properties of the enzyme is the one at Asn^{64}-Val^{65}. Subtilisin modifies Fru-P$_2$ase producing changes in the enzymatic properties similar to those of CE. In contrast to CE, cleavage by subtilisin takes place in one region only, extending from Tyr^{57} to Gly^{67} (Botelho and co-workers, 1977):

CE

↓

^{57} ^{67}

-Tyr-Gly-Ile-Ala-Gly-Ser-Thr-Asn-Val-Thr-Gly-

↑   ↑   ↑   ↑

Subtilisin

The multiple cleavages by subtilisin, which give tripeptides at various yields, indicate that this region of the subunit is exposed and contains an unstructured sequence, susceptible to proteolytic attack.

Rabbit liver Fru-P$_2$ase may also be partially modified, either in the cell or during the isolation procedure, at the carboxy terminal giving a subunit with either two or four amino acids less (Sun and co-workers, 1980). So far no catalytic changes have been associated with this particular pattern of cleavage.

Since Fru-P$_2$ases from various sources are susceptible to limited proteolysis with resulting modification in their catalytic properties (Horecker and co-workers,1980), it would appear that the properties of the converted enzyme have a physiological function. Among these properties, the decrease in the sensitivity to AMP is in agreement with the role of the enzyme in gluconeogenesis (Tsolas, Annamalai and Horecker, submitted for publication).

REFERENCES

Botelho, L. H., H. A. El-Dorry, O. Crivellaro, D. K. Chu, S. Pontremoli, and B. L. Horecker (1977). Digestion of rabbit liver fructose 1,6-bisphosphatase with subtilisin: Sites of cleavage and activity of the modified enzyme. Arch. Biochem. Biophys., 184, 535-545.

Crivellaro, O., P. S. Lazo, O. Tsolas, S. Pontremoli, and B. L. Horecker (1978). Properties of a fructose 1,6-bisphosphatase converting enzyme in rat liver lysosomes. Arch. Biochem. Biophys., 189, 490-498.

Hems, R., B. D. Ross, M. N. Berry, and H. A. Krebs (1966). Gluconeogenesis in the perfused rat liver. Biochem. J., 101, 284-292.

Horecker, B. L., J. S. MacGregor, V. N. Singh, O. Tsolas, S. C. Sun, O. Crivellaro, and S. Pontremoli (1980). Partial amino acid sequence of rabbit liver fructose 1,6-bisphosphatase (Fru-P$_2$ase, EC 3.1.3.11) and sites of cleavage by proteinases. In P. Mildner and B. Ries (eds.), Enzyme Regulation and Mechanism of action, Pergamon Press, Oxford. pp. 3-14.

Lazo, P. S., O. Tsolas, S. C. Sun, S. Pontremoli, and B. L. Horecker (1978). Modification of fructose bisphosphatase by a proteolytic enzyme from rat liver lysosomes. Arch. Biochem. Biophys., 188, 308-314.

Nakashima, K., and K. Ogino (1974). Regulation of rabbit liver fructose-1,6-diphosphatase. J. Biochem., 75, 355-365.

Sun, S. C., A. G. Datta, E. Hannappel, V. N. Singh, O. Tsolas, E. Melloni, S. Pontremoli, and B. L. Horecker (1980). The carboxy-terminal amino-acid sequence of rabbit liver fructose 1,6-bisphosphatase. Arch. Biochem. Biophys., in press.

Taketa, K., and B. M. Pogell (1963). Reversible inactivation and inhibition of liver fructose-1,6-diphosphatase by adenosine nucleotides. Biochem. Biophys. Res. Commun., 12, 229-235.

Tejwani, G. A., F. O. Pedrosa, S. Pontremoli, and B. L. Horecker (1976). The purification and properties of rat liver fructose 1,6-bisphosphatase. Arch. Biochem. Biophys., 177, 255-264.

Traniello, S., E. Melloni, S. Pontremoli, C. L. Sia, and B. L. Horecker (1972). Rabbit liver fructose 1,6-diphosphatase. Properties of the native enzyme and their modification by subtilisin. Arch. Biochem. Biophys., 149, 222-231.

SYNTHESIS OF PLASMINOGEN ACTIVATOR BY CYCLING CELLS

J. Sorić and B. Brdar

Central Institute for Tumors and Allied Diseases, Zagreb, Yugoslavia

ABSTRACT

The purpose of this study was to characterize the synthesis of the plasminogen activator by synchronized normal and Rous sarcoma virus-transformed chick embryo
fibroblasts in culture. The plasminogen activator activity of the cell lysates was
assayed on ^{125}I-fibrin-coated Petri dishes and was expressed as the radioactivity
released from the plates. It was found that normal cells only produce detectable
levels of plasminogen activator in the S-phase and late G_2-phase or mitosis of the
mitotic cycle. In contrast, transformed cultures synthesize this activator throughout the entire cell cycle although its rate of synthesis fluctuates and reaches a maximum in the G_2-M periods. We have also found that the appearance of plasminogen
activator activity following fibroblasts infection with Rous sarcoma virus depends on
both the synthesis of cellular DNA and cell division.

KEYWORDS

Plasminogen activator; chick embryo fibroblasts; Rous sarcoma virus; cell cycle.

INTRODUCTION

There is evidence that increased synthesis of the plasminogen activator is a pheno-
typic property of both malignantly transformed (Unkeless et al., 1973; Reich, 1974)
and normal mitotic cells (Aggeler et al., 1978; Hatcher et al., 1976; Bossman,
1974). The similarity in the plasminogen activator synthesis between normal mitotic
and cancer cells suggests that the expression of this phenotype might be programmed
through a series of events that are directed by cycles of DNA synthesis (Lungren and
Ross, 1976). Consequently, the expression of the cancer cell phenotype is associated with some transformation event that results in unregulated DNA synthesis. It
seems, therefore, that the synthesis of the plasminogen activator is probably uncontrolled within the cell cycle of transformed cells as compared to their normal coun-
terparts. To test this assumption we have studied the intracellular synthesis of plasminogen activator by synchronized normal and Rous sarcoma virus (RSV)-transfor-

med chick embryo fibroblasts in culture. Consideration has also been given to the appearance of the plasminogen activator activity in relation to the DNA synthesis or mitosis of cycling fibroblasts after virus infection.

In this communication, we report that the synthesis of the plasminogen activator within the cell cycle of normal fibroblasts is discontinious in contrast to the RSV-transformed cultures which synthesize this activator throughout the mitotic cycle. Moreover the synthesis of DNA and mitosis were both required for the appearance of the plasminogen activator in the "de novo" virus infected fibroblasts.

MATERIAL AND METHODS

Primary cultures of chick embryo fibroblasts were prepared from eleven day embryos and secondary cultures were transformed by infection at high multiplicity with a Schmidt-Ruppin D-type virus (RSV) or its temperature sensitive mutant for transformation (ts-68) as previously described (Brdar et al., 1973). Cultures were maintained in Dulbecco's medium supplemented with 10% fetal bovine serum. Exponentially growing normal or RSV-transformed fibroblasts (5×10^5 cells per 60 mm Petri dish) were synchronized by incubation with hydroxyurea (1 mM) for 12 h, followed by washing the cell monolayers free of the drug. This was considered 0 h postrelease. The synchrony of cell growth was determined both by DNA-labeling with 30 min pulses of ^{3}H-thymidine (2μCi/ml; sp. act. 20 Ci/mmol) and cell counting.

The plasminogen activator was assayed as previously described (Unkeless et al., 1974; Nagy et al., 1977). Cell monolayers were thouroughly washed with ice-cold phosphate-buffered saline (0.15 M N aCl; 0.005 M Sodium phosphate, pH 7.8) and scraped off the plate with a teflon policeman in the same buffer. After centrifugation at 700 g for 5 min at 4°C, the cell-pellet was washed once again with the same buffer and centrifugation was repeated. The drained cells were resuspended in 0.3 ml of 0.1% Triton X-100 in 0.1 M Tris, pH 8.1 and lysed by intensive vortexing for about 2 min. After centrifugation at 300-500 g for 10 min the nuclear pellet was discarded and the supernatant was assayed for the plasminogen activator activity. The lysate aliquots containing different concentrations of protein were added to ^{125}I-fibrin-coated Petri dishes and incubated at 37°C for different lengths of time in the presence of 4 μg/ml of chicken plasminogen; the reaction was expressed as the radioactivity released from the plates.

RESULTS

Figure 1 illustrates that chick embryo fibroblasts in culture transformed with Rous sarcoma virus synthesize high levels of plasminogen activator in contrast to their normal counterpart cells. It also shows the kinetics of plasminogen activator assay which is a function of cell lysate protein concentration. At a concentration of 50 ug protein, the difference in the enzyme activity between transformed (RSV-CEF) and normal (CEF) cells is about twenty fold. We have established with these measurements satisfactory conditions for assaying the plasminogen activator in other experiments designed to study the enzyme synthesis in synchronized normal and transformed fibroblast cultures.

As shown in Fig. 2, normal fibroblast cultures synchronized by hydroxyurea synthesize very low levels of plasminogen activator and its synthesis is discontinuous during the cell cycle. It could only be detected in the DNA-synthesizing or the dividing cells.

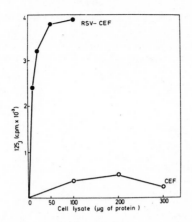

Fig. 1. Kinetics of plasminogen activator activity of lysate
         of transformed (RSV-CEF) and normal (CEF) chick
         fibroblast cultures as a function of protein concen-
         tration.

Since this result has been reproduced repeatedly, we believe that the plasminogen activation by non-transformed fibroblasts coordinates with their cell cycle. Unlike normal fibroblasts, RSV-transformed cultures produce the plasminogen activator at a high rate throughout the cell cycle, which is demonstrated in Fig. 3. However, this

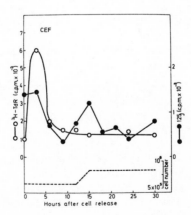

Fig. 2. Synthesis of plasminogen activator by synchronized
         normal fibroblast cultures.

activity fluctuates and it is lower in the S-phase than in the late G_2-phase or mitosis.
During the G_1-phase, the production of the plasminogen activator gradually declines
as the cells approach the second round S-phase. The difference in plasminogen activator activity between its lowest point in the S-phase and its maximum in the late
G_2 or mitosis is about three fold. A similar result was obtained with synchronized
SV-40 transformed hamster cells (H50) (data not shown) suggesting the general
pattern of plasminogen activation in the cell cycle of virally transformed fibroblasts.

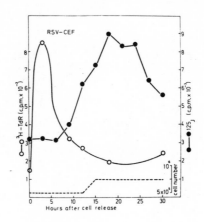

Fig. 3.   Synthesis of plasminogen activator by synchronized
RSV-transformed fibroblasts in culture.

We confirmed, in an experimental system of chick fibroblasts transformed by a temperature sensitive mutant for transformation of RSV (ts-68), that the difference in
the plasminogen activator synthesis between normal and RSV-transformed fibroblasts
is real. The data, which will be reported elsewhere,[1] show that the patterns of plasminogen activator synthesis in synchronized ts-68-transformed cultures at either
non-permissive (41°C) or permissive (36°C) temperatures are identical to those
in normal and RSV-transformed fibroblasts, respectively.

Since the synthesis of plasminogen activator is associated with oncogenic transformation, it is important to determine the stage of mitotic cycle this synthesis first
appears in following cell infection with a transforming virus. We did this by synchronizing cell cultures of chick fibroblasts in a usual way and infected them with a high
multiplicity of RSV at the S, G_2 and G_1-phases of the cell cycle. Following virus
infection the plasminogen activator activity was measured at intervals and this is
shown in Fig. 4. The enzyme activity appeared first in the cultures which were infected in the S-phase; also the level of its activity was the highest in these cells. It is
interesting that the cell cultures which were infected with RSV in the S-phase did not
initiate the synthesis of plasminogen activator as long as they were not permitted to
divide.[1] This suggests that the synthesis of cellular DNA and mitosis are both required for the appearance of plasminogen activator activity in the virally infected
fibroblast cultures.

---

[1] Sorić, J. and Brdar, B., manuscript in preparation.

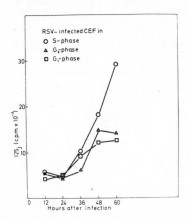

Fig. 4. Appearance of plasminogen activator in fibroblast
cultures after infection with RSV at different stages
of cell cycle.

DISCUSSION

The data presented above show that the synthesis of plasminogen activator is a function of the cell cycle of the normal but not of the RSV-transformed chick fibroblasts
in culture. The normal fibroblasts synthesize plasminogen activator transiently during the mitotic cycle in such a way that the measurable enzyme activity can only be
found in the cells that are committed to DNA synthesis and in those in the late G_2-
phase or mitosis. In contrast the transformed cultures produce high amounts of
plasminogen activator throughout the cell cycle, although its rate of synthesis somewhat fluctuates. All this might have some relevance to a generally accepted view
that the plasminogen activation by cells and tissues has an important function in pro-
viding local proteolysis. Namely, the zymogen plasminogen, which represents a
source of potential protease activity in the microenvironment of the cell, could be
converted to the enzymatically active plasmin and used for any cellular activity
requiering proteolysis. In this respect, a good correlation between the synthesis of
the plasminogen activator by cells in culture and their morphology, motility
(Ossowski et al., 1973) or invasiveness (Strickland et al., 1976) has been established. Moreover, the proteolysis could be part of a mechanism by which a growing
cell distroys surrounding connective tissue to provide space for its progeny cells
(Reich, 1975). Our results indicate that there is an essential difference between the
normal and RSV-transformed cells in generating plasminogen dependent proteolysis.
While this plasminogen activation seems to be regulated by normal cells, the synthesis of the plasminogen activator in transformed cells is apparently continuous
and uncontrolled and this observation needs additional comments. In view of previously reported data (Aggeler et al., 1978; Hatcher et al., 1976), we expected the
synchronized normal fibroblast cultures to produce the enzyme at its highest rate
during or before mitosis. However, in addition to this, we have repeatedly found
another peak of the enzyme activity in the cells confined to the S-phase of cell cycle.

We do not know whether these two enzyme peaks represent identical molecules and we can only speculate about their function in cellular growth. The synthesis of the plasminogen activator by cells in the G_2-M periods could be visualized through the inferred mechanism of proteolysis control which is important for the mitotic cell phenotype. With regard to plasminogen activator synthesized by the S-phase cells and its biological significance, the following points may be considered: a) enzyme synthesis by the S-phase cells could be due to the effect of the cell synchronizing agent, hydroxyurea which blocks the synthesis of DNA to give rise to the concomitant unbalanced synthesis of RNA. It should be stressed, however, that hydroxyurea does not cause a similar increase in plasminogen activator synthesis of RSV-transformed fibroblast cultures. b) it has been recently reported (Miskin and Reich, 1980) that chick fibroblasts in culture can be induced to synthesize high levels of plasminogen activator by factors causing DNA damage. This plasminogen activator was tentatively regarded as an eucaryot homolog of a bacterial protease (rec A gene product) which initiates events that lead to DNA repair. A probable candidate for this type of protease might be the plasminogen activator which is synthesized by the S-phase cells. This is solely based on the observation that the stimulation of the plasminogen activator synthesis associated with DNA damage is limited to the cells that retain the capacity for DNA replication; the post-mitotic cells such as myotubes are not capable of inducing plasminogen activator synthesis after DNA damage (Miskin and Reich, 1980). It remains to be seen, however, whether the synthesis of plasminogen activator by the S-phase cells can be enhanced with factors that cause DNA damage before drawing a final conclusion about the association of these two proteases.

Unlike normal fibroblasts, the RSV-transformed cultures synthesize the plasminogen activator continuously throughout the cell cycle and reach a peak during or before mitosis. The permanent production and shedding of this activator by transformed cells can be related to the loss of their growth regulating ability (Ossowski et al., 1973) which involves the loss of growth contact inhibition and the increased motility, invasiveness or affinity to grow in soft agar etc. This is compatible with a view (Reich, 1975) that malignant cells migrate and metastasize as a result of uncontrolled proteolysis of surrounding connective tissue, all independently of cell division or growth.

The appearance of the plasminogen activator in synchronized fibroblast cultures which are infected in the S-phase of cell cycle indicate that the cellular DNA synthesis is required for the initiation of enzyme synthesis. The cells infected with virus in the G_1 or G_2-phases also synthesize the plasminogen activator but after a delay that lasts longer than after infection of the S-phase cells. This increase in the plasminogen activator production can be correlated with morphological changes typical of transformed cells that are observed in the corresponding synchronious cultures. This all suggests a close relationship between the expression of transformed phenotype and plasminogen activator synthesis. Moreover, the requirement of mitosis for the appearance of plasminogen activator synthesis following S-phase cell infection with RSV also supports this notion.

Although the data presented in this communication indicate that the synthesis of the plasminogen activator by synchronized normal and RSV-transformed fibroblasts coordinates differently to the cell cycles of either cell type, this phenomenon should be verified in a variety of normal and neoplastic cell types before drawing general conclusions.

REFERENCES

Aggeler, J., L.N. Kapp, and Z. Werb (1978).Cell cycle specific secretion of
plasminogen activator by Chinese hamster fibroblasts. J. Cell Biol., 79, 10 a.

Bossman, H.B. (1974). Release of specific protease during mitotic cycle of
L5178 Y murine leukemic cells by sublethal autolysis. Nature, 249, 144-145.

Brdar, B., D.B. Rifkin, and E. Reich (1973). Studies of Rous sarcoma virus:
Effect of nucleoside analogs on virus synthesis.J. Biol. Chem., 248, 2397-2408.

Hatcher, V.B., M.S. Wertheim, C.Y. Rhee, G. Tsien, and P.G. Burk (1976).
Relationship between cell surface protease activity and doubling time in various
normal and transformed cells. Bioch. Biophys. Acta, 451, 499-510.

Lundgren, E., and G. Roos (1976). Cell surface changes in HeLa cells as an indication of cell cycle events. Cancer Res., 36, 4044-4051.

Miskin, R., and E. Reich (1980).Plasminogen activator: Induction of synthesis by
DNA demage. Cell, 19, 217-224.

Nagy, B., J. Ban, and B. Brdar (1977). Fibrinolysis associated with human neo-
plasia: Production of plasminogen activator by human tumors. Int. J. Cancer,
19, 614-620.

Ossowski, L., J.P. Quigley, G.M. Kellerman, and E. Reich (1973).Requirement
of plasminogen for correlated changes in cellular morphology, colony formation
in agar and cell migration. J. Exp. Med., 138, 1056-1064.

Ossowski, L., J.P. Quigley, and E. Reich (1974). Fibrinolysis associated with
oncogenic transformation. Morphological correlates. J. Biol. Chem., 249,
4312-4320.

Reich, E. (1974).Tumor-associated fibrinolysis. In Control of Proliferation in
Animal Cells, (eds. B. Clarkson and R. Baserga). Cold Spring Harbor Laboratory, 351-355.

Reich,E. (1975). Plasminogen activator: Secretion by neoplastic cells and macrophages. In Proteases and Biological Control (eds. E. Reich, D.B. Rifkin and
E. Shaw). Cold Spring Harbor Laboratory, 333-343.

Strickland, S., E. Reich, and M.I. Sherman (1976). Plasminogen activator in
early embryogenesis: enzyme production by trophoblast and parietal endoderm.
Cell, 9, 231-240.

Unkeless, J.C., A. Tobia, L. Ossowski, J.P. Quigley, D.B. Rifkin, and E. Reich
(1973). An enzymatic function associated with transformation of fibroblasts by
oncogenic viruses. I. Chick embryo fibroblast cultures transformed by avian
RNA tumor viruses. J. Exp. Med., 137, 85-111.

Unkeless, J.C., K. Danø, G.M. Kellerman, and E. Reich (1974). Fibrinolysis
associated with oncogenic transformation: Partial purification and characteri-
sation of the cell. J. Biol. Chem., 249, 4295-4305.

Properties and Structure
of Enzymes

STRUCTURES AND BIOLOGICAL FUNCTIONS OF CATHEPSIN B AND L

N. Katunuma, T. Towatari, E. Kominami and S. Hashida

Dept. of Enzyme Chem., Inst. for Enzyme Res., School of Medicine, Tokushima Univ., Tokushima City, Tokushima, Japan

ABSTRACT

Characterization of rat liver cathepsin B and cathepsin L including physicochemical properties, mechanism of inhibition by specific inhibitor, substrate specificity and partial amino acid sequence is described. On the basis of those results, roles of cathepsin B and cathepsin L in degradation of a selective enzyme liver aldolase is discussed.

KEYWORDS

Cathepsin B; cathepsin L; leupeptin; E-64; degradation of enzyme; aldolase.

INTRODUCTION

There is growing evidence for involvement of lysosomal proteases in intracellular protein degradation. In order to understand the role and mechanism of enzyme action in the catabolic processes it is necessary to know their characteristics. The lysosomal proteases were reviewed recently (Barrett, 1977). The most investigated thiol protease is cathepsin B (EC 3.4.22.1). The work from our laboratory has shown crystallization of cathepsin B from rat liver (Towatari and coworkers, 1978, 1979). Cathepsin B was subjected to the amino acid sequence study. Present study describes also characterization of much less investigated thiol protease, cathepsin L (EC 3.4.22.-). For study of the function of cathepsin B and cathepsin L, the specific and strong protease inhibitor, which is effective *in vivo*, is desired. Our results indicated that E-64 and its derivatives and leupeptin were useful in analysis of the possible role of cathepsin B and cathepsin L. Present reports describes that on administration of those protease inhibitors turnover rate of rat liver aldolase is reduced.

83

RESULTS AND DISCUSSION

Characterization of Cathepsin B and Cathepsin L from Rat Liver

Cathepsin B was purified from the lysosomal fraction of rat liver to a crystalline form by a series of step including acetone treatment, Sephadex G-75 chromatography, DEAE-Sephadex (A-50) chromatography, CM-Sephadex (C-50) chromatography and crystallization with ammonium sulfate (Towatari and coworkers, 1978, 1979). The spindle-shaped crystals of cathepsin B is shown in Fig. 1. Cathepsin L was also

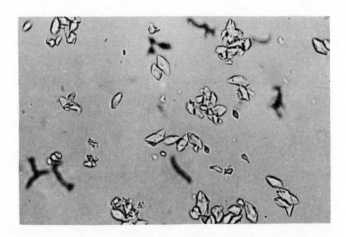

Fig. 1.   Crystals of cathepsin B from rat liver.

purified from rat liver lysosome by steps including acetone treatment, Sephadex G-75 chromatography, DEAE-Sephadex (A-50) chromatography and CM-Sephadex (C-50) chromatography (Towatari and colleagues, 1976, 1978). Crystalline cathepsin B was homogenous on ultracentrifugal analysis. However, the enzyme showed one major protein band (mol. wt. 29,000) and two minor protein bands (mol. wt. 25,000 and mol. wt. 5,000) on sodium dodecyl sulfate polyacrylamide gel electrophoresis. When stocked -20°C for more than 1 month, the major band disappeared and conversely two minor bands increased. Limited proteolytic cleavage probably due to autolysis might occur during storage. Purified preparation of cathepsin L also showed a single protein band coincided with that of activity on polyacrylamide gel electrophoresis in the absence of sodium dodecyl sulfate. Molecular weight of cathepsin L was estimated as 24,000 with gel filtration on Sephadex G-75 chromatography. Cathepsin B consists of four isozymes with isoelectric points between pH 4.9 and 5.3. The presence of pI isozymes of cathepsin B was also found in the human organs (Barrett, 1977) and the rabbit lung enzyme (Singh and Kalnitsky, 1978). Cathepsin B is found to be a glycoprotein (Towatari and colleagues, 1979) but analysis of sugar moiety of the enzyme is not determined. Antiserum against rat liver cathepsin B revealed the precipitin lines with highly purified cathepsin B and partially purified lysosomal fraction but no precipitin line with purified cathepsin L and cathepsin H (Fig. 2). Partially purified preparation of lysosome used here contained both cathepsin B and cathepsin H. Anti-rat cathepsin H gave a precipitin

line against both purified cathepsin H and partially purified lysosomal fraction. No precipitin line, however, was formed between
purified cathepsin B and cathepsin L. Those results indicate that
cathepsin B and cathepsin H are immunologically different each other,
and that cathepsin L is distinct from cathepsin B and cathepsin H,
respectively.

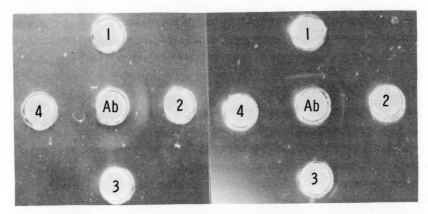

Fig. 2. Ouchterlony double diffusion analysis of
cathepsin B, L and H. Antiserum against cathepsin B
(left) and antiserum against cathepsin H (right)
were placed in the central well. Cathepsin H,
lysosomal extract, cathepsin B and cathepsin L
were placed in wells 1-4, respectively.

Mechanism of Inhibition of Cathepsin B and Cathepsin L by E-64

Cathepsin B and cathepsin L were inhibited by common thiol reagents
like monoiodoacetate, p-chloromerculibenzoate and activated by 2-
mercaptoethanol. Phenylmethylsulfonylfluoride, a serine protease
inhibitor had little effect on both cathepsins. It is remarkable
that protease inhibitors of microbial origin like leupeptin, E-64,
antipain and chymostatin inhibited cathepsin B and cathepsin L at
very low concentration. E-64 was isolated from Aspergillus japonicus
and very specific for thiol proteases (Hanada and coworkers a)b),
1978). Then, mechanism of inhibition of cathepsin B and cathepsin L
by E-64 was investigated (Hashida and others, 1980). Kinetic studies
indicated that E-64 was an irreversible inhibitor of these enzymes.
This inhibition was not recovered by dialysis or by passing through a
column of Sephadex G-25. Titration of 1 of the total 10 SH groups of
native cathepsin B with 2,2'-dithiopyridine resulted in complete loss
of enzyme activity. As shown in Table 1, proportional decrease of
titrable SH groups and activity of cathepsin B to concentrations of
added E-64 was observed, indicating that E-64 binds equimolorly to
active -SH residue of cathepsin B.

TABLE 1   Titratable SH-group and Remaining Activity of
Cathepsin B after Treatment with Different Amounts of E-64

E-64 added	Titratable SH-group	Remaining activity
(mol/mol of enzyme)	(mol/mol of enzyme)	(%)
0	1.0	100
0.4	0.6	45
1.0	0	2

Reaction mixtures containing 5.0×10^{-6} M of
cathepsin B, the indicated amount of E-64 and
0.1 M potassium phosphate buffer, pH 6.0 were
incubated at 25°C.   After 10 min, titration of
SH groups with 2,2'-dithiopyridine and assay of
remaining activity were performed.

Substrate Specificity of Cathepsin B and Cathepsin L

TABLE 2   Activities of Cathepsin B and Cathepsin L on Protein Substrates

Substrate	Source	Protease activity	
		Cathepsin B	Cathepsin L
		units/mg	
Casein		0.16	0.86
Acid denatured hemoglobin	bovine	0.24	0.76
Bovine serum albumin		0.10	0.53
Glucose-6-phosphate dehydrogenase	yeast	0.05	6.68
Glucose-6-phosphate dehydrogenase	rat liver	0	1.57
Ornithine aminotransferase (apo-)	rat liver	0.14	3.56
Tyrosine aminotransferase (apo-)	rat liver	0.05	4.63
Cystathionase (apo-)	rat liver	0.28	14.77
Glucokinase	rat liver	0.87	23.39
Glyceroaldehyde-3-phosphate dehydrogenase	rabbit muscle	0.04	3.24
Malate dehydrogenase	pig heart	0.04	7.57
Aldolase	rat liver	21.01	16.70
Aldolase	rabbit musice	11.56	5.90
Lactic dehydrogenase	rat liver	0	0
Glutamate dehydrogenase	beef liver	0	0
Alcohol dehydrogenase	yeast	0	0

Cathepsin B is very active on synthetic derivatives like Benzoyl-DL-
arginine p-nitroanilide and Benzoyl arginine 2-naphthylamine, whereas
cathepsin L is much less active on those synthetic substrates except
Benzoyl-L-arginine amide (Kirschke and colleagues, 1977; Towatari and
coworkers, 1979). However, protein substrates such as casein, acid-
denatured hemoglobin and bovine serum albumin are good substrates for
cathepsin L, but not for cathepsin B (Table 2). The rates of inactivation of various enzymes by cathepsin L were also much higher than
those by cathepsin B. Exception is aldolase. Either muscle aldolase
or liver aldolase was inactivated by not only cathepsin L but
cathepsin B. It seems to be due to that aldolase contains an exposed
peptide region around carboxyterminal that is susceptible to attack
by both cathepsins. Nakai and coworkers (1978) and recently, Bond
and Barrett (1980) showed that cathepsin B caused the release of
dipeptide, alanyltyrosine from carboxyterminal of rabbit muscle aldolase with which loss of enzyme activity was associated.
When specificity of cathepsin L was studied with a hexapeptide, Leu-
Try-Met-Arg-Phe-Ala, the enzyme cleaved between Met-Arg and further
splitting of peptide bonds was not observed (Fig. 3).

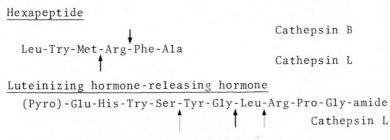

Fig. 3. The specificity of cathepsin L and cathepsin B toward Hexapeptide and Luteinizing hormone-
releasing hormone. The unbroken arrows indicate the
major sites of action of the enzyme and broken arrows
indicate the minor sites of action of the enzyme.

Cathepsin B hydrolysed the bond at Arg-Phe. On hydrolysis of the
luteinizing hormone releasing hormone by cathepsin L, the peptide
bond between Gly-Leu was the main cleavage point with minor hydrolysis
of the bond between Ser-Tyr and between Leu-Arg. Studies on specificity of cathepsin L with insulin B chain was recently reported
(Kärgel and colleagues, 1980). They demonstrated a clear specificity
for hydrobic amino acids at least in P_2 and P_3 position. But their
hypothesis could not be always applied to our results with peptide
substrates.

Partial Primary Structure of Rat Liver Cathepsin B

No mammalian thiol protease has yet been sequenced or subjected to a
detailed study of its catalytic mechanism. We have recently demonstrated the amino-terminal sequence analysis of rat liver cathepsin B
(Takio and coworkers, in preparation). As described previously,
stored crystalline cathepsin B was separated into two polypeptide
chains, namely heavy chain (mol. wt. 25,000) and light chain (mol.wt.
5,000) when subjected to sodium dodecyl sulfate polyacrylamide gel

electrophoresis. Then, after separation of the reduced and carboxy-methylated chains of the enzyme on Sephacryl S200 column, amino-terminal sequence analysis of both heavy chain and light chain was performed. Sequence of light chain which contains active site was completed. As shown in Fig. 4,

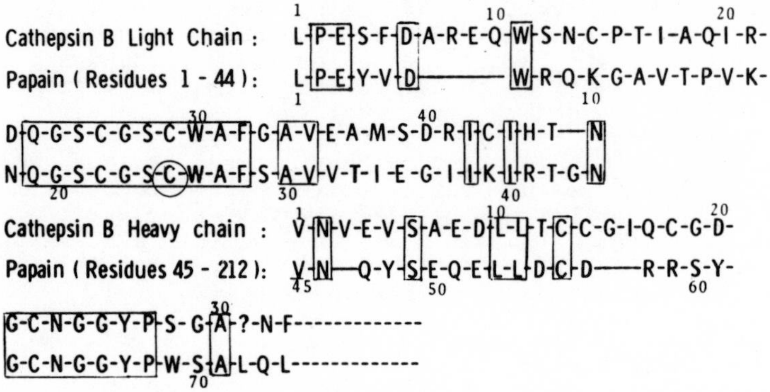

Cathepsin B Light Chain : L-P-E-S-F-D-A-R-E-Q-W-S-N-C-P-T-I-A-Q-I-R-

Papain (Residues 1 - 44): L-P-E-Y-V-D————————W-R-Q-K-G-A-V-T-P-V-K-

D-Q-G-S-C-G-S-C-W-A-F-G-A-V-E-A-M-S-D-R-I-C-I-H-T——N

N-Q-G-S-C-G-S-C-W-A-F-S-A-V-V-T-I-E-G-I-I-K-I-R-T-G-N

Cathepsin B Heavy chain : V-N-V-E-V-S-A-E-D-L-L-T-C-C-G-I-Q-C-G-D-

Papain (Residues 45 - 212): V-N——Q-Y-S-E-Q-E-L-L-D-C-D————R-R-S-Y-

G-C-N-G-G-Y-P-S-G-A-?-N-F————————

G-C-N-G-G-Y-P-W-S-A-L-Q-L————————

Fig. 4. Comparison of amino acid sequences of cathepsin B with that of papain. The sequences are aligned so as to maximize identities. Identical residues are indicated by enclosures. Gaps are indicated by long solid lines. X indicates that the residue was not identified. Circled Cys 25 in papain is at the active site.

the partial amino acid sequence of cathepsin B reveals an astonishingly high resemblance to that of papain, suggesting a common evolutionary origin. The ten residue surrounding the active site cysteine (residue 25) in papain (residues 19-28) is completely preserved in the light chain of the rat enzyme (residues 23-32). In addition, the seven residue sequence (residues 62-68) which forms a major part of the active site groove in papain is present in the heavy chain of cathepsin B (residue 21-27).

Inhibition of Cathepsin B and Cathepsin L by Leupeptin and E-64 and Its Derivative in Vivo

One of available useful methods for approach to the mechanism of intracellular protein degradation, especially to proteases involved in the process is the use of protease inhibitors. Actually, the roles of intracellular protein catabolism have been studied with help of proteinase inhibitors (Dean, 1975, Libby and Goldberg, 1978, Seglen and coworkers, 1979). We chosen leupeptin, E-64 and its derivatives for the study of role of cathepsin B and cathepsin L in intracellular degradation. Various derivatives of E-64 in which terminal agmatine in E-64-(L) is replaced inhibited cathepsins and their concentration for 50% inhibition were similar in vitro (Hashida and coworkers, 1980) (Table 3).

TABLE 3   Inhibitory Activity of E-64 Derivatives on Cathepsin B
and Cathepsin L in Vitro

$$HOOC\underset{O}{\overset{H}{C}}-\overset{}{C}-CO-NH-CH\overset{CO-R}{\underset{CH_2-CH}{<}}\overset{CH_3}{\underset{CH_3}{<}}$$

Compound	R	ID$_{50}$	
		Cathepsin B	Cathepsin L
		(M)	(M)
1. E-64 (L)	$-NH-(CH_2)_4-NH-C\overset{NH}{\underset{NH_2}{<}}$ $\frac{1}{2}H_2O$	2.4×10^{-7}	1.1×10^{-7}
2. E-64 (D)	$2H_2O$	5.4×10^{-7}	1.7×10^{-7}
3. E-64-a-z (SS)	$-NH-(CH_2)_4-NH-COOCH_2-C_6H_5$	2.4×10^{-7}	1.0×10^{-7}
4. E-64-a-z (RR)		3.0×10^{-7}	1.3×10^{-7}
5. E-64-a (RR)	$-NH-(CH_2)_4-NH_2$	2.4×10^{-7}	1.4×10^{-7}
6. E-64-b (SS)	$-NH-CH-COOH$	2.8×10^{-7}	1.4×10^{-7}
7. E-64-b (RR)	$\underset{CH}{\overset{CH_2}{|}}$	3.5×10^{-7}	1.4×10^{-7}
8. E-64-b (SS)	$CH_3\overset{CH}{<}CH_3$	2.4×10^{-7}	1.3×10^{-7}
9. E-64-c (RR)	$-NH-CH_2-CH_2-CH\overset{CH_3}{\underset{CH_3}{<}}$	2.4×10^{-7}	0.9×10^{-7}

We have first examined whether those inhibitors reduce the activities
of liver cathepsin B and cathepsin L when injected *in vivo*. The
animals were killed 1 h after the injection of inhibitors and the
cathepsin B and cathepsin L activities in the lysosomal fraction were
assayed. As shown in Table 4, leupeptin, E-64 and its derivatives re

TABLE 4   Inhibition of Cathepsin B and L by E-64 derivatives in Vivo

Compound	Cathepsin B		Cathepsin L	
	mu/g liver	% Inhibition	mu/g liver	% Inhibition
Control	12.3 ± 1.30	0	60.7 ± 10.4	0
Leupeptin	1.9 ± 0.30	85.0	24.3 ± 4.30	40.0
E-64-(L)	2.15 ± 0.52	82.5	31.9 ± 3.23	47.4
E-64-(D)	3.89 ± 0.49	68.4	36.7 ± 1.65	39.5
E-64-a-z (RR)	6.34 ± 0.26	48.5	46.0 ± 2.19	24.2
E-64-a-z (SS)	6.08 ± 1.53	50.6	37.2 ± 3.76	38.7
E-64-a (SS)	6.76 ± 0.77	44.9	41.0 ± 4.32	32.4
E-64-b (RR)	12.44 ± 0.92	0	61.4 ± 7.82	0
E-64-b (SS)	4.88 ± 0.91	60.3	48.2 ± 4.37	28.3
E-64-c (RR)	6.56 ± 0.68	46.5	54.0 ± 3.29	11.0
E-64-c (SS)	1.12 ± 0.25	90.9	35.9 ± 6.45	40.9

-vealed inhibitory activities cathepsin B and cathepsin L *in vivo*.
E-64-b (RR) was as active as E-64 (L) *in vitro* but was completely
inactive *in vivo* (Table 4). The discrepancy between the effects of
some E-64 derivative *in vitro* and *in vivo* may be due to differences
in their permeabilities. Inhibition of cathepsin H and cathepsin D
activities by those inhibitors were not observed. Time course of

N. Katunuma et al.

inhibition of cathepsin B and cathepsin L by E-64-(L), E-64-c (SS) and leupeptin was shown in Fig. 5. The activity of cathepsin B was

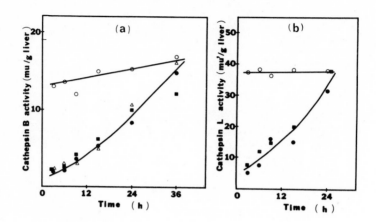

Fig. 5. Time course of change in cathepsin B (a) and cathepsin L (b) activities in isolated lysosomal fractions from rat liver treated with E-64, Leupeptin and E-64-c (SS). Rats weighing 100 to 120 g were given a single i.p. injection of 1.0 mg/100 g body weight of protease inhibitors. At the times indicated, rats were killed, the lysosomal fraction was isolated and cathepsin B and cathepsin L activities were determined. Without inhibitor (O——O), Leupeptin (△——△), E-64 (L) (●——●), E-64-c (SS) (■——■).

inhibited within 1 h after injection of these compounds and the inhibition persisted for at least 6 h, but gradually disappeared within 24-36 h. The activity of cathepsin L was inhibited in a similar fashion to that of cathepsin B but returned to the control level more rapidly. One possible explanation for the difference in the rate of return of cathepsin B and cathepsin L after injection of E-64 is due to differences in the degradation rate of both enzymes. Thus, leupeptin and E-64 should be useful in analysis of the mechanism of protein turnover.

Turnover Rates of Liver Aldolase in Control and Protease Inhibitors-treated Rats

Rats were injected with leupeptin or E-64 once every 8 h. Injections of the inhibitors was started by 1 h after injection of $[C^{14}]NaHCO_3$. The results in Fig. 6 show that the apparent turnover rate of aldolase was reduced significantly by leupeptin or E-64 (Kominami and colleagues, 1980). In E-64-treated animals the half lives of the proteins and especially aldolase were longer. The half lives of aldolase, lactic dehydrogenase and total soluble proteins in control animals

were estimated as 2.4 days, 3.5 days and 2.1 days, respectively.
Stronger effect of E-64 than that of leupeptin to reduce turnover
rates of marker proteins was observed.

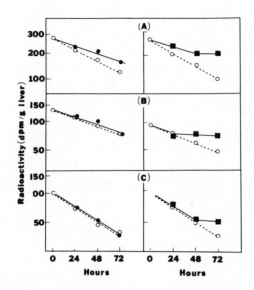

Fig. 6. Effects of protease inhibitors on the
rates of degradation of fructose biphosphate
aldolase (A), lactic dehydrogenase (B) and total
soluble proteins (C). Rats weighing 140 to 160 g
were given a single injection of 100 μCi of
[^{14}C]NaHCO$_3$ and then injected with an inhibitor
and every 8 h starting one h later. Dose of 0.8
mg/rat of leupeptin and 2.0 mg/rat of E-64 were
given. Leupeptin was injected intraperitoneally
in 0.9% saline and E-64 was injected subctaneously
as an emulsion with oil. Rats were killed at the
indicated times after the first injection of
inhibitor, and aldolase and lactic dehydrogenase
were isolated from the pooled livers from 3 rats.
Results are shown as total counts of radioactivity
in enzyme protein or protein fractions/g liver.
(O----O), specific radioactivities of control;
(●——●), specific radioactivities after treatment
with leupeptin; (■——■), specific radioactivities
after treatment with E-64.

Injection of those inhibitors did not cause any effect on protein
synthesis. Hopegood and coworkers (1977), Seglen (1978) and Libby
and Goldberg (1978) also found that leupeptin had no effect on protein
synthesis. These results indicate that proteases, which are sensi-
tives to inhibition by E-64 or leupeptin, especially cathepsin L and
cathepsin B may be important in degradation of aldolase.

Analysis of amino acid sequences of cathepsin B was performed in cowork with Dr. H. Takio and Dr. K. Titani (Haward Huges Med. Inst. at the Dept. of Biochem., SJ-70, Univ. of Washington, Seattle, Washington 98195, U.S.A.)

REFERENCES

Barrett, A.J. (1977) Cathepsin B and other thiol proteinases. In A.J. Barrett (Ed.), Proteinases in Mammalian Cells and Tissues, North-Holland, Amsterdam, pp.181-208.
Bond, T.S., and A.J. Barrett (1980) Biochem. J., 189, 17-25.
Dean, R.T. (1975) Nature, 257, 414-416.
Hanada, K., M. Takami, M. Yamagishi, S. Ohmura, J. Sawada, and I. Tanaka (1978) Agr. Biol. Chem., 42, 523-528.
Hanada, K., M. Takami, Ohmura, S., J. Satada, T. Seki, and I. Tanaka (1978) Agr. Biol. Chem., 42, 529-536.
Hashida, S., T. Towatari, E. Kominami, and N. Katunuma (1980) J. Biochem., Accepted for publication.
Hopegood, M.F., M.G. Clark, and F.J. Ballard (1977) Biochem. J., 164, 399-407.
Kärgel, H.J., R. Dettmer, G. Etzold, H. Kirschke, P. Bohley, and J. Langner (1980) FEBS Lett., 114, 257-260.
Kirschke, H., J. Langner, B. Wiederanders, S. Ansorge, and P. Bohley (1977) Eur. J. Biochem., 74, 293-301.
Kominami, E., S. Hashida, and N. Katunuma (1980) Biochem. Biophys. Res. Commun., 93, 713-719.
Libby, P., and A.L. Goldberg (1978) Science, 199, 534-536.
Nakai, N., K. Wada, K. Kobashi, and J. Hase (1978) Biochem. Biophys. Res. Commun., 83, 881-885.
Seglen, P.P. (1978) Biochem. J., 174, 469-474.
Seglen, P.O., B. Grinde, and A.E. Solheim (1979) Eur. J. Biochem., 95, 215-225.
Singh, H., and G. Kalnitsky (1978) J. Biol. Chem., 253, 4319-4326.
Towatari, T., and N. Katunuma (1978) Biochem. Biophys. Res. Commun., 83, 513-520.
Towatari, T., Y. Kawabata and N. Katunuma (1979) Eur. J. Biochem., 102, 279-289.
Towatari, T., K. Tanaka, D. Yoshikawa, and N. Katunuma (1976) FEBS Lett., 67, 284-288.
Towatari, T., K. Tanaka, D. Yoshikawa, and N. Katunuma (1978) J. Biochem., 84, 659-671.

CATHEPSIN L

H. Kirschke*, H.–J. Kärgel**, S. Riemann* and P. Bohley*

**Physiologisch–chemisches Institut der Martin–Luther–Universität Halle,
Wittenberg, DDR–402 Halle (Saale), GDR*

***Zentralinstitut für Molekularbiologie der Akademie der Wissenschaften der DDR,
DDR–1115, Berlin–Buch, GDR*

ABSTRACT

A view is given on the distribution of cathepsin L. This enzyme has
been identified in each tissue of all species so far tested.
Cathepsin L has a marked ability to hydrolyse proteins. It has an
action on soluble collagen by conversion of ß- and higher components to mainly α-chains.
The specificity of cathepsin L has been revealed using insulin B
chain and glucagon as substrates: those peptide bonds are hydrolysed which have an apolar amino acid in subsite position P_2.
Cathepsin L has been found to be a glycoprotein which binds to concanavalin A-Sepharose.

KEYWORDS

Cathepsin L; cathepsin B; cathepsin H; cysteine proteinase; substrate specificity

INTRODUCTION

The purpose of this paper is to give details of some new properties
and distribution of cathepsin L rather than a full review of this
enzyme.
While studying the in-vitro degradation of rat liver cytosol pro-

teins by proteinases in different cell organelles we detected a proteinase in rat liver lysosomes characterized by its particularly
high activity in degrading short-lived cytosol proteins as well as
its ability to degrade other proteins (Bohley and colleagues, 1971).
This proteinase, now known as cathepsin L, is one of the five lysosomal cysteine proteinases: cathepsins B, H, L, N and S. We have
shown that cathepsin L and H are ubiquitous in mammals comparable
in this respect to cathepsin B.

The special physiological role of each of the cysteine proteinases
is at present unknown. All that we can attempt at this stage is a
description of the substrate specificity and susceptibility to specific inhibitors in the hope that we shall be able to get some insight into their physiological functions.

DISTRIBUTION

Cathepsin L has been isolated from rat liver lysosomes (Kirschke
and colleagues, 1972, 1974, 1977; Towatari and colleagues, 1976,
1978; DeMartino, Doebber and Miller, 1977; Lynen, Sedlaczek and
Wieland, 1978), rat kidney (Strewler and Manganiello, 1979) and
rabbit skeletal muscle (Okitani and colleagues, 1980). Barrett
(1980) reported on the presence of cathepsin L in human liver.
Using an inhibitor of cathepsin L (Z-Phe-Phe-diazomethyl ketone,
kindly contributed by E. Shaw) which is specific for cathepsin L
under appropriate conditions (Bohley and colleagues, unpublished
results), we have identified cathepsin L in different organs of the
rat, rabbit, ox, pig, pigeon and carp. Cathepsin L also occurs in
L-cells, Jensen sarcomas, Ehrlich ascites tumor cells and in every
human tumor and metastasis so far tested. Preliminary results indicate the existence of this enzyme in Euglena gracilis (Krauspe,
unpublished results). An enzyme isolated from slime moulds Dictyostelium discoideum and characterized by Gustafson and Thon (1979)
seems to have the same properties as cathepsin L.

These data suggest that cathepsin L evolved at an early stage in
phylogeny.

The question then arises why this enzyme had not been detected
earlier. The most probable reason is that intracellular proteins
which are the best substrates of cathepsin L (especially short
lived cytosol proteins; Bohley and colleagues, 1972) usually had

not been used as substrates to identify intracellular proteinases.
Two further points are the instability of cathepsin L on storage,
even in tissues and its similarity to cathepsin B in regard to molecular weight, isoelectric point and catalytical properties such
as its ability to hydrolyse benzoyl arginine amide and proteins
(including native enzymes).

SUBSTRATES AND PROPERTIES

All cysteine proteinases can be identified by their action on different substrates, though we do not yet have selective substrates
for cathepsins H, L, N and S.

TABLE 1   Specific Activities of Cathepsins L and B

Substrate	Cathepsin L	Cathepsin B
	units x mg^{-1}	
Azocasein, pH 6.0	5,500	205
Collagen, insoluble, pH 6.0	0.60	0.14
Collagen, insoluble, pH 3.5	5.30	0.74
Z-Arg-Arg-2-NNap, pH 6.0	0.65	78.00
Bz-Arg-2-NNap, pH 6.0	0.08	2.50
Bz-Arg-NH$_2$, pH 6.0	2.00	2.70
Z-Lys-OPhNO$_2$, pH 5.4	45.00	12.00

1 unit: for azocasein: 1 μg substrate degraded in
1 min at 37° C
for collagen: 1 μMol Hypro released in 1 min
at 24° C
for synthetic substrates: 1 μMol substrate
degraded in 1 min at 37° C

Abbreviations: -2-NNap, -2-naphthylamide;
-OPhNO$_2$, -4-nitrophenyl ester

Benzoyl arginine amide and benzyloxycarbonyl lysine 4-nitrophenyl
ester are, so far, the only low molecular weight substrates hydrolysed by cathepsin L (Kirschke and colleagues, 1977; Strewler and
Manganiello, 1979), but these are also substrates of cathepsins B
and H. It is now wellknown that cathepsin L has a marked ability to
hydrolyse proteins. The specific activity of cathepsin L with proteins as substrates exceeded that of all other cysteine proteinases.

This has been shown for a mixture of various cytosol proteins (Bohley and colleagues, 1972), for collagen (Kirschke, Bohley and Barrett, in preparation), actin, myosin, myofibrils (Sohar and colleagues, 1979) and others.

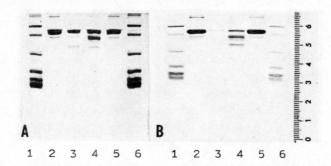

Fig. 1.   Degradation of actin by cathep-
             sins L, B and H.

Samples of actin and enzymes were incubated 1 h (A) and 2 h (B) and then subjected to SDS electrophoresis (12.5% gel).

(1 and 6) molecular weight standards –
albumin, ovalbumin, chymotrypsinogen,
soya bean trypsin inhibitor, myoglobin,
cytochrom c; (2) actin standard; (3) actin and cathepsin L (1.2 μg); (4) actin
and cathepsin B (2.5 μg); (5) actin and
cathepsin H (2.5 μg)

The action of cathepsin L on soluble collagen was shown to be similar to that of cathepsin B by the conversion of ß- and higher components to mainly α-chains (Kirschke, Bohley and Barrett, in preparation). The degradation of insoluble collagen by cathepsin L at pH 3.5
was about ten times faster than at pH 6.0 (TABLE 1) and the specific
activity at pH 3.5 exceeded that of cathepsin B by a factor of ten
and that of cathepsin N from bovine spleen (Ducastaing and Etherington, 1978) by a factor of six (Kirschke and colleagues, 1980).
The degradation of chymotrypsinogen by cathepsin L has been shown to
produce some active chymotrypsin (Maercker, unpublished results),
but on the whole cathepsin L causes an overall degradation of several proteins so far tested as it is shown for actin in Fig. 1. (Actin was kindly contributed by I. Sohar).
Cathepsin L degrades proinsulin and perhaps only fortuitously gene-
rates some insulin (Ansorge, Kirschke and Friedrich, 1977).

The same may be true for inactivation of several enzymes by cathepsin L whereby the inactivation occurs during the overall degradation and perhaps not by a limited and specific proteolytic attack.
All these results imply a very broad specificity of cathepsin L.
We have determined the bonds cleaved in insulin B chain and in glucagon by cathepsin L so as to get some insight in its specificity.

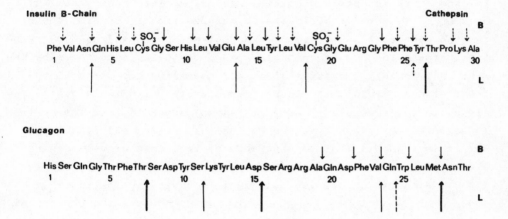

Fig. 2.   Specificity of cathepsin L and cathepsin B on insulin B
          chain and glucagon.

The data for the action of cathepsin L on insulin B chain (Kärgel
and colleagues, 1980) and glucagon (Kärgel, unpublished results)
are compared with the cleavage points by cathepsin B on insulin B
chain (Otto, 1971) and glucagon (Aronson and Barrett, 1978)

In summary these experiments show that cathepsin L hydrolyses those
peptide bonds of insulin B chain and glucagon, which have an apolar
amino acid such as Val, Phe, Leu and Tyr in the position P_2 (using
the nomenclature introduced by Schechter and Berger (1967) . A further hydrophobic residue in P_3 enhances the susceptibility to the
enzyme. The amino acid in P_1 appears to be of little importance to
the attack of cathepsin L, since we found Met, Glu, Asp and Thr in
this position. Nor can we recognize a requirement for a special amino acid in P' position, or any sequence of amino acids adjacent to
the peptide bond to be cleaved.
The studies with insulin B chain as substrate have proved to be in-

sufficient to demonstrate the different specificities of cathepsin L
and cathepsin B (Fig. 2.). In this respect glucagon has been found
to differentiate the action of the two enzymes. Cathepsin B hydrolyses glucagon in an unusual manner by a peptidyldipeptidase action
(Aronson and Barrett, 1978), whereas cathepsin L cleaves those peptide bonds with a hydrophobic amino acid in position P_2. Both enzymes act with a different specificity - but, by accident, there are
two identical split positions.

In general it is not been possible to extrapolate from specificity
studies of cysteine proteinases on polypeptides to predict the attack on synthetic substrates. This contrasts with the good correlation between the action on proteins and synthetic substrates by serine proteinases, such as trypsin and chymotrypsin. Cathepsin B,
for instance, hydrolyses synthetic substrates with two arginine residues in the molecule (Z-Arg-Arg-2-NNap) several times faster than
those which contain only one arginine (Bz-Arg-2-NNap) (TABLE 1).
But it is evident, that the sequence of two consecutive arginine residues of glucagon does not enhance the susceptibility by cathepsin B. Cathepsin L acts very slowly in cleaving synthetic substrates
with a sequence of three hydrophobic amino acids although work with
polypeptides suggested that such compounds would be good substrates.
For example, benzyloxycarbonyl-Phe-Leu amide is hydrolysed very
slowly (Salama, unpublished results), whereas the enzyme has a
strong action on benzoyl arginine amide and benzyloxycarbonyl lysine
4-nitrophenyl ester.

However, the action of cathepsin L on polypeptides with hydrophobic
amino acids in P_1, P_2 and P_3 has correlated surprisingly well with
the inhibitory action of diazomethyl ketone derivatives of Z-Phe-
Phe and Z-Phe-Ala on the enzyme (Kirschke and colleagues, 1980).
Cathepsin L is a glycoprotein. We determined carbohydrate in the molecule by the Anthron-method. Cathepsin H is a glycoprotein, too, and
its purification by adsorption to Concanavalin A-Sepharose has been
described (Kirschke and colleagues, 1979; Schwartz and Barrett, 1980)
But all experiments of binding active cathepsin L to Con A-Sepharose
resulted a loss of activity of about 90% whereas protein has been recovered almost completely. We found that the mercury derivative of ca
thepsin L can be eluted from Con A-Sepharose without inactivation.
There is reason to suggest that binding of cathepsin L to Con A-Sepha
rose by its carbohydrate residue enhances the autolysis of the enzyme

REFERENCES

Ansorge, S., H. Kirschke, and K. Friedrich (1977). Conversion of proinsulin into insulin by cathepsins B and L from rat liver lysosomes. Acta biol. med. germ., 36, 1723-1727.

Aronson, N. N., and A. J. Barrett (1978). The specificity of cathepsin B. Biochem. J., 171, 759-765.

Barrett, A. J. (1980). Human lysosomal thiol proteinases. In Proceedings FEBS Special Meeting on Enzymes, Dubrovnik-Cavtat.
In press.

Bohley, P., H. Kirschke, J. Langner, S. Ansorge, B. Wiederanders,
and H. Hanson (1971). Intracellular protein breakdown. In A. J.
Barrett, and J. T. Dingle (Eds.), Tissue Proteinases. North-
Holland Publishing Co., Amsterdam and London. Chap. 9, pp.
187-219.

Bohley, P., C. Miehe, M. Miehe, S. Ansorge, H. Kirschke, J. Langner,
und B. Wiederanders (1972). Intrazellulärer Proteinabbau. V. Be-
vorzugter Abbau kurzlebiger Zytosolproteine durch Lysosomenendo-
peptidasen aus Rattenleber. Acta biol. med. germ., 28, 323-330.

DeMartino, G. N., T. W. Doebber, and L. L. Miller (1977). Pepstatin-
insensitive proteolytic activity of rat liver lysosomes. J. Biol.
Chem., 252, 7511-7516.

Ducastaing, A., and D. J. Etherington (1978). Purification of bovine
spleen collagenolytic cathepsin (cathepsin N). Biochem. Soc.
trans., 6, 938-940.

Gustafson, G. L., and L. A. Thon (1979). Purification and characterization of a proteinase from Dictyostelium discoideum. J. Biol.
Chem., 254, 12471-12478.

Kärgel, H.-J., R. Dettmer, G. Etzold, H. Kirschke, P. Bohley, and
J. Langner (1980). Action of cathepsin L on the oxidized B-chain
of bovine insulin. FEBS Lett., 114, 257-260.

Kirschke, H., J. Langner, B. Wiederanders, S. Ansorge, und P. Bohley
(1972). Intrazellulärer Proteinabbau. IV. Isolierung und Charakterisierung von Peptidasen aus Rattenleberlysosomen. Acta biol.
med. germ., 28, 305-322.

Kirschke, H., J. Langner, B. Wiederanders, S. Ansorge, P. Bohley,
and H. Hanson (1974). Cathepsin L and proteinases with cathepsin B 1-like activity from rat liver lysosomes. In H. Hanson,
and P. Bohley (Eds.), Intracellular Protein Catabolism. Wissen-

schaftliche Beiträge 1974/6 (R 27), Martin-Luther-Universität Halle-Wittenberg, Halle (Saale). pp. 210-217.

Kirschke, H., J. Langner, B. Wiederanders, S. Ansorge, and P. Bohley (1977). Cathepsin L. A new proteinase from rat-liver lysosomes. Eur. J. Biochem., 74, 293-301.

Kirschke, H., J. Langner, B. Wiederanders, S. Ansorge, and P. Bohley (1979). The role of cathepsins L and H from rat liver lysosomes in protein degradation. In S. Rapoport, and T. Schewe (Eds.), Processing and Turnover of Proteins and Organelles in the Cell, Vol. 53. Pergamon Press, Oxford and New York. pp. 107-115.

Kirschke, H., J. Langner, S. Riemann, B. Wiederanders, S. Ansorge, and P. Bohley (1980). Lysosomal cysteine proteinases. In D. Evered, and J. Whelan (Eds.), Protein Degradation in Health and Disease. Excerpta Medica, Amsterdam, Oxford and New York. pp. 15-35.

Lynen, A., E. Sedlaczek, and O. H. Wieland (1978). Partial purification and characterization of a pyruvate dehydrogenase-complex-inactivating enzyme from rat liver. Biochem. J., 169, 321-328.

Okitani, A., U. Matsukura, H. Kato, and M. Fujimaki (1980). Purification and some properties of a myofibrillar protein-degrading protease, cathepsin L, from rabbit skeletal muscle. J. Biochem., 87, 1133-1143.

Otto, K. (1971). Cathepsins B 1 and B 2. In A. J. Barrett, and J. T. Dingle (Eds.), Tissue Proteinases. North-Holland Publishing Co., Amsterdam and London. Chap. 1, pp. 1-28.

Schechter, I., and A. Berger (1967). On the size of the active site in proteases. I. Papain. Biochem. Biophys. Res. Commun., 27, 157-162.

Schwartz, W. N., and A. J. Barrett (1980). Human cathepsin H. Biochem. J., in press.

Sohar, I., Ö. Takacs, F. Guba, H. Kirschke, and P. Bohley (1979). Degradation of myofibrillar proteins by cathepsins from rat liver lysosomes. In H.-J. Hütter, und P. Bohley (Hrsg.), Proteinstoffwechsel, Teil 2. Wissenschaftliche Beiträge 1979/23 (R. 45), Martin-Luther-Universität Halle-Wittenberg, Halle (Saale). pp. 121-125.

Strewler, G. J., and V. C. Manganiello (1979). Purification and characterization of phosphodiesterase activator from kidney. J. Biol. Chem., 254, 11891-11898.

Towatari, T., K. Tanaka, D. Yoshikawa, and N. Katunuma (1976). Separation of a new protease from cathepsin B 1 of rat liver lysosomes. FEBS Lett., 67, 284-288.

Towatari, T., K. Tanaka, D. Yoshikawa, and N. Katunuma (1978). Purification and properties of a new cathepsin from rat liver. J. Biochem., 84, 659-671.

HUMAN CATHEPSINS B AND H: ASSAY, ACTIVE—SITE TITRATION AND SELECTIVE INHIBITION

A.J. Barrett

*Biochemistry Department, Strangeways Laboratory,
Cambridge CB1 4RN, England*

ABSTRACT

Methylcoumarylamides of Z-Phe-Arg- and Arg- have proved to be excellent assay substrates for cathepsins B and H, respectively. The inhibitor, E-64, reacts stoichiometrically with cysteine proteinases, and often the reaction is rapid enough to permit accurate active-site titration. Synthetic analogues of E-64 differ greatly in their reactivities toward papain, cathepsin B and cathepsin H, and there is the prospect of designing new, highly specific inhibitors.

Keywords: Cathepsin B; cathepsin H; aminomethylcoumarin; E-64.

INTRODUCTION

The purpose of this paper is to report results of some recent studies with new substrates and inhibitors for human cathepsins B and H.

The enzymes were isolated from human liver as described previously (Barrett, 1973; Schwartz & Barrett, 1980). Arg-NMec (arginine 4-methyl-7-coumarylamide) and Z-Phe-Arg-NMec (benzyloxycarbonyl-phenylalanyl-arginine 4-methyl-7-coumarylamide) were obtained from either the Peptide Research Foundation, Osaka, Japan, or Bachem Feinchemikalien A.G., CH-4416 Bubendorf, Switzerland. (We favour the abbreviation '-NMec' for these compounds as indicating '-N-linked methyl-coumarin'. The relationship of the fluorogenic ester methylumbelliferones is clarified by the use of '-OMec' for these).

Other chemicals were from B.D.H. Chemicals, Poole, Dorset, England. E-64 (L-trans-epoxysuccinyl-leucyl-agmatine), and related synthetic compounds were kindly provided by, and used in collaboration with, Dr. K. Hanada, Taisho Pharmaceutical Co. Ltd., Tokyo. Benzyloxycarbonyl-phenylalanyl-alanyl-diazomethane (Z-Phe-Ala-CH$_2$N$_2$) was kindly given by Dr. Elliott Shaw, Brookhaven National Laboratory, New York.

ASSAY OF THE ENZYMES

Much work has been done with the lysosomal cysteine proteinases by use of assays with benzoyl-arginine 2-naphthylamide (Bz-Arg-NNap), often by the method that I

103

developed several years ago (Barrett, 1972, 1976). This substrate lacks
specificity, however, being cleaved by cathepsin B and cathepsin H (and there may
be other 'BANA-hydrolases'). Some other naphthylamides are more selective (see
Barrett, 1980), but the carcinogenicity of 2-naphthylamine makes all of these
compounds undesirable in the laboratory. Fortunately, the 4-methyl-7-coumarylamide
leaving group is proving a more than adequate substitute. We tested a range of
commercially available aminomethylcoumaryl substrates that had been designed for
serine proteinases for sensitivity to the cysteine proteinases. As I have reported
(Barrett, 1980), Z-Phe-Arg-NMec proved to be an excellent substrate for cathepsin B
(Table 1), and Arg-NMec was good for cathepsin H. Each enzyme was over 1000-fold
more active on its appropriate substrate than was the other. The substrates are
equally suitable for continuous rate or stopped assays. The assay for cathepsin B
was extremely sensitive, easily being adapted to the quantification of 1 ng or less
of the enzyme.

Table 1   Kinetic constants of some substrates of cathepsin B

	$k_{cat.}$ (s^{-1})	$10^3 K_m$ (M)	$k_{cat.}/K_m$ ($s^{-1}.M^{-1}$)
Bz-Arg-2-NNap	9	4.3	2.1×10^3
Z-Arg-Arg-2-NNap	61	0.19	3.2×10^5
Z-Val-Lys-Lys-Arg-2-NNapOMe	68	0.14	4.9×10^5
Z-Phe-Arg-NMec	91	0.285	3.2×10^5

Note: $k_{cat.}$ values in the above table (taken from Barrett, 1980) were calculated
on the assumption that the enzyme preparation contained one mole of active
sites/25,000 g of protein. In fact, titration with E-64 (see below) has now shown
that apparently pure enzyme contains only one mole of active sites/50,000 g, so
the values shown should probably be doubled for fully active enzyme.

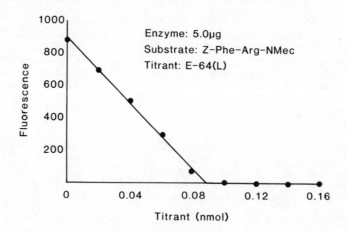

Fig. 1. Titration of a solution of cathepsin B with E-64

INHIBITION BY E-64

Hanada and his collaagues have described the inhibition of papain, bromelain and ficin by E-64, a compound produced by a species of Aspergillus (Hanada et al., 1978 a-c). The inhibition of rat cathepsin B and L, and the calcium-dependent cysteine proteinase have been reported by others (Towatari et al., 1978; Sugita et al., 1980), and the compound seems to be very selective for the cysteine proteinases. Inhibition is irreversible, the active site thiol group attacking the epoxide ring and forming a stable thioether.

E-64 reacts stoichiometrically with cysteine proteinases, even in the presence of low molecular weight thiols. Titration of very dilute solutions of the enzyme with the inhibitor, and assay of residual activity with Z-Phe-Arg-NMec, has been found to provide an excellent method of determining the absolute molarity of active enzyme, and of calibrating the assay (Fig. 1).

RATES OF REACTION FOR E-64 AND SOME ANALOGUES

The rate of inactivation of papain (and cathepsin B) by E-64 is so rapid as to have been impossible to measure by the conventional methods. What was needed was an assay method capable of working with extremely dilute enzyme solutions, and a computational method that took into account the depletion of the very dilute inhibitor during the reaction, i.e. gave a second order rate constant directly. The assay procedures with the methylcoumarylamides were very suitable, and we developed an appropriate technique and computer program to obtain the data (A. J. Barrett, C. G. Knight and K. Hanada, unpublished results).

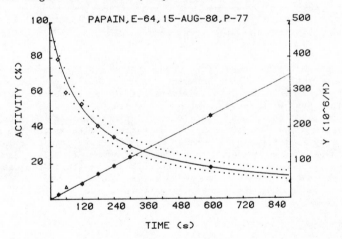

Fig. 2. Computer plot used to determine rate of reaction of cathepsin B with E-64

We found it necessary to work with enzyme and inhibitor concentrations down to 2×10^{-8}M. After reaction periods of 30 seconds to 10 minutes, samples were further diluted 40-fold in the 10 minute assay with Z-Phe-Arg-NMec. A second order rate constant for the inactivation was obtained by use of the computer program, which plots activity and the estimate of $k_2 \cdot t$, against time (t) for each sample (Fig. 2). Out-lying points are eliminated by inspection, and a best estimate of k_2 is obtained by linear regression analysis of the others. This value is used to draw a curve through the experimantal points, and further curves corresponding to values of k_2 20% above and below the 'best' value. These are expected to enclose all of the experimental points. Some of the results obtained are shown in Table 2.

Table 2   Rates ($M^{-1}.s^{-1}$) of inactivation of cysteine proteinases by some E-64 analogues

	Compound	Papain	Cathepsin B	Cathepsin H
E-64	HO-Eps-Leu-Agm	388,000	64,700	4,000
E-64(D)	HO-D-Eps-Leu-Agm	60,900	1,700	65
Ep-459	HO-Eps-Leu-amido-(4-amino)butane	10,000	69,500	-
Ep-475	HO-Eps-Leu-amido-(3-methyl)butane	197,400	87,200	-
Ep-434	EtO-Eps-Leu-Leu-OMe			
(a)	-L-L-L-	2,200	17,230	25
(b)	-L-L-D-	870	376	325
(c)	-D-D-L-	0.8	44	2.7
Z-Phe-Ala-CH$_2$N$_2$		35,000	1,220	-

Note: the data are taken from the unpublished work of A. Kembhavi, A. J. Barrett and K. Hanada. Abbreviations: Eps, trans-epoxysuccinic acid; -, not determined.

It will be noted that the naturally occuring L-isomer of trans-epoxysuccinic acid reacted far more rapidly than the D-isomer, especially with the human enzymes. The compounds Ep-459, Ep-434a and Ep-434c were interesting for their greater reactivity with cathepsin B than with papain, which was generally the more reactive. The highest rate of reaction we have found so far for cathepsin B has been that with Ep-475. Cathepsin H has not proved nearly as reactive with any of the new compounds as with E-64 itself, but compound Ep-434b was distinctive for its relatively good rate with cathepsin H, as compared to the other enzymes. Finally, it was found that Z-Phe-Ala-CH$_2$N$_2$, the excellent selective inhibitor of cysteine proteinases described by Watanabe, Green & Shaw (1979) has distinctly lower rates of reaction with papain and cathepsin B than E-64 and several of the analogues.

CONCLUSIONS

The methylcoumarylamide substrates have great potential for use with the cysteine proteinases; they can exhibit a high degree of sensitivity and selectivity.

E-64 reacts rapidly and specifically enough with some cysteine proteinases to be used as an active-site titrant for the determination of the molar concentration of active sites, and when it is used in conjunction with the new substrates, the determination can be made with very small amounts of enzyme indeed.

Synthetic analogues of E-64 show widely differing reaction rates with papain, cathepsin B and cathepsin H, and there is the prospect of designing highly selective inhibitors.

ACKNOWLEDGEMENTS

I am grateful to Dr. K. Hanada, Dr. C. Graham Knight and Mrs. Asha Kembhavi for permission to mention preliminary results of our collaborative studies. I also thank Mrs. Molly A. Brown for her excellent technical assistance.

REFERENCES

Barrett, A. J. (1972). A new assay for cathepsin B1 and other thiol proteinases. Anal. Biochem. 47, 280-293.

Barrett, A. J. (1973). Human cathepsin B1. Purification and some properties of the enzyme. Biochem. J. 131, 809-822.

Barrett, A. J. (1976). An improved color reagent for use in Barrett's assay of cathepsin B. Anal. Biochem. 76, 374-376.

Barrett, A. J. (1980). Fluorimetric assays for cathepsin B and cathepsin H with methylcoumarylamide substrates. Biochem. J. 187, 909-912.

Hanada, K., Tamai, M., Morimoto, S., Adachi, T., Ohmura, S., Sawada, J. & Tanaka, I. (1978). Inhibitory activities of E-64 derivatives on papain. Agr. Biol. Chem. 42, 537-541.

Hanada, K., Tamai, M., Ohmura, S., Sawada, J., Seki, T. & Tanaka, I. (1978). Structure and synthesis of E-64, a new thiol protease inhibitor. Agr. Biol. Chem. 42, 529-536.

Hanada, K., Tamai, M., Yamagishi, M., Ohmura, S., Sawada, J. & Tanaka, I. (1978). Isolation and characteriztion of E-64, a new thiol protease inhibitor. Agr. Biol. Chem. 42, 523-528.

Schwartz, W. N. & Barrett, A. J. (1980). Human cathepsin H. Biochem. J. in the press.

Sugita, H., Ishiura, S., Suzuki, K. & Imahori, K. (1980). Inhibition of epoxide derivatives on chicken calcium-activated neutral protease (CANP) in vitro and in vivo. J. Biochem. 87, 339-341.

Towatari, T., Tanaka, K., Yoshikawa, D. & Katunuma, N. (1978). Purification and properties of a new cathepsin from rat liver. J. Biochem. 84, 659-671.

Watanabe, H., Green, G. D. J. & Shaw, E. (1979) A comparison of the behaviour of chymotrypsin and cathepsin B towards peptidyl diazomethyl ketones. Biochem. Biophys. Res. Commun. 89, 1354-1360.

THE BOVINE CYSTEINE PROTEINASES, CATHEPSIN B, H AND S

P. Ločnikar, T. Popović, T. Lah, I. Kregar, J. Babnik, M. Kopitar
and V. Turk

*Department of Biochemistry, J. Stefan Institute, E. Kardelj University,
61000 Ljubljana, Yugoslavia*

ABSTRACT

Cathepsin B, H and S were purified simultaneously from bovine spleen by a method involving acid extraction, acetone fractionation, gel chromatography on Sephadex G-50, covalent chromatography on thiol Sepharose 4B and chromatography on CM - cellulose. All three enzymes were isolated in electrophoretically pure forms. The enzymes exist in multiple forms. Molecular weight, substrate specificity, effect of inhibitors and some other enzymatic properties are presented.

KEYWORDS

Cathepsin B; cathepsin S; cathepsin H; cysteine proteinase; thiol proteinase; lyso-somal proteinase.

INTRODUCTION

During the last few years, important progress has been made in the study of intra-cellular cysteine (thiol) proteinases. Cathepsin B, the most studied enzyme of this group, was recently crystallized and its partial amino acid sequence shows a high resemblance to the papain molecule (Towatari and co-workers, 1979; Katunuma and co-workers, in this book). Beside cathepsin B, the number of known lysosomal cys-teine proteinases has increased: cathepsins L, S, H and some other similar enzymes, present in different tissues and species were found (Kirschke and co-workers, 1977a, 1977b, 1980, in this book; De Martino and co-workers, 1977; Turk and co-workers, 1980). The cysteine proteinases so far known were isolated by various purification procedures. Their molecular weights are in the range of about 21,000 - 30,000.Their isoelectric points are similar, and leupeptin and other known inhibitors also inhibit cathepsin B, H, L, and S, although to a different extent. Therefore the task of separa-ting them is a rather difficult one.

In this communication we present a procedure for the simultaneous isolation of bovi-ne spleen cathepsin B, S and H and the properties of these enzymes.

METHODS

Fresh bovine spleens were used as the source of the enzymes. The purification procedure is based on the method for the isolation of bovine lymph node cathepsin B and
H (Zvonar and co-workers, 1979) with some modifications. The purification steps
include acid extraction at pH 4.0, concentration by ultrafiltration and acetone fractionation (45–65%), gel chromatography on Sephadex G–50, covalent chromatography
on thiol Sepharose, and as the last step, CM-cellulose chromatography which results
in the separation of cathepsin B, S and H. The details of the purification procedure
and the enzymic properties will be published elsewhere (submitted for publication).

The activity of cathepsin B against α-N-benzoyl-D,L-arginine- β -naphthylamide
(BANA) was assayed as described by Barrett (1972). Cathepsin H activity was measured against L-Leu- β -naphthylamide (Leu- β -NA) and the amount of released
β -naphthylamine was determined as previously mentioned. Cathepsin S activity was
determined according to the Anson method (1939) using hemoglobin as substrate in
the presence of 10^{-7} M pepstatin. Polyacrylamide gel electrophoresis under non-
denaturing conditions was carried out on 7% gels, at pH 9.5 by the method of Davis
(1964). The molecular weight was determined by sodium dodecyl sulphate (SDS)
disc gel electrophoresis as described by Weber and Osborn (1969) using 10% gels.
Gels were stained for protein with Coomassie blue. The native molecular weight of
enzymes was determined by gel filtration (Andrews, 1965). Isoelectric focusing was
performed on a Desaga apparatus (Desaga, FRG) as described by the manufacturer.
Protein hydrolyzates were analyzed for amino acids on a Beckman model 118 CL
analyzer. Circular dichroism spectra of the enzymes were measured in a Jobin-
Yvon Mark III dichrograph. Antisera were prepared in New Zealand white rabbits by
repeated intramuscular injections of purified cathepsin S emulsified with complete
Freund's adjuvant. The Ouchterlony double immunodiffusion test and immunoelectrophoresis were performed.

RESULTS AND DISCUSSION

The general scheme summarizing the procedure for purification of cathepsin B, S,
and H is shown in Fig. 1

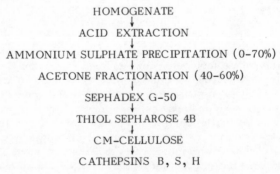

HOMOGENATE

ACID EXTRACTION

AMMONIUM SULPHATE PRECIPITATION (0–70%)

ACETONE FRACTIONATION (40–60%)

SEPHADEX G–50

THIOL SEPHAROSE 4B

CM–CELLULOSE

CATHEPSINS B, S, H

Fig. 1. Purification scheme of cysteine proteinases.

The last purification step (Fig. 2) shows the stepwise elution profiles from CM-cellulose of cathepsin B, S and H. The first two fractions from this procedure correspond
to cathepsin B activity (two multiple forms), whereas the last two proteolytically
active peaks are cathepsin S and H, respectively.

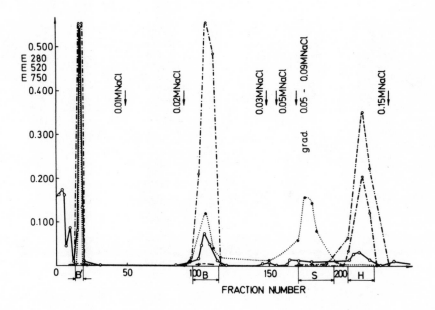

Fig. 2. Separation of cysteine proteinases on CM-cellulose:
B'and B represent multiple forms of cathepsin B; S is cathepsin S and H is cathepsin H activity. Starting elution buffer:
0.02 M Na-acetate, 1 mM EDTA, pH 5.0. Column: 1 x 14.5cm,
flow rate 9.6 ml/hr, fraction/20 min. ———A_{280}: protein;
-.-. A_{520}: activity on BANA; -..- A_{520}: activity on Leu-β-NA;
.... A_{750}: activity on hemoglobin.

Analytical polyacrylamide gel electrophoresis of each of the separated cysteine proteinases is presented in Fig. 3a. A single protein band was obtained for cathepsin B,
S and H, when a high protein concentration was applied. Fig. 3b shows the SDS polyacrylamide gels of three enzymes with a single polypeptide chain. This is the first

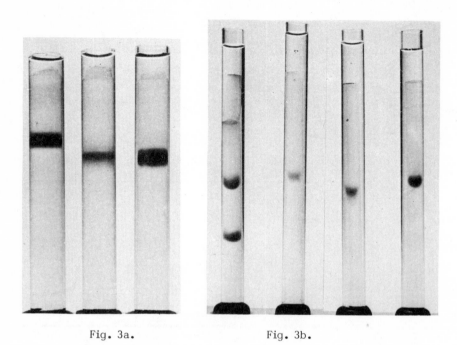

Fig. 3a.   Fig. 3b.

Fig. 3. Polyacrylamide gel electrophoresis in the absence (3a)
and in the presence of 1% SDS; from left to right: cathepsin B,
S and H; (3b) polypeptide composition: standards, cathepsin B,
S and H.

report of the simultaneous purification of three cysteine proteinases, isolated in a
pure form. Covalent chromatography on thiol Sepharose 4B represents a very efficient purification step, with a high increase in the specific activity, and where only
cathepsin B, S and H were bound without other contaminating proteins. The bound
enzymes were eluted from thiol Sepharose by the addition of cysteine to the elution
buffer. Covalent chromatography was first introduced for the purification of papain
from Papaya latex (Brocklehurst and co-workers, 1973), and recently applied in our
laboratory for the purification of cathepsin B and H (Zvonar and co-workers, 1979,
1980). In other known purification procedures chromatography on an organomercurial adsorbent was incorporated as an essential step. The method was introduced by
Barrett (1973) in the purification procedure for cathepsin B. Recently the same method was also used for the purification of cathepsin H (Kirschke and co-workers,
1977b; Singh and Kalnitsky, 1978, 1980).

Cathepsin B (EC 3.4.22.1)

The molecular weight of native cathepsin B determined by gel chromatography is about 26,000. A single polypeptide chain of the same molecular weight was obtained by polyacrylamide gel electrophoresis in the presence of SDS and (or without) β-mercaptoethanol. Although cathepsin B is homogeneous by the last mentioned method, heterogeneity must be present in the protein, since on isoelectric focusing it gives two bands for each cathepsin B active peak, indicating four multiple forms within the range pH 4.7-5.5. Cathepsin B is most stable in the narrow pH range between pH 5.5-6.5 and unstable above pH 7. A good synthetic substrate for cathepsin B is BANA (pH optimum is 6.0), whereas Leu-β-NA is not cleaved at all. Protein substrates are azocasein (pH 6.0), insoluble collagen (pH 6.0), and hemoglobin (pH 4.0). Cathepsin B inactivates aldolase but not lactate dehydrogenase, glutamate dehydrogenase (in agreement with the results of Towatari and co-workers, 1978), urease and galactose dehydrogenase. Cathepsin B activity was completely inhibited by $HgCl_2$, 4-chloromercuribenzoate, iodoacetic acid (all in 1 mM concentration), Tos-Phe-CH_2Cl (0.1 mM) and leupeptin (1 μM). Preparing immobilized leupeptin on agarose resin we found reversible binding of cathepsin B to the inhibitor (unpublished results).

Cathepsin H

Bovine cathepsin H hydrolyzed substrates Leu-β-NA and BANA, thus exhibiting both aminopeptidase and endopeptidase activities. This enzyme also shows proteolytic activity on azocasein as substrate (pH 6.0), although the specific activity is lower compared to cathepsin B. This substrate specificity studies can be compared with other reported data (Kirschke and co-workers, 1977b; Singh and Kalnitsky, 1978; Schwartz and Barrett, 1980). The molecular weight of the enzyme obtained by gel chromatography and the SDS polyacrylamide electrophoresis method is 28,000. Isoelectric focusing shows two major multiple forms with pI values of 7.1 and 7.3. The enzyme is most stable around pH 6.0. Compared to cathepsin D, this enzyme is more stable in the acid pH range and even shows activity above pH 7.0.

The tested inhibitors 4-chloromercuribenzoate, iodoacetic acid and iodoacetamide (all in 1 mM concentration) completely inhibit cathepsin H activity, whereas Tos-Phe-CH_2Cl (0.01 mM) inhibits 67% of its activity. With the microbial inhibitor leupeptin we found that 1 mM concentration has practically no effect on cathepsin H activity. We obtained the same results for bovine lymph node cathepsin H (Zvonar and co-workers, 1980), and it is also in agreement with other reported data (Singh and Kalnitsky, 1980; Schwartz and Barrett, 1980). It is interesting to add that under our experimental conditions cathepsin H was not bound to immobilized leupeptin agarose resin, whereas cathepsin B was bound. The explanation is that K_i of leupeptin for cathepsin H (6.9×10^{-6} M) is lower than for cathepsin B (1.4×10^{-8} M) as already described (Schwartz and Barrett, 1980).

Cathepsin S

This cysteine proteinase was originally first isolated in our laboratory from calf lymph nodes (Turnšek and co-workers, 1975) and can also be found in bovine spleen (Turk and co-workers, 1978, 1980). The enzyme hydrolyses hemoglobin as substrate in acid pH, similarly to cathepsin D, but its activity was not affected at all by pepstatin. For routine enzyme assay, the Anson method was used (Anson, 1939).

The enzyme was isolated in an electrophoretically pure form. SDS polyacrylamide gel electrophoresis and ultracentrifugation analysis give a molecular weight of about 23,000 - 25,000, whereas gel filtration showed a lower value of about 20,000. The enzyme displays several multiple forms with pI in the range 6.3 - 6.9. Synthetic substrates such as BANA and Leu- β -NA were not hydrolyzed by cathepsin S. On the

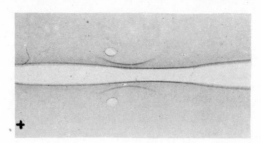

Fig. 4. Immunoelectrophoretic analysis of cathepsin S. Wells contain 5 μl of pure enzyme, trough contain 100 μl of anti-cathepsin S serum.

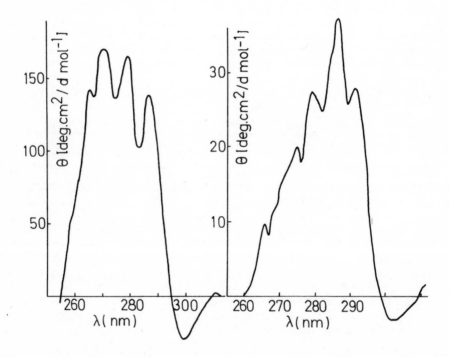

Fig. 5. Near ultraviolet CD spectra of papain (left) and cathepsin S (right).

contrary, cathepsin S prefers the hydrolysis of hemoglobin, azocasein, histones, collagen and other protein substrates. The enzyme inactivates aldolase, whereas the activity of urease, lactate dehydrogenase and glutamatedehydrogenase was unaffected.

The enzyme shows maximal proteolytic activity between pH 3.5 to 6.0, depending on substrate. The enzyme is completely inhibited by $HgCl_2$, iodoacetic acid, iodoacetamide, 4-chloromercuribenzoate (all in 1 mM concentration), and by 1 μM leupeptin. On immunoelectrophoresis, the antibody produced a single arc with the purified protein (Fig. 4), which is also a test for apparent homogeneity.

The structure of cathepsin S was studied by circular dichroism (CD). The CD bands in the near ultraviolet spectral region are presented in Fig. 5. The spectrum exhibits strong broad positive band, representing aromatic acid residues. The spectrum obtained is very similar to that of papain (Su and Jirgensons, 1977), indicating some structural similarities with this enzyme.

In many of its properties cathepsin S is similar to those reported for cathepsin L (Kirschke and co-workers, 1977a; Okitani and co-workers, 1980) and some other cysteine proteinases (De Martino and co-workers, 1977; Evans and Etherington, 1978). Further experiments will show whether this enzyme is different or identical with cathepsin L.

CONCLUSIONS

At least three cysteine proteinases, cathepsin B, S and H are present in bovine spleen and were isolated simultaneously by a new purification procedure which includes co-valent chromatography on thiol Sepharose resin. All three enzymes were isolated in the pure form and characterized. Whereas cathepsin B is well investigated, cathepsin H and especially cathepsin S need more studies for their detailed classification.

ACKNOWLEDGEMENT

We thank Mrs. A. Burkeljc and Mr. K. Lindič for their excellent technical assistance. This work was supported by the Research Community of Slovenia and in part by the National Science Foundation, USA.

REFERENCES

Andrews, P. (1965). The gel filtration behaviour of proteins related to their molecular weights over a wide range. Biochem.J., 96, 595-606.

Anson, M.L. (1939). The estimation of pepsin, trypsin, papain and cathepsin with hemoglobin. J.Gen.Physiol., 22, 79-89.

Barrett, A.J. (1972). A new assay for cathepsin B1 and other thiol proteinases. Analyt.Biochem., 47, 280-293.

Barrett, A.J. (1973). Human cathepsin B1. Purification and some properties of the enzyme. Biochem.J., 131, 809-822.

Brocklehurst, K., J. Carlsson, M.P.J. Kierstan, and E.M. Crook (1973). Covalent chromatography. Preparation of fully active papain from dried Papaya latex. Biochem.J., 133, 573-584.

Davis, B.J. (1964). Disc electrophoresis. 2. Method and application to human serum protein. Ann.New York Acad.Sci., 23, 404-427.

De Martino, G., T.W. Doebber, and L.L. Müller (1977). Pepstatin-insensitive proteolytic activity of rat liver lysosomes. J.Biol.Chem., 252, 7511-7516.

Evans, P., and D.J. Etherington (1978). Characterization of cathepsin B and collagenolytic cathepsin from human placenta. Europ.J.Biochem., 83, 87-97.

Katunuma, N., T. Towatari, E. Kominami, and S. Hashida.   Structures and biological functions of cathepsin B and L. In this book.

Kirschke, H., J. Langner, B. Wiederanders, S. Ansorge, and P. Bohley,(1977a). Cathepsin L. A new proteinase from rat liver lysosomes. Europ.J.Biochem., 74, 293-301.

Kirschke, H., J. Langner, B. Wiederanders, S. Ansorge, P. Bohley, and H. Hanson (1977b). Cathepsin H: an endoaminopeptidase from rat liver lysosomes. Acta Biol.Med.Germ., 36, 185-199.

Kirschke, H., J. Langner, S. Rieman, B. Wiederanders, S. Ansorge, and P. Bohley (1980). Lysosomal cysteine proteinases. In Protein degradation in health and disease. Ciba Foundation Symposium 75, Elsevier/North Holland, Amsterdam, pp, 15-35.

Kirschke, H., H.J. Kärgel, S. Rieman,and P. Bohley. Cathepsin L. In this book.

Okitani, A., U. Matsukura, H. Kato, and M. Fujimaki (1980). Purification and some properties of a myofibrilar protein degrading protease, cathepsin L, from rabbit skeletal muscle. Biochem.J., 87, 1133-1143.

Schwartz. W.N., and A.J. Barrett (1980). Human cathepsin H. Biochem.J., 191, 487-497.

Singh, H., and G. Kalnitsky (1978). Separation of a new α-N benzoylarginine- β - -naphthylamide hydrolase from cathepsin B1. J.Biol.Chem., 253, 4319-4326.

Singh, H., and G. Kalnitsky (1980). α-N-benzoylarginine- β -naphthylamide hydrolase, an aminoendopeptidase from rabbit lung. J.Biol.Chem., 255, 369-374.

Su, Y.T., and B. Jirgensons,(1977). Further studies on detergent-induced conformational transitions in proteins. Arch.Biochem.Biophys., 181, 137-146.

Towatari, T., K. Tanaka, D. Yoshikawa, and N. Katunuma (1978). Purification and properties of a new cathepsin from rat liver. J. Biochem., 84, 659-671.

Towatari, T., Y. Kawabata, and N. Katunuma (1979). Crystallization and properties of cathepsin B from rat liver. Europ.J.Biochem., 102, 279-289.

Turk, V., I. Kregar, F. Gubenšek, and P. Ločnikar (1978). Bovine spleen cathepsin B and S: purification, characterization and structural studies. In H.L. Segal and D.J. Doyle (Eds.) Protein turnover and lysosome function. Academic Press, New York, pp. 353-361.

Turk, V., I. Kregar, F. Gubenšek, T. Popović, P. Ločnikar, and T. Lah (1980). Carboxyl and thiol intracellular proteinases. In P. Mildner and B. Ries (Eds.), Enzyme regulation and mechanism of action. Trends in Enzymology, Vol. 60, Pergamon Press, Oxford, pp. 317-330.

Turnšek, T., I. Kregar, and D. Lebez (1975). Acid sulfhydryl protease from calf lymph nodes. Biochim.Biophys.Acta, 403, 514-520.

Weber, K., and M. Osborn (1969). The reliability of molecular weight determination by dodecyl sulphate polyacrylamide gel electrophoresis. J. Biol.Chem., 244, 4406-4412.

Zvonar, T., I. Kregar, and V. Turk (1979). Isolation of cathepsin B and α-N-benzoylarginine- β -naphthylamide hydrolase by covalent chromatography on activated thiol Sepharose. Croat. Chem. Acta, 52, 411-416.

Zvonar-Popović, T., T. Lah, I. Kregar and V. Turk (1980). Some characteristics of cathepsin B and α-N-benzoylarginine- β -naphthylamide hydrolase from bovine lymph nodes. Croat.Chem.Acta, 53, 509-517.

STRUCTURAL AND DENATURATION STUDIES OF CATHEPSIN D AND THE EXISTENCE OF ITS POSSIBLE PRECURSOR

V. Turk, V. Puizdar, T. Lah, F. Gubenšek and I. Kregar

Department of Biochemistry, J.Stefan Institute, University E.Kardelj,
61000 Ljubljana, Yugoslavia

ABSTRACT

Autolysis studies of cathepsin D show that this process occurs at acid pH during an incubation period of several days. As a result the polypeptide chain of molecular weight of 42,000 was partially cleaved immediately to chains of molecular weight of 28,000 and 14,000. Prolonged incubation resulted in the disappearance of 42,000 and 14,000 polypeptide chains. The enzyme activity was also decreased. Structural and denaturation studies using circular dichroism and fluorescence show similarities with pepsin. There is an indication of the possible existence of a precursor of cathepsin D, or an endogenous inhibitor – cathepsin D complex.

KEYWORDS

Cathepsin D; intracellular proteinases; denaturation; circular dichroism; fluorescence; secondary structure; precursor; aspartic proteinase.

INTRODUCTION

Aspartic (carboxyl) proteinases represent a group of extracellular and intracellular proteolytic enzymes with diverse species origins and biological functions. The most investigated are mammalian and microbial extracellular enzymes with quite well established common features, structure and catalytic mechanism. Thus, the amino acid sequences show high homology, the active centers are formed by two aspartic acids, they display similar specificity with preference for cleavage between hydrophobic residues, the polypeptide chain form two distinct globular lobes, etc. (Tang and co-workers, 1978; Blundell and co-workers, 1979; Andreeva and Gustchina, 1979; Tang, 1979; Fruton, 1980). Among the mammalian intracellular aspartic proteinases cathepsin D is the main enzyme which plays an important role in the protein degradation in the normal metabolism and in the pathological processes of tissue breakdown. Cathepsin D was isolated in the pure form and characterized (reviewed by Barrett, 1977).

117

Now we are at the beginning of structural studies. In this report we present autolysis studies, as well as structural and denaturation studies. In addition we present an evidence for the existence of bovine spleen cathepsin D in a precursor form.

RESULTS AND DISCUSSION

Cathepsin D has been purified for several decades using classical procedures which include autolysis as a purification step, usually for 24 hrs. When the enzyme was purified in this way, we found that cathepsin D consists of several polypeptide chains of different molecular weights (Turk and co-workers, 1974; Smith and Turk, 1974), which we have explained by partial degradation of the enzyme. This degradation of the native molecule can be due to proteolysis or autolysis. To avoid proteolysis, we added in the first purification step after homogenization specific inhibitors of thiol and serine proteinases, yet again in final enzyme preparation, cathepsin D was composed of several polypeptide chains.

A new rapid purification method which avoids the autolysis step was developed in our laboratory (Smith and Turk, 1974; Kregar and co-workers, 1977). This method included affinity chromatography on haemoglobin or pepstatin-agarose resin and resulted in cathepsin D with a single polypeptide chain. Therefore we decided to study more carefully the process of its possible autolytic degradation. The enzyme was incubated at 37 °C in acetate buffer, pH 3.5, for several days. The results are presented in Fig. 1.

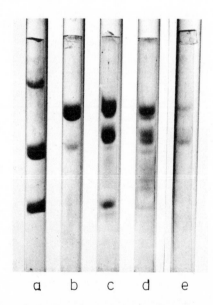

$$a \quad b \quad c \quad d \quad e$$

Fig. 1. SDS- gel electrophoretic presentation of autolytic degradation of cathepsin D. a) standard proteins: bovine serum albumin, chymotrypsinogen A,lysozyme; b) cathepsin D in phosphate buffer pH 8.4; c) cathepsin D after acidification to pH 3.5; incubated cathepsin D d) for 4 hrs, e) for 5 days.

Immediately after acidification additional polypeptide chains appeared with molecular weights of about 28,000 and 14,000. Further incubation resulted in the disappearance of the low molecular weight chain. On prolonged incubation the 42,000 polypeptide chain nearly disappeared, whereas 28,000 polypeptide chain persisted. A lowering of proteolytic activity during the incubation period was observed. Kinetic studies of autolysis will elucidate the mechanism of degradation of cathepsin D (in preparation).

Structural studies were followed by circular dichroism (CD) and fluorescence measurements. A very low content of α–helix and a moderate content of β–structure strongly support the suggestion that cathepsin D has the general structure of aspartic proteinases (Turk and co-workers, 1980). Slight differences were, however, observed in the near ultraviolet (UV) CD spectra of cathepsin D and pepsin (Turk and co-workers, 1980; Kozlov and co-workers, 1979; Lah and co-workers, 1980). Fluorescence spectra of cathepsin D and pepsin show that there are marked differences in the position of the aromatic amino acid residues. While pepsin exhibits strong tryptophan emission at a wavelength of about 348 (Nakatani and co-workers, 1976), several peaks at 345, 330 and 313 nm which belong to tryptophan and additional one at 301 nm which belongs to tyrosine, were observed for cathepsin D (Fig. 2).

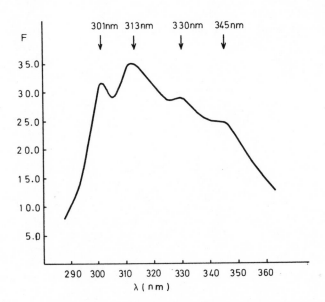

Fig. 2. Corrected fluorescence emission spectrum of cathepsin D (0.01% protein solution in 0.1 M 2-N-morpholino-ethan sulphonic acid buffer, pH 6.6); excitation at 285 nm.

The binding of pepstatin to cathepsin D was reflected in negligible changes of the fluorescence spectrum, while the near UV CD spectrum of the enzyme was affected. A strong increase of the bands at 292, 285 nm and an increase at 280 nm was observed.

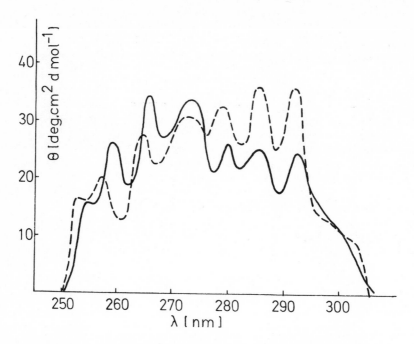

Fig. 3. Near ultraviolet CD spectrum of cathepsin D.
—— cathepsin D (0.12% protein solution in acetate buffer
pH 5.0; ---- cathepsin D and pepstatin in acetate buffer
pH 5.0.

These results show similarity with CD binding studies of pepstatin to porcine pepsin
(Kozlov and co-workers, 1979).

The native conformation of cathepsin D can be perturbed very effectively with polar
denaturants, such as urea and guanidinium hydrochloride (Gdn.HCl). We followed
the conformational changes – at pH 6.6 and room temperature – by fluorescence,
near and far UV CD, as well as by measurement of residual enzyme activity on haemoglobin substrate at pH 3.5 (Anson, 1939). A typical denaturation curve is presented in Fig. 4. We have observed a slight increase in proteolytic activity by the addition of about 0.5 M Gdn.HCl. This phenomenon has already been noticed by some
authors (Wojtowicz and Odense, 1970), and it could possibly be ascribed to the influence of the small molecule of the polar denaturant on the aromatic amino acids in
the active center which are in the primary or secondary subsite positions, resulting
in the increase in enzymatic activity. Subsequent addition of 1.0 to 2.5 M Gdn.HCl·
resulted in almost simultaneous loss of the ordered structure and activity. Gdn.HCl
at concentrations of 4.0 – 5.0 M still further unfolded the molecule leading to a ran-
dom coil at 6.0 M Gdn.HCl, although in the absence of reducing agents, the disulphide bonds in cathepsin D could still represent some strain in the molecule.

In refolding experiments, which were carried out after 24 hrs incubation of cathepsin

D in different concentrations of Gdn.HCl or urea, by dilution we failed to restore the native conformation or the activity. The irreversibility of the denaturation of pepsin has been reported by several authors (Ahmad and Mc Phie, 1978, 1979). The possible autolysis, disulphide rearrangements and/or the aggregation of unfolded pepsin were overwhelmed by the hypothesis that the irreversibility of denaturation is an intrinsic property of the pepsin molecule. Its active conformation should be determined in the precursor molecule of pepsinogen, which, however, could be refolded (Ahmad and Mc Phie, 1978).

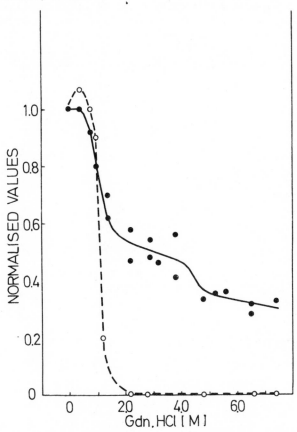

Fig. 4. Denaturation curve of cathepsin D. ——— relative negative ellipticity at 220 nm; ―――― relative cathepsin D activity; the enzyme solution (0.3% in 0.1 M phosphate buffer pH 6.6) was incubated with increasing concentrations of Gdn.HCl

On the basis of the present results we can say that cathepsin D is unfolded in more than one step leading to the irreversible loss of its activity and structure after prolonged incubation in the denaturant solution. Further experiments on the kinetics of unfolding would tell us more about the complex mechanism of denaturation and the reason for the lack of restoration of enzyme native conformation.

The results presented suggest the possibility of the existence of the cathepsin D molecule in a zymogen form. Until very recently, it was believed that lysosomal enzymes, and particularly cathepsins, do not exist in the form of precursors. The first suggestion of a possible precursor of cathepsin D was published by Firfarova and Orekhovich (1971). Later, when we developed the affinity chromatography method for the purification of cathepsin D, we found that besides cathepsin D another high molecular weight protein was bound to haemoglobin or pepstatin agarose resin (Smith and Turk, 1974; Kregar and co-workers, 1977) and was separated by gel chromatography from cathepsin D in an electrophoretically pure form. First we tried to activate this protein with cysteine proteinases, following the idea of activation of a possible precursor of cathepsin D (Huang and co-workers, 1979). Under our experimental conditions the activation was not successful. Therefore, we decided to activate the inactive protein by lowering the pH of the enzyme solution to 2.5. The results are presented in Fig. 5.

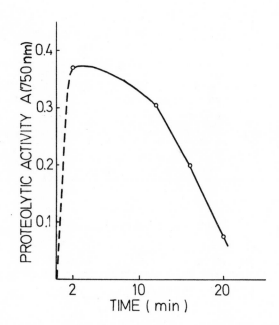

Fig. 5. Activation of the high molecular weight inactive protein with time.

It is evident that the activation occurs in a very short time – about 1 to 2 min – after the addition of HCl. The activation is transient, and after 20 min the enzyme activity almost disappeared. This can be explained by the instability of cathepsin D in this pH region. Polyacrylamide gel electrophoresis of the activated protein shows several bands, their position indicating the presence of cathepsin D.

Very recently the existence of α-glucosidase, β-hexosaminidase and cathepsin D in the form of a precursor was reported in cultures of human skin fibroblasts (Hasilik and Neufeld, 1979; Hasilik, 1980). They found that polypeptide chains of these lysosomal enzymes are first synthesized in a form of larger precursors and then processed to smaller active forms of enzymes and other fragments. These results and our activation studies suggest the existence of cathepsin D in the form of a precursor or in the endogenous inhibitor – enzyme complex.

ACKNOWLEDGEMENT

We greatly appreciate the helpful advice and enthusiastic interest of Dr. Roger Pain, Biochemistry Department, University of Newcastle upon Tyne, UK, where fluorimetric experiments were performed. This work was supported by the Research Council of Slovenia, and in part by the NSF grant no. F7F030Y.

REFERENCES

Ahmad, F., and P. Mc Phie (1978). The denaturation of covalently inhibited swine pepsin. Int.J.Peptide Prot.Res., 12, 155-163.

Ahmad, F., and P. Mc Phie (1979). Characterization of a stable intermediate in the unfolding of diazoacetylglycine ethyl ester – pepsin by urea. Canad.J.Biochem., 57, 1090-1092.

Andreeva, N.S., and A.E. Gustchina (1979). On the supersecondary structure of acid proteinases. Biochem.Biophys.Res.Comm., 87, 32-42.

Anson, M.L. (1939). The estimation of pepsin, trypsin, papain, and cathepsin with hemoglobin. J.Gen.Physiol., 22, 79-89.

Barrett, A.J. (1977). Cathepsin D and other carboxyl proteinases. In A.J.Barrett (Ed.) Proteinases in Mammalian Cells and Tissues, North Holland, Amsterdam, pp.209-248.

Blundell, T.L., B.T. Sewell, and A.D. Mc Lachlan (1979). Four-fold structural repeat in the acid proteases. Biochim.Biophys.Acta, 580, 24-31.

Firfarova, K.F., and V.N. Orekhovich (1971). On inactive precursor of cathepsin D from chicken liver. Biochem.Biophys.Res.Comm., 45, 911-916.

Fruton, J.S. (1980). Fluorescence studies on the active sites of proteinases. Mol. Cell.Biochem., 32, 105-114.

Hasilik, A., and E.F. Neufeld (1979). Biosynthesis of lysosomal enzymes in fibroblast. J.Biol.Chem., 255, 4937-4950.

Hasilik, A. (1980). Biosynthesis of lysosomal enzymes. TIBS, 5, 237-240.

Huang, J.S., S.S. Huang, and J. Tang (1979). Structure and function of carboxyl proteases: a comparison of the intracellular and extracellular enzymes. Symp. on Frontiers in Protein Chemistry (in press).

Kozlov, L.V., E.A. Meshcheryakova, L.L. Zavada, E.S. Efremov, and L.G. Rashkovetskij (1979). Dependence of pepsin conformational states on pH and temperature. Biokhimiya, 44, 338-349.

Kregar, I., I. Urh, H. Umezawa, and V. Turk (1977). Purification of cathepsin D by affinity chromatography on pepstatin-Sepharose resin. Croat.Chim.Acta, 49, 587-592.

Lah, T., V. Turk, and R.H. Pain (1980). The influence of denaturing agents on pepsin. Vest.Slov.Kem.Drus., 27, 237–250.

Nakatani, H., K. Kitagishi, and K. Hiromi (1976). Spectroscopic studies of pepsin and its complex with Streptomyces pepsin inhibitor. Biochim.Biophys.Acta, 452, 521–524.

Smith, R., and V. Turk (1974). Cathepsin D: rapid isolation by affinity chromatography on haemoglobin-agarose resin. Eur.J.Biochem., 48, 245–254.

Tang, J., M.N.G. James, I.N.Hsu, J.A. Jenkins, and T.L. Blundell (1978). Structural evidence for gene duplication in the evolution of the acid proteases. Nature, 271, 618–621.

Tang, J., (1979). Evolution in the structure and function of carboxyl proteases. Mol. Cell.Biochem., 26, 93–109.

Turk, V., I. Kregar, F. Gubenšek, R. Smith, and S. Lapanje. (1974). Some properties of cathepsin D isolated from different organs. In H.Hanson and P. Bohley (Eds.) Intracellular Protein Catabolism, J.A. Barth, Leipzig, pp. 260–278.

Turk, V., I. Kregar, F. Gubenšek, T. Popović, P. Ločnikar, and T. Lah (1980). Carboxyl and thiol intracellular proteinases. In P. Mildner and B. Ries (Eds.), Enzyme Regulation and Mechanism of Action. Trends in Enzymology, Vol. 60, Pergamon Press, Oxford, pp. 317–330.

Wojtowicz, M.B., and P. Odense (1970). The effect of urea upon the activity measurement of cod muscle cathepsin with hemoglobin substrate. Canad.J.Biochem., 48, 1050–1053.

CHICKEN PEPSIN: STRUCTURE AND HOMOLOGY
WITH OTHER ACID PROTEINASES

V. Kostka, H. Keilová and M. Baudyš

Institute of Organic Chemistry and Biochemistry, Czechoslovak Academy
of Sciences, 166 10 Prague, Czechoslovakia

ABSTRACT

The amino acid sequence of chicken pepsin (CP) has been studied
to provide a basis for the elucidation of (a) the catalytic mechanism of CP at the molecular level, and (b) the evolutionary relationship of CP with other carboxyl proteinases. For more efficient
preparation of the enzyme CP was isolated directly from acid extracts of chicken proventriculi mucosa by covalent and affinity
chromatography; these procedures were based on the coupling of CP
either via its SH-group to mercurial Sepharose or to Sepharose with
immobilized A. lumbricoides pepsin inhibitor as affinant. The
latter support was used to advantage also for affinity chromatography purification of other carboxyl proteinases. As starting material for sequence studies the zymogen (CPG) was isolated by anion
exchange chromatography of alkaline extracts of the mucosa. The
process of CPG activation to CP starts (at the amino acid sequence
level) by scission of the bond between Phe (26) and Leu (27). The
26-residue N-terminal activation peptide released inhibits to different degrees CP, calf chymosin, and hog pepsin. Sequence studies
carried out by automatic degradation of CPG and CP derivatives and
by the classical approach based on overlapping peptides permitted
a partial structure of CPG to be derived. The latter accounts for
322 residues arranged in 19 segments of interchangeable order. The
alignment of this structure with those of hog pepsinogen and calf
prochymosin shows sequential homologies in the terminal parts of
the molecule, around half-cystines and the reactive aspartic acid
residues.

KEYWORDS

Chicken pepsin(ogen); affinity chromatography; zymogen activation;
primary structure.

INTRODUCTION

The past decade has witnessed increasing interest in the investigation of carboxyl (acid) proteinases at the molecular level. The profound learning of the catalytic mechanism of these enzymes, reviewed before (cf., e.g., Fruton, 1977; Antonov, 1978) and also in this book, has been paralleled by the examination of the amino acid sequences of the main representatives of these enzymes isolated from various sources. The knowledge inferred from these studies, recently reviewed by Foltmann and Pedersen (1977), stimulated a search for relatedness among these enzymes and their first tentative classification. The carboxyl proteinases sequentially studied so far have been isolated from organisms very distant from the evolutionary viewpoint, ranging from molds to mammals. There are no data, however, on the avian enzymes which may contribute to a better understanding of the evolution of species. For this reason we considered interesting to study the primary structure of chicken[1] pepsin. Our interest in this source was enhanced by the fact that this enzyme has successfully been employed as a rennet substitute (cf., e.g., Gutfeld and Rosenfeld, 1975). We assumed that the knowledge of the primary structure of chicken pepsin may contribute to a better understanding of its milk-clotting mechanism and, later on, also to the elucidation of its three-dimensional structure, also because of its free SH-group, offering the possibility of isomorphous replacements.

STARTING MATERIAL

Preparation of Chicken Pepsin

The isolation of pepsin by purification of acid extracts of the gastric mucosa leads mostly to inhomogeneous products contaminated with products of autolysis (Rajagopalan and coworkers, 1966b). Chicken pepsin bears a free thiol group (Bohak, 1969) which unlike in its native zymogen, can be titrated and chemically modified. It seemed reasonable to make use of this group for rapid isolation of the enzyme from crude preparations by coupling it via an S-S bond to a support bearing thiol groups. In the experiment (Keilová and Kostka, 1975) illustrated by Fig. 1a crude pepsin was immobilized on mercurial Sepharose ("Hg-Sepharose") (Sluyterman and Wijdenes, 1970), equilibrated in 50 mM acetate at pH 5. After elution of inactive material CP was displaced by addition of 20 mM mercaptoethanol to the same buffer. The yield of the operation was 90% and the specific activity of the enzyme increased 6 times. When subjected to disc electrophoresis at pH 8.6 the presence of two main zones and at least 4 minor zones was observed. By anion exchange chromatography the preparation was resolved accordingly into 5 fractions (Fig. 2), the main bulk of activity being eluted in the most negative fraction. Fractions II-V showed the same specificity when assayed with the B-chain of oxidized insulin as substrate.

The failure to obtain a homogeneous pepsin preparation by a rapid single step operation led us to examine another procedure of affinity chromatography based on the interaction of pepsin with its naturally occurring inhibitor from the roundworm Ascaris lumbricoides.

---

[1] In this study the word "chicken" is used for "common domestic fowl" and not for the young of it.

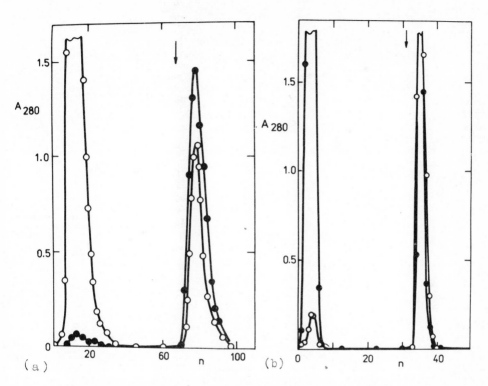

Fig. 1   Chromatography of chicken pepsin on mercurial Sepharose 4B
         (a) and on Ascaris Sepharose 4B (b)

(a) Crude CP (500 mg) was adsorbed to a 2.5 x 16 cm column equilibrated in 50 mM acetate (pH 5), containing 0.2 M KCl,   1 mM EDTA,
and 0.5% of n-butanol. Flow rate 90 ml/h. CP was eluted by 20 mM
2-mercaptoethanol in the acetate buffer added as marked by the
arrow. o, absorbance at 280 nm; •, proteolytic activity; n, tube
number.

(b) Crude CP (50 mg) was adsorbed to a 1 x 6 cm column equilibrated
in acetic acid (pH 3). Flow rate 60 ml/h. CP was eluted by 10 mM
phosphate (pH 6.8) added as marked by the arrow. •, absorbance at
280 nm; o, proteolytic activity; n, tube number.

The inhibitor, obtained by a modification of the method of Peanasky
and Abu-Erreish (1971), was coupled to Sepharose 4B at pH 6.8
(Keilová and coworkers, 1975). Crude CP was adsorbed to the support
("Ascaris-Sepharose") at pH 3 and displaced by 10 mM phosphate
(pH 6.8) as shown in Fig. 1b. The yield and increase of specific
activity were comparable to the results obtained with Hg-Sepharose.
Affinity chromatography on Ascaris Sepharose, however, appears
to be slightly more specific: preparations of CP obtained from Hg-
-Sepharose had a specific activity by ca. 15% higher after additional chromatography on Ascaris Sepharose. CP preparations obtained by this procedure, however, showed an electrophoretic and

chromatographic behavior almost identical to preparations obtained
from Hg-Sepharose, i.e. represented a mixture of pepsins. The interaction of the A. lumbricoides pepsin inhibitor is not limited to CP.
We found complex formation and inhibition of hog pepsin (like Peanasky and Abu-Erreish, 1971),　　　　gastricsin and rabbit bone marrow
cathepsin E (Keilová and Tomášek, 1977) and made an effort to purify these enzymes on Ascaris Sepharose (Salvetová, 1975). The
latter may therefore be used to advantage for rapid isolation of
various carboxyl proteinases. CP is stable over a wider pH-range
(up to pH 8) and that is very useful for its immobilization on Sepharose. Such "CP-Sepharose" columns can then be used for the isolation and purification of other pepsin inhibitors which form a
complex with the enzyme in acid solutions. The procedure was
successfully employed for the preparation of the A. lumbricoides
pepsin inhibitor.

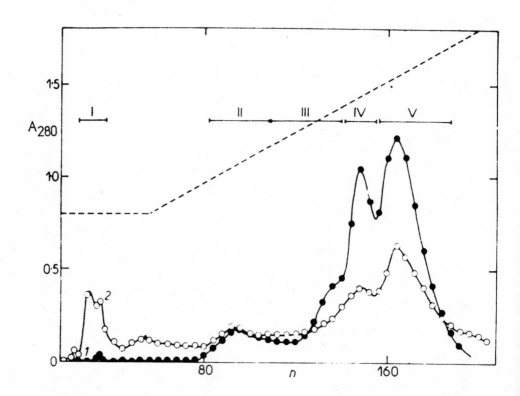

Fig. 2　Chromatography of chicken pepsin on DEAE-cellulose

The 1.5 x 25 cm column was eluted first by 20 mM phosphate
(pH 6.9) 0.2 M in NaCl, then (starting from fraction No 50)
by linear gradient (300 + 300 ml) of 0.2 - 0.4 M NaCl
(marked ----). Flow rate 9 ml/h. ●, absorbance at 280 nm;
o, proteolytic activity; n, tube number.

Preparation of Chicken Pepsinogen

The crude CPG preparation was obtained by alkaline extraction of
the mucosa and repeated acetone precipitation essentially by the
method of Bohak (1969). DEAE-cellulose chromatography of this preparation yielded one major active fraction and little activity in
two additional minor fractions. The major fraction was rechromato-
graphed (Kostka and coworkers, 1977) to serve as starting material
for the sequence studies.

Pepsin from Pepsinogen

CP was prepared by 20-min activation of CPG at pH 2.5 and room temperature and the activation peptides separated by gel chromatography in dilute acid.

Molecular Characteristics of CPG and CP

Both preparations gave single zones on disc electrophoresis at pH
8.6 and behaved as homogeneous compounds in the ultracentrifuge.
A value of $s_{20, w} = 3.95$ S corresponding to a mol. wt. of $42,000 \pm
1,000$ was found for CPG whereas 3.7 S and 33,400 were ob-
tained for CP. These data are in good agreement with the results of
amino acid analysis showing 379 residues (total mol. wt. 43,879)
in CPG and 304 (34,220) in CP. Controversial data were reported by
various authors on the carbohydrate moiety (Bohak, 1979; Green and
Llewellin, 1973). From our analyses carried out both with the proteins and isolated peptide fragments (see below) CPG and CP contain
both 3 mannose residues and 7 glucosamines. The homogeneity of both
proteins was further evidenced by the existence of single terminal
groups: serine and threonine are N-terminal in CPG and CP, respectively, and the same single C-terminal sequence ...-Leu-Ser.COOH
was detected in both proteins.

ZYMOGEN ACTIVATION

The original scheme of pepsin formation from its precursor derived
from the experiments of Van Vunakis and Herriott (1956), who studied this process as the first ones quantitatively at the molecular
level, has become subject to change in view of the results obtained
recently by various groups (Kay and Dykes, 1975; Sanny and co-
workers, 1975). The transformation of CPG into CP was studied first
by Herriott and coworkers (1938) who also isolated its intermediary
product, named pepsin inhibitor, and examined its activity toward
pepsins from various species (Herriott, 1941). The kinetics of complex formation between this inhibitor and CP studied later Bohak
(1973). Differences shown to exist in the activation process between hog pepsinogen (Dykes and Kay, 1976) and prochymosin (Kay
and Dykes, 1977; Pedersen and coworkers, 1979) stimulated our
interest in a similar investigation of CPG intended to show (a) the
size of the activation peptide released, and (b) its interspecies
inhibitory activity.

First Step in the Activation of CPG

The activation of CPG at pH 2.5 in the presence of a 2.2 molar excess of pepstatin (Aoyagi and coworkers, 1971) and the separation
of the activation peptide from the remaining protein moiety were
carried out as described in detail by Keilová and coworkers (1977).
The peptide was purified by gel filtration and subjected to automated sequencing. The sequence found was Ser-Ile-His-Arg-Val-Pro-
-Leu-Lys-Lys-Gly-Lys-Ser-Leu-Arg-Lys-Gln-Leu-Lys-Asp-His-Gly-Leu-
-Leu-Glu-Asp-Phe and corresponded to the N-terminal 26 residues
in CPG determined from sequencing of the protein and from ancillary
data (see below). This suggests that CPG activates to release in
the first instance a peptide containing 26 residues.

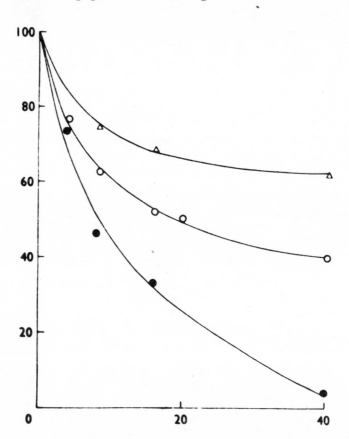

Fig. 3 Inhibition of milk-clotting activity
by the chicken activation peptide

Inhibition of the activity of chicken pepsin (o), hog pepsin (●),
and calf chymosin (Δ) by various amounts of the chicken 1-26 activation peptide. Each enzyme was preincubated for 10 min at 25°C
with the solution of the peptide (0.5 mg in 1 ml of 0.2 M acetate
(pH 5.3) before the determination of the residual activity. Ordin-
ate, residual activity in %; abscissa, molar peptide to enzyme ratio.

Interspecies Inhibitory Activity of the Chicken Activation Peptide

Early data (Herriott, 1941) showed that crude preparations of the chicken activation peptide were inactive toward CP. We studied the inhibition of the milk-clotting activity of CP, hog pepsin and calf chymosin by the purified 26-residue activation peptide. The results of this investigation (Keilová and coworkers, 1977) are illustrated in Fig. 3 and show that the peptide inhibits all three enzymes yet to various degrees.

SEQUENCE STUDIES

Two lines of approach were followed, namely (a) automated sequencing of CPG and CP derivatives and (b) "classical" analysis of overlapping peptides derived from different enzymatic digests of both proteins. Experiments under (a) afforded the sequence of 17 and 7 residues at the N-terminus of CPG and CP, respectively. The CPG sequence was extended to 26 residues by an analysis of the mixture of peptides liberated during the activation of the zymogen in the absence of pepstatin (Kostka and coworkers, 1977). The experiments based on overlapping peptides were undertaken with the tryptic digest of S-sulfo-CPG and citraconylated S-sulfo-CPG (Baudyš and coworkers, 1980) and with the chymotryptic digest of alkali-denaturated CP (Baudyš and coworkers, 1980) and of S-sulfo-CPG (Kostka, unpublished observations). The thermolytic digest of native and on Hg-Sepharose immobilized CP was used for the isolation of the disulfide peptides (Keilová and coworkers, 1979), elucidation of the three S-S bridges and as a source of links for peptides derived from other digests (Baudyš and coworkers, 1981). The neighborhood of the aspartic acid residue reacting with diazoacetyl-D,L-norleucine methyl ester (Rajagopalan and coworkers, 1966a) was characterized by selective isolation of the corresponding peptide from the thermolytic digest of the labeled protein (Keilová and coworkers, 1980). The results of these studies permitted us to derive the sequence of 322 out of 379 present in CPG according to amino acid analysis. These are contained in a 28-residue N-terminal sequence, a 21-residue C-terminal sequence, and in 19 segments of interchangeable order, representing the middle part of the protein. These segments range in size from 2 to 58 residues. The results of the sequence studies are summarized in a partial formula of CPG shown in Fig. 4.

DISCUSSION

Purity and Identity of CP and CPG Preparations

The process of acid extraction of pepsin from the mucosa is likely to be paralleled by autolysis of the protein. Crude CP preparations are moreover contaminated by inactive proteins. Covalent chromatography of CP on Hg-Sepharose removes a large bulk of inactive material (Fig. 1a), yet the protein itself is inhomogeneous as judged by its electrophoretic and chromatographic (Fig. 2) behavior. Chromatographic CP fractions I to V all have the same N-terminal group, threonine, and show the same specificity when assayed with the B-chain of oxidized insulin as substrate. They necessarily all bear a free SH-group since they bind to Hg-Sepharose. From these data

```
                                         10                              20
Ser-Ile-His-Arg-Val-Pro-Leu-Lys-Lys-Gly-Lys-Ser-Leu-Arg-Lys-Gln-Leu-Lys-Asp-His-Gly-Leu-Leu-Glu-Asp-Phe-Leu-

         30                         40                              50
Lys•/Lys-Ser-Pro-Tyr-Asn-Pro-Ala- - -Lys/          /Thr-Ala-Thr-Glu- - -Ser-Tyr-Glu-Pro-Met-Thr-Asn-Tyr/

                  60                        70
Met- - -Asp-Ala-Ser/   /Tyr- - - - -Gly-Thr-Ile-Ser-Ile-Gly-Thr-Pro-Gln-Gln-Asp/Phe-Thr/Val-Ile-Phe-Asp-Thr-

  80                               90                        100
Gly-Ser-Ser-Asn-Leu-Trp-Val-Pro-Ser/Ile-Tyr-Cys- - -Lys- - -Ser-Ser-Ala-Cys-Ser-Asn-His-Lys/Arg-Phe-Asp-Pro-
                                                carb.            120                        130
Ser-Lys-Ser-Ser-Thr-Tyr/Val-Ser-Thr-Asn-Glu-Thr-Val-Tyr-Ile-Ala-Tyr-Gly-Thr-Gly-Ser-Met- - -Ser-Gly/Ile-Leu-
                                                 110

Gly-Tyr-Asp-Thr/                                    Ile-Phe-Gly-Leu-Ser-Glu-Thr-Glu-Pro-Gly-Ser-Phe/
Ala                                                                          150
    160                    140

             170                                              180
   /Phe-Asp-Gly/   /Leu-Gly-Leu-Ala(Tyr,Pro)Ser-Ile-Ser-Ser-Gly-Gly-Ala-Thr-Pro-Val-Phe-Asp-
         190                         200                              210
Asn-Met-Met-Asn-Gln-His-Leu-Val-Ala-Gln-Asp-Leu-Phe-Ser-Val-Tyr-Leu-Ser-Asp-Gly-Glu-Thr-Gly-Ser-Phe-Val-

                      220                              230
Leu-Phe-Gly-Gly-Ile-Asp-Pro-Asn-Tyr-Thr-Thr-Lys-Gly/        /Glu-Thr-Tyr- X -Gln-
        240                              250                       260 +
Ile-Thr(Met,Asx,Glx)Val-Thr-Val-Gly-Asn-Lys-Tyr-Val-Ala-Cys-Phe-Phe-Thr-Cys-Gln-Ala-Ile-Val-Asp-Thr-Gly-Thr-

Ser-Leu-Leu-Val-Met         /Ile-Ile-Lys- - -Asp-Leu-Gly-Val-Ser-Ser-
             Phe-Pro-Asp-Gly/                       290
                  270          280
                                     300                              310
Asp-Gly-Glu-Ile-Ser-Cys-Asp-Asp-Ile-Ser-Lys-Leu-Pro-Asp-Val-Thr-Phe-His/Ile-Asn-Gly/
                                                              330
     320   /Val-Leu-Asn-Glu-Asp-Gly-Ser- - -Cys-Met-Leu-Gly-Phe-Glu-Asn-Met-Gly-Thr-Pro-Thr-Glu-Leu-Gly-Glu-
                                                                            340
                                          350                        360                              370
Gln-Trp/Ile-Leu-Gly-Asp-Val-Phe/Ile-Arg-Glu-Tyr-Tyr-Val-Ile-Phe-Asp-Arg-Ala-Asn-Asn-Lys-Val-Gly-Leu-Ser-Pro-

Leu-Ser-
```

Fig. 4   Partial amino acid sequence of chicken pepsinogen

the differences in their chromatographic behavior may indicate the presence of different enzyme or isoenzyme forms or merely reflect charge differences which have occurred during the separation procedure and could well be ascribed e.g. to differences in the carbohydrate moiety (discussed below). The existence of three CP types (or classes, cf. Foltmann and Pedersen, 1977), named A, D, and C in analogy to hog pepsin has been reported by Donta and Van Vunakis (1970a). These enzymes, prepared by activation of the corresponding zymogens resolved chromatographically, showed, however, a high degree of electrophoretic inhomogeneity caused most likely by autolysis during their prolonged dialysis. They were not studied chromatographically and the data necessary for a comparison with our samples are missing. It is unlikely though that CP C, as defined by Donta and Van Vunakis (1970a), is present in our preparation; according to the latter authors this enzyme does not contain a free SH-group and should not therefore bind to Hg-Sepharose. Unlike the active enzyme, CPG prepared by us behaves as a homogeneous protein as judged by various criteria, predominantly by N-terminal sequence analysis, and equally homogeneous is the pepsin derived from it. We have not observed in our experiments the two major components found by Foltmann and Axelsen (1979) in the mucosal extracts of hen and giving rise to two pepsins of different alkali stability. Our results, like Bohak's (1969) strongly suggest the existence of one major zymogen only in chicken gastric mucosa. This is not in agreement with the data of Donta and Van Vunakis (1970a) who chromatographed chicken mucosal extracts under the conditions originally employed by Ryle (1970) for the isolation of the minor components of hog pepsin and pepsinogen. These authors observed three pepsinogens named A, D, and C, existing in chicken mucosa in relative concentrations of 3:2:1. They postulate the existence of two main CPG types, of the A+D type and the C type and have been able to demonstrate (Donta & Van Vunakis, 1970b) significant differences (such as electrophoretic, in amino acid composition and immunochemical properties) between the A+D type on the one hand and the C type on the other. The nomenclature itself, however, rests on uncertain grounds: according to common usage (Enzyme Nomenclature 1979) pepsin D (EC 3.4.23.3), and accordingly its zymogen, is defined as dephosphorylated pepsin A (EC 3.4.23.1). This is hardly the case with chicken pepsin shown by two groups (Levchuk and Orekhovich, 1963; Green and Llewellin, 1973) to lack phosphate. From our data the differences in chromatographical behavior between CPG A and D (practically the only differences, since the other ones, such as in amino acid composition, lie within limits of experimental error) could be accounted for by variations in carbohydrate moiety. In our sequence studies we have been able to isolate (Baudyš, unpublished observations) from the thermolytic and chymotryptic digest of CPG three peptides of identical amino acid sequence (Val-Ser-Thr-Asn--Glu-Thr-Val-Tyr, res. 111 to 118, Fig. 4) markedly differing in electrophoretic mobility. The electrophoretic behavior of these peptides can be accounted for by varying carbohydrate moiety attached to the asparagine residue: the fastest peptide was found to contain 3 mannose residues and 7 glucosamines (of their number 5 bearing a sulfo group), the peptide of medium mobility 3 mannoses and 6 glucosamines (4 sulfonated), and the slowest peptide 2 mannoses and 5 glucosamines (2 sulfonated). These variations, regardless of their origin, would then necessarily affect the charge of the proteins and could probably be responsible for the observed difference in the chromatographic behavior of CPG A and D. Clearly, however, other characteristics, preferentially sequential, will be needed for a more sound classification of CPG and CP in the system of carboxyl proteinases.

Conversion of CPG into CP

Bohak (1969), studying as the first one the process of CPG activation since the early experiments of Herriott and coworkers (1938), postulated that the number of peptide bonds hydrolyzed during this process is much smaller than in hog pepsin (one or possibly two bonds). A similar conclusion was made by Green and Llewellin (1973) assuming the scission of a bond near the N-terminus, leading to the liberation of a peptide of mol. wt. about 2,000. The analysis of the peptide mixture formed by 20-min CPG activation at pH 2.5 and room temperature in our Laboratory (Kostka and coworkers, 1977) showed that both these assumptions are not correct: we were able to isolate and characterize sequentially peptides reconcilable with 26-residue N-terminal sequence of CPG and indicating cleavage of 7 bonds at least. The remaining peptides could not be identified beyond doubt since the mixture clearly contained also autolytical products. To circumvent this difficulty and to gain an insight into the primary steps of the activation process at the sequence level we performed CPG activation in a manner similar to that used by Kay and Dykes (1975) to detect the first bond in the hog pepsinogen to pepsin conversion. The finding of the 26-residue peptide, corresponding to the N-terminal sequence of CPG and the lack of evidence for any other shorter peptide seem to suggest that the first bond split during CPG activation is that between Phe(26) and Leu(27). The chicken activation peptide is comparable in size to the peptide liberated from calf prochymosin (Pedersen and coworkers, 1979) and longer than the peptide liberated from hog or ox pepsinogen. In the latter two zymogens the first activation peptide is formed by cleavage of the Leu(17)-Ile(18) bond. These positions are occupied by Leu-Lys in CPG; we have hypothesized earlier (Keilová and coworkers, 1977) that the higher resistance of this bond to peptic cleavage may be the reason why the initial split is deferred to a subsequent susceptible bond. The complete understanding of the primary steps in CPG activation would require also the knowledge of the remaining protein moiety, named "pseudopepsin" in the case of the hog enzyme. Our efforts to obtain this protein in homogeneous state have been unsuccessful so far.

In contrast to Herriott's (1941) early data we find inhibition of CP by the activation peptide. We ascribe this discrepancy to the little purity of the chicken inhibitor preparations then used for the assays which moreover were not carried out quantitatively. The chicken activation peptide inhibits also hog pepsin and calf chymosin, yet to varying degrees, depending on the peptide to enzyme ratio (Fig. 3). Under conditions where CP is inhibited by 50%, hog pepsin is inhibited by 75% whereas calf chymosin still retains two thirds of its original activity. Kay and Dykes (1977) examined the inhibition of hog pepsin by the hog activation peptide and the chicken activation peptide studied in our Laboratory. They found almost complete inhibition in both cases, yet at different molar ratios. It remains to be shown how this difference can be accounted for by the 9-residue difference in the length of the two peptides.

Partial Amino Acid Sequence of CPG

The results of our sequential studies are summarized in the form of
a partial amino acid sequence of CPG shown in Fig. 4. The residues
are numbered according to prochymosin, as suggested by Foltmann and
Pedersen (1977). Peptides in the middle part of the chain were
assigned their positions with regard to the complete amino acid sequence of hog pepsin, determined earlier in our Laboratory (Morávek
and Kostka, 1974), and to sequences of other carboxyl proteinases,
as summarized first by Foltmann and Pedersen (1977). The aspartic
acid residue which we found to react with diazoacetylnorleucine
ethyl ester is marked by a cross; the site of attachment of the
carbohydrate moiety is indicated by "carb". The formula presented
postulates 373 amino acid residues in the polypeptide chain of CPG.
This value is in fair agreement with the analytical data on the
protein (379 residues, cf. Kostka and coworkers, 1977). The size of
the propart liberated during CPG activation should contain (from
differences in amino acid composition and mol. wt. determination of
CPG and CP) approximately 75 residues, leaving 304 residues for CP.
These data are clearly irreconcilable with our present sequential
information. According to the tentative partial formula arranged
to fit best into the known sequences of other carboxyl proteinase
zymogens, the propart should represent roughly 45 residues leaving
about 330 residues for the pepsin part of the molecule. The solution of this discrepancy should result from the knowledge of a sequence overlapping the terminal portions of the propart and of the
enzyme.

An alignment of the partial amino acid sequence of CPG with the
complete primary structure of hog pepsinogen (Ong and Perlmann,
1968; Pedersen and Foltmann, 1973; Morávek and Kostka, 1974) and
calf prochymosin (Foltmann and coworkers, 1977) is shown in Fig. 5.
The positions occupied in all three zymogens by the same residues
are set in capital letters. The comparison reveals several sections
where the peptide chains are especially homologous. There are 8
highly conservative sites where five or more neighboring positions
are occupied by identical amino acids in all three zymogens. Two
of these sites (77-82; 261-266) contain the reactive aspartic acid
residues (marked +). Extensive homologies are observed in the C-terminal part of the zymogens (346-354; 360-365). Homologies in the
middle part of the chain are found around half-cystine residues,
especially around those which form the two short loops (91-96;
279-283). Like other carboxyl proteinase zymogens, CPG also shows
a cumulation of basic amino acid residues in the N-terminal portion
of the chain. This seems to provide additional piece of evidence
in favor of the hypothesis voiced by Foltmann (1966), postulating
a stabilization of the inactive zymogen conformation through electrostatic interactions of the basic residues localized in the propart with negative charges of residues situated in the enzyme
moiety.

The homologies demonstrated to exist between the primary structures
of hog pepsinogen and calf prochymosin and the partial amino acid
sequence of CPG shown here, especially homologies in those parts of
molecules which are important for enzymatic function, seem to suggest the existence of a common ancestor of these zymogens. Complete
sequential information on CPG will be needed, however, to verify
this postulate and to permit the definition of the type (class) of
CP relative to other carboxyl proteinases.

```
                    10                          20                                    50
HPG  Leu-Val-Lys-Val-PRO-LEU-Val-Arg-Lys-LYS-SER-LEU-ARG-Gln-Asn-LEU-Ile-Lys-Asp-GLY-Lys-LEU-Lys-
CPG  Ser-Ile-His-Arg-Val-PRO-LEU-Lys-Lys-Gly-LYS-LYS-SER-LEU-ARG-Lys-Gln-LEU-Lys-Asp-His-GLY-Leu-LEU-Glu-
CPC  Ala-Glu-Ile-Ile-Thr-Arg-Ile-PRO-LEU-Tyr-Lys-GLY-LYS-LYS-SER-LEU-ARG-Lys-Ala-LEU-Lys-Glu-His-GLY-His-GLY-Leu-LEU-Glu-

                    30                          40                                    70
HPG  ASP-PHE-LEU-Lys-Thr-His-Lys-His-Asn-Pro-Ala-Ser-LYS-Tyr-Phe-Pro-Glu-Ala-Ala-Leu-Ile-Gly-Asp-Glu-
CPG  ASP-PHE-LEU-Lys-Lys-Ser-Pro-Tyr-Asn-Pro-Ala- - -LYS/                 /Thr-Ala-Thr-Glu- - -Ser-Tyr-Glu
CPC  ASP-PHE-LEU-Gln-Lys-Gln-Tyr-Gly-Ile-Gly-Ser-LYS-Tyr-Ser- - - - -Gly-Phe-Gly-Glu-Val-Ala-Ser-Val

                                                60                          80
HPG  PRO-Leu-Glu-ASN-TYR-Leu- - - -ASP-Thr-Glu-Tyr-Phe- - - - -GLY-Thr-ILE-GLY-Ile-GLY-THR-PRO-Ala-GLN-Asp-
CPG  PRO-Met-Thr-ASN-TYR/Met- - -ASP-Ala-Ser/    /Tyr- - - -GLY-Thr-ILE-Ser-Ile-GLY-THR-PRO-Gln-Gln-Asp/
CPC  PRO-Leu-Thr-ASN-TYR-Leu- - - -ASP-Ser-Gln-Tyr-Phe- - - - -GLY-Lys-Ile-Tyr-Leu-GLY-THR-PRO-Pro-GLN-Glu-

                                                90                          120
HPG  PHE-THR-VAL-Ile-PHE-ASP-THR-GLY-SER-SER-Asn-Leu-TRP-VAL-PRO-SER-Val-TYR-CYS-Ser- - -SER-Leu-ALA-CYS-
CPG  PHE-THR/VAL-Ile-PHE-ASP-THR-GLY-SER-SER-Asn-Leu-TRP-VAL-PRO-SER/Ile-TYR-PRO-SER-Lys- - -SER-Ser-ALA-CYS-
CPC  PHE-THR-VAL-Leu-PHE-ASP-THR-GLY-SER-SER-Asp-Phe-TRP-VAL-PRO-SER-Ile-TYR-CYS-Lys- - -SER-Asn-ALA-CYS-

                                               110                          140
HPG  Ser-Asp-HIS-Asn-Gln-PHE-Asn-PRO-Asp-Asp-SER-SER-THR-Phe-Glu-Ala-Thr-Ser-Gln-Glu-Leu-Ser-ILE-Thr-TYR-
CPG  Ser-Asn-HIS-Lys/Arg-PHE-Asp-PRO-Asp-Lys-SER-SER-THR-Tyr/Val-Ser-Thr-Asn-Glu-Thr-Val-Tyr/ILE-Ala-TYR-
CPC  Lys-Asn-HIS-Gln-Arg-PHE-Asp-PRO-Arg-Lys-SER-SER-THR-Phe-Gln-Asn-Leu-Gly-Lys-Pro-Leu-Ser-ILE-His-TYR-

                                               130                          160
                                                                    Gly
                                                                    Ala
HPG  GLY-THR-GLY-SER-MET- - - -Thr-GLY-ILE-LEU-Gly-TYR-ASP-THR-Val-Gln-Val-Gly-Gly-Ile-Ser-Asp-Thr-Asn-Gln-
CPG  GLY-THR-GLY-SER-MET- - - -Ser-GLY/ILE-LEU-LEU-TYR-ASP-THR/
CPC  GLY-THR-GLY-SER-MET- - - -Gln-GLY-ILE-LEU-GLY-TYR-ASP-THR-Val-Ser-Asn-Ile-Asp-Ile-Gln-Gln-

                                               150                          170
HPG  Ile-Phe-GLY-LEU-SER-Glu-Thr-GLU-PRO-GLY-Ser-Phe-Leu-Tyr-Tyr-Ala-Pro-PHE-ASP-GLY-Ile-LEU-GLY-Leu-ALA-
CPG  Ile-Phe-GLY-LEU-SER-Glu-Thr-GLU-PRO-GLY-Ser-Phe/                 /PHE-ASP-GLY/  /LEU-GLY-Leu-ALA-
CPC  Thr-Val-GLY-LEU-SER-Thr-Gln-GLU-PRO-GLY-Asp-Val-Phe-Thr-Tyr-Ala-Glu-PHE-ASP-GLY-Ile-LEU-GLY-Met-ALA-

                                               180                          190
HPG  Tyr-Pro-SER-Ile-Ser-Ala-Thr-Gly-Ala-Thr-PRO-VAL-PHE-ASP-ASN-Leu-Trp-Asp-Gln-Gly-LEU-VAL-Ser-GLN-ASP-
CPG  (Tyr,Pro)SER-Ile-Ser-Ser-Gly-Gly-Ala-Thr-PRO-VAL-PHE-ASP-ASN-Met-Met-Asn-His-LEU-VAL-Ala-GLN-ASP-
CPC  Tyr-Pro-SER-Leu-Ala-Ser-Glu-Gln-Tyr-Ser-Ile-PRO-VAL-PHE-ASP-ASN-Met-Met-Asn-Arg-His-LEU-VAL-Ala-GLN-ASP-
```

```
                          200                                     210                                      220
HPG  LEU-PHE-SER-VAL-TYR-Leu-Ser-Asn-Asp-Asp-Ser-Gly-SER-Val-Val-Leu-Leu-GLY-Gly-ILE-ASP-Ser-Ser-TYR-
CPG  LEU-PHE-SER-VAL-TYR-Leu-Ser-Lys-Asp-Gly-Glu-Thr-Gly-SER-Phe-Val-Leu-Phe-Gly-ILE-ASP-Pro-Asn-TYR-
CPC  LEU-PHE-SER-VAL-TYR-Met-Asp-Arg-Asp-Gly-Gln-Glu- -   -SER-Met-Leu-Thr-Leu-GLY-Ala-ILE-ASP-Pro-Ser-TYR-

                                                230                                    240
HPG  Tyr-THR-Gly-Ser-Leu-Asn-Trp-Val-Pro-Val- -  - -Ser-Val-Glu-Gly-TYR-Trp-GLN-Ile-THR-Leu-Asp-Ser-Ile-THR-
CPG  Thr-THR-Lys-Gly/                              /Glu-Thr-TYR- X -GLN-Ile-THR(Met,Asx,Glx)Val-THR-
CPC  Tyr-THR-Gly-Ser-Leu-His-Trp-Val-Pro-Val- -  - -Thr-Val-Gln-Gln-TYR-Trp-GLN-Phe-THR-Val-Asp-Ser-Val-THR-

                          250                                      260 +                                          Met
HPG  Met-Asp-Gly-Glu-Thr-Ile-ALA-CYS-Ser-Gly-Gly-CYC-GLN-ALA-ILE-ASP-THR-GLY-THR-SER-Leu-LEU-Thr-Gly-
CPG  Val-Gly-Asn-Lys-Tyr-Val-ALA-CYS-Phe-Thr-CYS-GLN-ALA-ILE-ASP-THR-GLY-THR-SER-Leu-LEU-Val-Met-       Phe
CPC  Ile-Ser-Gly-Val-Val-ALA-CYS-Glu-Gly-Gly-CYS-GLN-ALA-ILE-Leu-ASP-THR-GLY-THR-SER-Lys-LEU-Val-Gly-

     270                                    280                                     290
HPG  PRO-Thr-Ser-Ala-Ile-Ala-  - -Asn-Ile-Gln-Ser-Asp-ILE-Gly-Ala-  - -Ser-Glu-Asn-Ser-Asp-Gly-Glu-Met-Val-
CPG  PRO-Gln-Gly/             /ILE-Ile-Lys-  - -Asp-Leu-Gly-Val-Ser-Ser-Asp-Gly-Glu-
CPC  PRO-Ser-Asp-Ile-Leu-  - -Asn-Ile-Gln-Ala-ILE-Gly-Ala-  - -Thr-Gln-Asn-Gln-Tyr-Gly / Glu-Phe-Asp-
                                                                                    Asp

                          300                                    310
HPG  ILE-Ser-CYS-Ser-Ser-Ile-Asp-Ser-Leu-PRO-Asp-Ile-Val-PHE-Thr-ILE-ASN-GLY-Val-Gln-Tyr-Pro-Leu-Ser-Pro-
CPG  ILE-Ser-CYS-Asp-Asp-Ile-Ser-Lys-Leu-PRO-Asp-Val-Thr-PHE-His/ILE-ASN-GLY/
CPC  ILE-Asp-CYS-Asp-Asn-Leu-Ser-Tyr-Met-PRO-Thr-Val-PHE-Glu-ILE-ASN-GLY-Lys-Met-Tyr-Pro-Leu-Thr-Pro-

          320                                    330                                    340
HPG  Ser-Ala-Tyr-Ile-Leu-Gln-Asp-Asp-Asp-Ser-  - -CYS-Thr-Ser-GLY-PHE-Glu-Gly-Met-Asp-Val-Pro-Thr-Ser-Ser-
CPG  /Val-Leu-Asn-Glu-Asp-Gly-Ser-  - -CYS-Met-Leu-GLY-PHE-Glu-Asn-Met-Gly-Thr-Pro-Thr-Glu-Leu-
CPC  Ser-Ala-Tyr-Thr-Ser-Glu-Asp-Gln-Gly-Phe-  - -CYS-Thr-Ser-GLY-PHE-Gln-Ser-Glu-Asn-  -  -  -  -His-Ser-

                          350                                    360
HPG  Gly-Glu-Leu-TRP-ILE-LEU-GLY-ASP-VAL-PHE-ILE-LEU-GLY-ASP-VAL-PHE-ILE-ARG-Gln-TYR-Thr-Val-PHE-ASP-ARG-ALA-ALA-ASN-ASN-Lys-VAL-
CPG  Gly-Glu-Gln-TRP/ILE-LEU-GLY-ASP-VAL-PHE/ILE-ARG-Glu-TYR-TYR-Ile-PHE-ASP-ARG-ALA-ASN-ASN-Lys-VAL-
CPC  -  -Gln-Lys-TRP-ILE-LEU-GLY-ASP-VAL-PHE-ILE-ARG-Glu-TYR-TYR-Ser-Val-PHE-ASP-ARG-ALA-ASN-ALA-ASN-Leu-VAL-

HPG  GLY-LEU-Ala-Pro-Val-Ala
CPG  GLY-LEU-Ser-Pro-Leu-Ser
CPC  GLY-LEU-Ala-Lys-Ala-Ile
```

Fig. 5 Alignment of amino acid sequences of hog pepsinogen (HPG), chicken pepsinogen (CPG), and calf prochymosin (CPC)

REFERENCES

Antonov, V. A. (1978). Mechanism of acid proteinases. Proceedings 12th FEBS Meeting, Dresden, abstr. No 17.

Aoyagi, T., S. Kunimoto, H. Morishima, T. Takeuchi, and H. Umezawa (1971). Effect of pepstatin on acid proteases. J. Antibiot., 24, 687-694.

Baudyš, M., H. Keilová, and V. Kostka (1980). C-Terminal amino acid sequence of chicken pepsinogen and its homology with Sequences of other acid proteases. Collect. Czech. Chem. Commun., 45, 1144-1154.

Baudyš, M., V. Kostka, and H. Keilová (1981). Thermolytic hydrolysate of chicken pepsin. Collect. Czech. Chem. Commun., 46, in the press.

Bohak, Z. (1969). Purification and characterization of chicken pepsinogen and chicken pepsin. J. Biol. Chem., 244, 4638-4648.

Bohak, Z. (1973). The kinetics of the conversion of chicken pepsinogen to chicken pepsin. Eur. J. Biochem., 32, 547-554.

Donta, S. T., and H. Van Vunakis (1970a). Chicken pepsinogens and pepsins. Their isolation and properties. Biochemistry, 9, 2791-2797.

Donta, S. T., and H. Van Vunakis (1970b). Immunochemical relationships of chicken pepsinogens and pepsins. Biochemistry, 9, 2798-2802.

Dykes, C. W., and J. Kay (1976). Conversion of pepsinogen into pepsin is not a one-step process. Biochem. J., 153, 141-144.

Enzyme Nomenclature (1979). Recommendations (1978) of the Nomenclature Committee of the International Union of Biochemistry on the Nomenclature and Classification of Enzymes. Academic Press, New York, pp. 332-333.

Foltmann, B. (1966). A review on prorennin and rennin. C. R. Trav. Lab. Carlsberg., 35, 143-231.

Foltmann, B., and N. H. Axelsen (1979). Gastric proteinases and their zymogens. Phylogenetic and developmental aspects. In P. Mildner and B. Ries (Eds), Enzyme Regulation and Mechanism of Action, Pergamon Press, New York, pp. 272-280.

Foltmann, B., and V. B. Pedersen (1977). Comparison of the primary structures of acidic proteinases and of their zymogens. In J. Tang (Ed.), Acid Proteases, Structure, Function, and Biology, Plenum Press, New York, pp. 3-22.

Foltmann, B., V. B. Pedersen, H. Jacobsen, D. Kaufmann, and G. Wybrandt (1977). The complete amino acid sequence of prochymosin. Proc. Nat. Acad. Sci. USA, 74, 2321-2324.

Fruton, J. (1977). Specificity and mechanism of pepsin action on synthetic substrates. In J. Tang (Ed.), Acid Proteases, Structure, Function, and Biology, Plenum Press, New York, pp. 131-140.

Green, M. L., and J. M. Llewellin (1973). The purification and properties of a single chicken pepsinogen fraction and the pepsin derived from it. Biochem. J., 133, 105-115.

Gutfeld, M., and P. P. Rosenfeld (1975). The solution to Israel's rennet shortage. Dairy Industries, 40, 52-55.

Herriott, R. M. (1941). Isolation, crystallization and properties of pepsin inhibitor. J. Gen. Physiol., 24, 325-328.

Herriott, R. M., Q. R. Bartz, and J. H. Northrop (1938). Transformation of swine pepsinogen into swine pepsin by chicken pepsin. J. Gen. Physiol., 21, 575-582.

Kay, J., and C. W. Dykes (1975). Pepsinogen activation - the first split is not where you would expect it. Proceedings 10th FEBS

Meeting, Paris, abstr. No 758.

Kay, J., and C. W. Dykes (1977). First cleavage site in pepsinogen activation. In J. Tang (Ed.), Acid Proteases, Structure, Function, and Biology, Plenum Press, New York, pp. 103-127.

Keilová, H., and V. Kostka (1975). Isolation and some properties of chicken pepsin. Collect. Czech. Chem. Commun., 40, 574-579.

Keilová, H., and V. Tomášek (1977). Naturally occurring inhibitors of intracellular proteinases. Acta biol. et med. germanica, 36, 1877-1881.

Keilová, H., A. Salvetová, and V. Kostka (1975). Affinity chromatography of chicken pepsin using pepsin inhibitor from Ascaris lumbricoides as affinant. Collect. Czech. Chem. Commun., 40, 580-584.

Keilová, H., V. Kostka, and J. Kay (1977). The first step in the activation of chicken pepsinogen is similar to that of prochymosin. Biochem. J., 167, 855-858.

Keilová, H., M. Baudyš, and V. Kostka (1979). Primary structure of peptides which form the disulfide bonds of chicken pepsin. Collect. Czech. Chem. Commun., 44, 2284-2292.

Keilová, H., V. Kostka, and M. Baudyš (1980). Isolation and characterization of a peptide derived from the active site of chicken pepsin. Collect. Czech. Chem. Commun., 45, 2131-2134.

Kostka, V., H. Keilová, K. Gruner, and J. Zbrožek (1977). N-Terminal sequence analysis of chicken pepsinogen and pepsin. Collect. Czech. Chem. Commun., 42, 3691-3704.

Levchuk, T. P., and V. N. Orekhovich (1963). Poluchenie i nekotorie svojstva kurinnogo pepsina. Biokhimiya, 28, 1004-1011.

Morávek, L., and V. Kostka (1974). Complete amino acid sequence of hog pepsin. FEBS Lett., 43, 207-211.

Ong, E. B., and G. E. Perlmann (1968). The amino-terminal sequence of porcine pepsinogen. J. Biol. Chem., 243, 6104-6109.

Peanasky, R. J., and G. M. Abu-Erreish (1971). Inhibitors from Ascaris lumbricoides. Interactions with the host's digestive enzymes. In H. Fritz (Ed.), Proc. Int. Res. Conf. on Proteinase Inhibitors, W. de Gruyter, Berlin, pp. 281-293.

Pedersen, V. B., and B. Foltmann (1973). The amino acid sequence of a hitherto unobserved segment from porcine pepsinogen preceding the N-terminus of pepsin. FEBS Lett., 35, 255-256.

Pedersen, V. B., K. A. Christensen, and B. Foltmann (1979). Investigations on the activation of bovine prochymosin. Eur. J. Biochem., 94, 573-580.

Rajagopalan, T. G., W. H. Stein, and S. Moore (1966a). The inactivation of pepsin by diazoacetylnorleucine methyl ester. J. Biol. Chem., 241, 4295-4297.

Rajagopalan, T. G., S. Moore, and W. H. Stein (1966b). Pepsin from pepsinogen. Preparation and properties. J. Biol. Chem., 241, 4940-4950.

Ryle, A. P. (1970). The porcine pepsins and pepsinogens. Methods Enzymol., 19, 316-336.

Salvetová, A. (1975). Isolace a vlastnosti některých kyselých proteas. Strukturní studie vepřového pepsinu. CSc. thesis, Institute of Organic Chemistry and Biochemistry, Czechoslovak Academy of Sciences, Prague.

Sanny, C., J. A. Hartsuck, and J. Tang (1975). Conversion of pepsinogen to pepsin. Further evidence for intramolecular and pepsin catalyzed activation. J. Biol. Chem., 250, 2635-2639.

Sluyterman, L. A. Ae., and J. Wijdenes (1970). A agarose mercurial column for the separation of mercaptopapain from nonmercaptopapain. Biochim. Biophys. Acta, 200, 593–595.

Tang, J., P. Sepulveda, J. Marciniszyn Jr., K. C. Chen, W. Y. Huang, N. Tao, D. Liu, and J. P. Lanier (1973). Amino acid sequence of porcine pepsin. Proc. Nat. Acad. Sci. USA, 70, 3437–3439.

Van Vunakis, H., and R. M. Herriott (1956). Structural changes associated with the conversion of pepsinogen to pepsin. Biochim. Biophys. Acta, 22, 537–543.

NON—COVALENT INTERMEDIATES IN CATALYSIS BY PROTEOLYTIC ENZYMES

V.K. Antonov

Shemyakin Institute of Bioorganic Chemistry, USSR Academy of Sciences, 117988 Moscow, USSR

ABSTRACT

The methods which allow to establish the formation of the covalent or non-covalent intermediates between enzymes and the acyl fragments of substrates in the course of proteolytic cleavage of the amide bond have been put forward. It was found that in contrast to the chymotrypsin, such proteinases as pepsin, leucine aminopeptidase and thermolysine do not form acyl-enzymes, functioning by the mechanism of general base catalysis at the stage of the amide bond breaking. To explain the acyl transfer reactions (transpeptidation) catalyzed by these enzymes formation of the relatively stable non-covalent complexes between enzyme and substrate and/or its acyl fragment has been postulated. The role of such intermediate complexes in the specificity of the enzymes is discussed.

KEYWORDS

Proteinases; mechanism; oxygen exchange; transpeptidation; specificity.

INTRODUCTION

Enzyme-catalyzed reactions, as a rule if not always, are multistep processes. On the route from a substrate to the products, intermediate complexes are formed in which the substrate or its fragments is bound to the enzyme by sorption forces or through covalent bonds. The formation of covalent intermediates is corroborated, along with other data, by experiments in which intermediate forms are "trapped". Typical reactions of entrapment for proteolytic enzymes are those of acyl transfer and transpeptidation (Borsook, 1953).
The transpeptidation reactions of the acyl transfer type are catalyzed by various enzymes, such as serine (Dobry, 1952) and thiol (Johnston, 1950) proteinases, certain metal-containing peptidases (Hanson, 1967) and carboxyl proteinases (Takahashi, 1974). The latter enzymes catalyze, besides the reactions of acyl transfer, those of amino transfer (Fruton, 1961).

V.K. Antonov

All these facts indicate that the formation of the so-called "acyl-enzymes", i.e. covalent compounds of proteolytic enzymes and the acyl fragments of substrates, is typical of these enzymes. Furthermore, it is difficult to explain the mechanism of acyl transfer otherwise than in terms of intermediate covalent compounds being formed in the course of these reactions.

Is the formation of acyl-enzymes a necessary condition for hydrolytic enzyme-catalyzed reactions in which an amide bond is being cleaved? If not, what is the mechanism of transfer reactions? I shall try to answer the questions.

COVALENT OR GENERAL BASE MECHANISM?

During the recent decade, in our laboratory techniques have been developed that make it possible to discriminate between: (1) the covalent mechanism of enzyme-catalyzed proteinolysis, namely, a mechanism in which the carbonyl carbon of a substrate is attacked by the catalytically active group of the enzyme (X),

$$
\text{E-X}^{\ominus} \quad \begin{array}{c} R_1 \\ | \\ C=O \\ | \\ NHR_2 \end{array} \rightleftharpoons \begin{array}{c} R_1 \\ | \\ E-X-C-O^{\ominus} \\ | \\ NHR_2 \end{array} \rightleftharpoons \begin{array}{c} R_1 \\ | \\ E-X-C=O \end{array} \xrightarrow{H_2\bullet} \begin{array}{c} R_1 \\ | \\ E-X-C-\bullet^{\ominus} \\ | \\ OH \end{array} \rightleftharpoons \text{E-X}^{\ominus} + \begin{array}{c} R_1 \\ | \\ C=\bullet \\ | \\ OH \end{array}
$$

$$
\downarrow\uparrow R_3NH_2
$$

$$
\tag{1}
$$

$$
\begin{array}{c} R_1 \\ | \\ E-X-C-O^{\ominus} \\ | \\ NHR_3 \end{array} \rightleftharpoons \text{E-X}^{\ominus} + \begin{array}{c} R_1 \\ | \\ C=O \\ | \\ NHR_3 \end{array}
$$

and (2) the general base mechanism of cleavage in which the catalytically active group of the enzyme promotes the abstraction of a proton from the water molecule that directly attacks the carbonyl carbon (Antonov, 1978, 1979, 1980; Lyakisheva, 1973).

$$
\text{E-X}^{\ominus} \quad \begin{array}{c} H-\bullet \\ | \\ H \end{array} \begin{array}{c} R_1 \\ | \\ C=O \\ | \\ NHR_2 \end{array} \rightleftharpoons \begin{array}{c} R_1 \\ | \\ E-XH \cdot H\bullet-C-O^{\ominus} \\ | \\ NHR_2 \end{array} \rightleftharpoons \begin{array}{c} R_1 \\ | \\ E-X^{\ominus} \cdot C=\bullet \\ | \\ OH \end{array} \rightleftharpoons \text{E-X}^{\ominus} + \begin{array}{c} R_1 \\ | \\ C=\bullet \\ | \\ OH \end{array}
$$

$$
\downarrow\uparrow R_3NH_2
$$

$$
\tag{2}
$$

$$
\begin{array}{c} R_1 \\ | \\ E-XH \cdot {}^{\ominus}\bullet-C-NHR_3 \\ | \\ OH \end{array} \rightleftharpoons \text{E-X}^{\ominus} + \begin{array}{c} R_1 \\ | \\ C=\bullet \\ | \\ NHR_3 \end{array}
$$

These techniques are based on the fact that the above mechanisms differ in the step at which water is involved in the reaction. In covalent catalysis, water participates at the step when the acyl-enzyme is cleaved; in the case of general base mechanism, water is involved in the first step of chemical transformation of the substrate.

If the reaction of transfer or transpeptidation is conducted in

$H_2^{18}O$, heavy oxygen will not be incorporated into the amide bond of a transfer product in the case of covalent catalysis since the acyl-enzyme attacked by the acceptor does not contain ^{18}O. In general base catalysis, the transfer to the acceptor occurs apparently after the incorporation of ^{18}O into the acyl product, and a new peptide must contain heavy oxygen.

Two circumstances should be considered here. First, one has to be certain that, along with the transpeptidation, the direct resynthesis proceeds at a minimal rate or does not take place at all. Second, one should consider whether heavy oxygen might be incorporated into the product of transfer due to reversible hydration-dehydration of the intermediate acyl-enzyme.

There are several arguments against this supposition. For instance, model studies of ester and anhydride hydrolysis (Bender, 1960) indicate that intermediate tetrahedral compounds decompose much faster toward hydrolysis products than toward initial substrates, i.e. hydration-dehydration would make a very minor contribution to the enrichment of the acyl-enzyme with heavy oxygen. Direct experiments on the incorporation of ^{18}O into a substrate in chymotrypsin-cata-lyzed hydrolysis gave negative results (Bender, 1957; Lyakisheva, 1973).

Finally, the results of experiments on ^{18}O incorporation into the products of transfer may be verified independently in the following way (3). Many proteolytic enzymes catalyze the exchange of oxygen atoms in the carboxyl group of the acyl product of hydrolysis with the ^{18}O of heavy water (Breslow, 1977; Ginodman, 1966).

$$ \tag{3} $$

If the second (amine) product of hydrolysis is introduced in the system, then along with the reaction of exchange, the reversible reaction of synthesis-hydrolysis of the amide bond occurs; the reaction also results in the incorporation of oxygen into the carboxyl of the acyl product. Apparently, if the reaction proceeds via the formation of an acyl-enzyme, synthesis-hydrolysis of the peptide cannot increase the observed rate of oxygen exchange in the carboxyl of the acyl reaction product. In the case of the general base mechanism, if the rate of synthesis-hydrolysis is comparable with the rate of "genuine" exchange, the overall rate of exchange at the carboxyl of the acyl product would increase upon the addition of an adequate amine component. Obviously, the results of experiments on the exchange acceleration do not depend on the capability of an inter-

V.K. Antonov

TABLE 1   Incorporation of ^{18}O into the Transpeptidation Products

E n z y m e	Reaction studied	Form of the product analysed*	Per cent of exch-ange**
Chymotrypsin	$2LeuLeuNH_2 \rightarrow (Leu)_3 + LeuNH_2 + NH_3$	$Tfa(Leu)_3OMe$	0
Pepsin	$2LeuTyrNH_2 \rightarrow (Leu)_2 + 2TyrNH_2$	$TMS(Leu)_2OTMS$	56 ± 10
Leucine amino-peptidase	$2LeuNH_2 \rightarrow LeuLeuNH_2 + NH_3$	$LeuLeuNH_2$	46 ± 5
	$2\|^{18}O\|\text{-}LeuNH_2 \rightarrow \|^{18}O\|\text{-}LeuLeuNH_2 + NH_3$	$\|^{18}O\|\text{-}LeuLeuNH_2$	$55 \pm 5^{\neq}$
Thermolysine	$2LeuLeuNH_2 \rightarrow LeuLeu + 2LeuNH_2$	$Tfa(Leu)_2OMe$	47 ± 5

*
**Tfa - trifluoroacetyl, TMS - trimethylsilyl; Analyzed by mass-spectrometry. Water containing 62-80 at.% of ^{18}O was used. $^{\neq}$Reaction was carried out in normal water.

mediate acyl-enzyme for hydration-dehydration. The use of the both approaches may provide unambiguous information about the mechanism of cleavage of the amide bond.

Tables I and 2 show our data pertinent to the incorporation of heavy oxygen into the products of transpeptidation and to oxygen exchange in the carboxyl of acyl products for reactions catalyzed by chymotrypsin, pepsin, leucine aminopeptidase and thermolysine. As might be expected, ^{18}O was not incorporated into the product of transpeptidation in the case of chymotrypsin. However, it was incorporated with the other enzymes. Moreover, in the case of pepsin and leucine aminopeptidase, the exchange of oxygen in the carboxyl of acetylphenyl-

TABLE 2   Rates of Oxygen Exchange Catalyzed by Some Proteinases

E n z y m e	Substrates	Incub. time min.	% of exchange	Exptl. V_{max} of exchange $M\ min^{-1}$	Calcul.* rate of exch. due to synth. hydrol. $M.\ min^{-1}$
Pepsin	AcPhe	360	13 ± 2	2.5×10^{-5}	
	AcPhe+PheAlaAlaOMe	360	29 ± 3	5.6×10^{-5}	7.8×10^{-5}
Leucine amino-peptidase	Leu	8	21 ± 1	1.6×10^{-3}	
	Leu+NH$_4$Cl	8	19 ± 1	1.8×10^{-3}	8.3×10^{-6}
	Leu+LeuNH(CH$_2$)$_2$OH	8	40 ± 2	3.5×10^{-3}	1.2×10^{-3}

*Calculate by the formula given in (Breslow, 1977).

alanine and leucine, respectively, was accelerated when adequate amino products were added.

All these data clearly indicate that, for pepsin, thermolysine and leucine aminopeptidase, hydrolysis of substrates involves water at the first chemical step of the process and no intermediate covalent acyl-enzymes are formed. It is noteworthy that the data of X-ray diffraction analysis of penicillopepsin (James, 1977) and thermolysine (Weaver, 1977) also support this conclusion.

Therefore, the known types of proteolytic enzymes may be subdivided into two groups according to the mechanism of their action: (1) proteinases that operate by the covalent type of catalysis (serine and, apparently, thiol proteinases) and (2) proteinases that operate by general base mechanism (carboxyl and metal-dependent proteinases). It would be relevant to note that carboxyl proteinases cannot be found in prokaryotic organisms; it is possible therefore that the general base mechanism of hydrolysis appeared later in the course of evolution.

MECHANISM OF TRANSPEPTIDATION

The covalent mechanism of catalysis has been studied sufficiently well, at least for chymotrypsin, but not that of general base catalysis. Here, I would like to consider only one question. If no covalent intermediate compounds are formed with the acyl fragment of a substrate in the case of carboxyl and metal-dependent proteinases, what is the mechanism of acyl transfer or transpeptidation catalyzed by these enzymes? One must discriminate here between the thermodynamic and kinetic aspects of the problem.

Not infrequently, when the thermodynamics of the transpeptidation reaction is analyzed (see, for example, James, 1977; Knowles, 1970) only synthesis of the amide bond from free carboxyl and amino components is considered (4).

$$R-COO^{\ominus} + {}^{\oplus}H_3N-R_1 \rightleftharpoons RCONHR_1 + H_2O \qquad (4)$$

In this case, indeed, the yield of a peptide does not exeed 1-2% at concentrations of the reacting substances usually employed. Actually, however, the reaction of transpeptidation is an exchange reaction for which, regardless of pH, a change in free energy must be close to zero and, in the conditions of equilibrium, the yield of a peptide may reach 50% as it can be easily calculated (5):

$$RCONHR_1 + R_2COO^{\ominus} \rightleftharpoons R_2CONHR_1 + RCOO^{\ominus}$$
$$AB \qquad\quad C \qquad\qquad\quad BC \qquad\qquad A \qquad\qquad\qquad (5)$$

$$|BC|_{eq} = |AB|_o |C|_o \,/\, |AB|_o + |C|_o$$

The kinetic aspect of the problem is more complicated. The non-covalent enzyme complex with a free product must exist for a sufficiently long time for the acceptor to be bound and for a new amide bonde to be formed. The pepsin-catalyzed reaction of transpeptidation of the amino transfer type may be taken as an example; data are available about the initial rates of the amide bond formation for this reaction (Antonov, 1974). Using these data, one may calculate that the observed rate of transpeptidation can be reached if the rate constant for the dissociation of an enzyme-product complex will be only about 10 s^{-1} for the system:

$$AcPhe\text{-}Tyr + AcPhe(NO_2) \longrightarrow AcPhe(NO_2)\text{-}Tyr + AcPhe$$

(for the calculation equation 10 of the above cited paper has been used and assumptions $k_3 = k_{-4}$ and $k_4 = 1 \times 10^7$ $M^{-1}s^{-1}$ were made). This is at least by 2-3 orders of magnitude lower than the usually observed rates for the dissociation of such complexes.

In order to solve this contradiction, one should suppose that at least two enzyme-product complexes (6) exist on the route from an enzyme-substrate complex to products, and the equilibrium in the system is shifted toward the "external" (E_oP) complex:

$$\ldots \; \rightleftharpoons \; E_pP \underset{k-3}{\overset{k_3}{\rightleftharpoons}} E_oP \overset{K_i}{\rightleftharpoons} E + P$$

(6)

were $\dfrac{k_3}{k_{-3}} = K_q \gg I$ and $K_{i(app)} = \dfrac{K_i K_q}{1 + K_q} \approx K_i$

In this case, in the experiments on the inhibition by the product, the observed constant for dissociation of the enzyme-product complex, $K_{i(app)}$, would reflect only the formation of this "external" complex (K_i). Its conversion into the "internal" (E_pP) complex must be very slow, in above mentioned case slower than with $k \approx 10$ s^{-1}. Consequently, the "internal" complex which is responsible for transpeptidation can be formed from the substrate, but only to a very low extent from free reaction products and the enzyme.

One more consequence follows from the existence of relatively stable non-covalent complexes of the above type: any attempts, which are often made to prove the covalent character of intermediate compounds in enzymatic catalysis by trapping with an acceptor, actually prove nothing at all. Such an experiment might be successful both in the presence and in the absence of a covalent intermediate.

SPECIFICITY

If we suppose that special complexes exist between the enzyme and a reaction product, we must necesserily assume also the existence of similar enzyme-substrate complexes; in other words, we have to presume that interactions between the enzyme and a substrate are also stepwise. Indeed, the stepwise formation of enzyme-substrate and enzyme-inhibitor complexes was found experimentally in many cases (Goto, 1980; Harrison, 1978; Johannin, 1966; Kitagishi, 1980; Sachdev, 1975; Yon, 1976).

How can we describe an "internal" enzyme-substrate complex (E_pS, see eqn 7)? I think that it is this complex which is productive and undergoes chemical transformation. It is difficult to consider its formation as a one-step process controlled by diffusion since this formation requires the appearance of numerous specific bonds and, possibly, conformational changes. Therefore, the interaction between the enzyme and a substrate should commence with a rapid "anchoring" of the substrate through a sufficiently specific group, which is followed by rearrangements resulting in a productive complex.

The kinetic consequence of the two-step formation of a productive enzyme-substrate complex is the following (eqn. 7 and Table 3): the observed kinetic constants of the enzyme-catalyzed reaction would depend on the equilibrium between the "external" (E_oS) and productive complexes. I think that this is very important for understanding the specificity of proteolytic enzymes.

$$E + S \underset{k_{-1}}{\overset{k_1}{\rightleftharpoons}} E_oS \underset{k_{-2}}{\overset{k_2}{\rightleftharpoons}} E_pS \overset{k_3}{\longrightarrow} E + P$$

$$v = \frac{k_3 K_p |E|_o |S|_o}{K_s + (1 + K_p)|S|_o} \text{ , were } K_s = \frac{k_{-1}}{k_1} \text{ and } K_p = \frac{k_2}{k_{-2} + k_3} \qquad (7)$$

$$k_{cat} = \frac{k_3 K_p}{1 + K_p} \text{ and } K_{m(app)} = \frac{K_s}{1 + K_p}$$

It is well known that the specificity of proteinases may be manifested in the maximal rates of hydrolysis, in the binding of a substrate, or in the both, i.e. the "better binding - better catalysis" principle (Knowles, 1965) is obeyed. The two-step scheme for the formation of a productive enzyme-substrate complex makes it possible to interpret different manifestations of the specificity of proteinases in the same terms. According to the scheme, for all substrates possessing an identical, or similar, "anchoring" site (or a site of primary specificity), the dissociation constant for an "external" enzyme-substrate complex will be the same, and the observed substrate constant (or Michaelis constant) at first will not change (for substrates with $K_p<1$) and then will decrease (when $K_p>1$). Here, the values of k_{cat} will increase reaching a constant value equal to the rate constant for chemical transformation of the productive complex (k_3). The latter value must, apparently, be identical for a series of substrates of the same type and should reflect the maximal catalytical effectiveness of an enzyme for the given group of substrates. Indeed, the plots of log K_m versus log k_{cat} for a series of substrates of different proteinases (Fig. 1) are in good agreement with those predicted on the basis of the two-step scheme for the formation of a productive complex. It would be relevant to note that the above concepts are applicable not only to proteinases operating by the general base mechanism, but also to proteinases accomplishing

TABLE 3   Meaning of the Experimental Kinetic Constants

K_p	k_3	k_{cat}	$K_{m(app)}$
$<<1$	$<<k_{-2}$	$k_2 k_3/k_{-2}$	K_s
$<<1$	$>>k_{-2}$	k_2	K_s
$>>1$	$<<k_{-2}$	k_3	$K_s k_{-2}/k_2$
$>>1$	$>>k_{-2}$	k_3	$K_s k_3/k_2$

covalent catalysis.

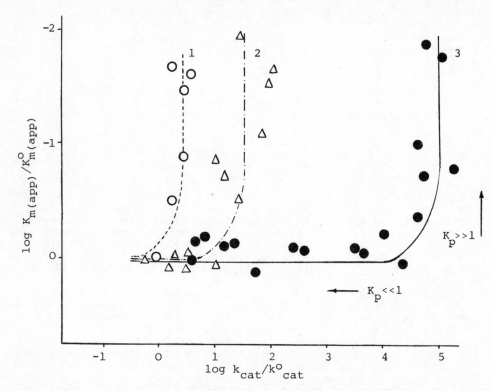

Fig. 1.   Plots of log K_m (relative) vs. log k_{cat} (relative)
          for series of substrates of carboxypeptidase A (1),
          chymotrypsin (2) and pepsin (3). Data for substrates
          of type AcX-Phe (o) are taken from Klyosov (1977),
          for AcX-Phe-Y (Δ) from Bauer (1976, 1976a) and Bau-
          mann (1973) and for ZX-Phe(NO_2)-PheY(●) from Zin-
          chenko (1977) and Medzihradzky (1970). X are amino
          acid or peptide residues, Y - C-protected amino
          acid or peptide residues.

CONCLUSION

We may conclude therefore that, on the route from a substrate to products, proteolytic enzymes form a series of non-covalent intermediate
complexes playing a crucial role in catalysis: (1) complexes formed
upon the "anchoring" of a substrate when the primary specificity of
the enzyme is realized; (2) productive complexes in which the secondary specificity of the enzyme is realized: and (3) complexes with
the reaction products whose existence account for the reactions of
transpeptidation. Apparently, besides the above complexes, there are
also enzyme complexes with unstable intermediate reaction products,
such as tetrahedral intermediates, but their consideration is beyond
the scope of this report.

ACKNOWLEDGEMENT

The author gratefully acknowledge the contributions of Drs. L.M.
Ginodman, L.D.Rumsh, A.A.Zinchenko, Yu.V.Kapitannikov, A.G.Gurova
and L.P.Yavashev.

REFERENCES

Antonov, V.K., L.D. Rumsh, and A.G. Tichodeeva (1974). FEBS Lett.,
    46, 29-33.
Antonov, V.K., L.M. Ginodman, Yu.V. Kapitannikov, T.N. Barshevskaya,
    A.G. Gurova, and L.D. Rumsh (1978). FEBS Lett., 88, 87-90.
Antonov, V.K., L.P. Yavashev, A.G. Gurova, V.A. Sadovskaya, and L.D.
    Rumsh (1979). Bioorg. Khim., 5, 1427-1429.
Antonov, V.K., L.M. Ginodman, L.D. Rumsh, Yu.V. Kapitannikov, T.N.
    Barshevskaya, L.P. Yavashev, A.G. Gurova, and L.I. Volkova (1980).
    Bioorg. Khim., 6, 436-446.
Bauer, C.-A., R.C. Thompson, and E.R. Blout (1976). Biochemistry, 15,
    1291-1295.
Bauer, C.-A., R.C. Thompson, and E.R. Blout (1976a). Biochemistry,
    15, 1296-1299.
Baumann, W.K., S.A. Bizzozero, and H. Datler (1973). Eur. J. Biochem.,
    39, 381-391.
Bender, M.L., and K.C. Kemp (1957). J. Am. Chem. Soc., 79, 111-116.
Bender, M.L. (1960). Chem. Rev., 60, 53-113.
Borsook, H. (1953). Peptide Bond Formation. In M.L. Anson, K. Bailey
    and J.T. Edsall (Eds.), Advances in Protein Chemistry, Vol. 8,
    Academic Press, New York. pp. 127-174.
Breslow, R., and D.L. Wernick (1977). Proc. Natl. Acad. Sci. USA,
    74, 1303-1307.
Dobry, A., J.S. Fruton, and J.M. Sturtevant (1952). J. Biol. Chem.,
    195, 149-155.
Fruton, J.S., S. Fujii, and M.H.Knappenberger (1961). Proc. Natl.
    Acad. Sci. USA, 47, 759-761.
Ginodman, L.M., N.I. Maltsev, and V.N. Orechovich (1966). Biokhimia,
    31, 1073-1078.
Goto, S. (1980). J. Biochem., 87, 399-406.
Hanson, H., and J. Lasch (1967). Hoppe-Seyler's Z. Physiol. Chem.,
    348, 1525-1539.
Harrison, L.W., B.L. Vallee (1978). Biochemistry, 17, 4359-4363.
James, M.N., I.N. Hsu, and L.I.J. Delbaere (1977). Nature, 267, 808-
    813.
Johannin, G., and J. Yon (1966). Biochem. Biophys. Res. Commun., 25,
    320-325.
Johnston, R.B., M.J.Mycek, and J.S. Fruton (1950). J. Biol. Chem.,
    187, 205-213.
Kitagishi, K., H. Nakatani, and K. Hiromi (1980). J. Biochem., 87,
    573-579.
Klyosov, A.A., and B.L. Vallee (1977). Bioorg. Khim., 3, 806-815.
Knowles, J.R. (1965). J. Theor. Biol., 9, 213-228.
Knowles, J.R. (1970). Phil. Trans. Roy. Soc. London, B 257, 135-146.
Lyakisheva, A.G., L.M. Ginodman, and V.K. Antonov (1973). Moleku-
    lyarnaya Biol., 7, 810-816.
Medzihradzky, K., I.M. Voynick, H. Medzihradszky-Schweiger, and
    J.S. Fruton (1970). Biochemistry, 9, 1154-1162.
Sachdev, G.P., and J.S.Fruton (1975). Proc. Natl. Acad. Sci. USA,
    72, 3424-3429.

Takahashi, M., T.T. Wang, and T. Hofmann (1974). _Biochem. Biophys. Res. Commun._, _57_, 39-46.

Weaver, L.H., W.R. Kester, and B.W.Matthews (1977). _J. Mol. Biol._, _114_, 119-132.

Yon, J.M. (1976). _Biochimie_, _58_, 61-69.

Zinchenko, A.A., L.D. Rumsh, and V.K. Antonov (1977). _Bioorg. Khim._, _3_, 1663-1670.

A DIFFERENT FORM OF THE Ca^{2+} –DEPENDENT PROTEINASE ACTIVATED BY MICROMOLAR LEVELS OF Ca^{2+}

A. Szpacenko, J. Kay[1], D.E. Goll, and Y. Otsuka

Muscle Biology Group, University of Arizona
Tucson, Az. 85721, U.S.A.

ABSTRACT

A new form of the calcium-activated proteinase has been isolated from beef heart muscle. This new form contains two subunits, 80,000 and 30,000 daltons, respectively, and is inhibited completely by iodoacetamide and 1mM EDTA. In these respects, it is identical to the originally isolated calcium-activated proteinase but, while the original enzyme required 2 to 5mM Ca^{2+} for maximal activity and was inactive below 0.2mM Ca^{2+}, the new form is completely activated at a Ca^{2+} concentration of 50μM. The new form is less negatively charged than the original form.

Two pathways have been suggested to be involved in the catabolism of intracellular proteins (Ballard, 1977); a lysosomal route that, in isolated hepatocytes, seems to account for about 70% of the total protein turnover (Seglen and colleagues 1979), and a nonlysosomal route that accounts for the remainder and seems to be responsible for degradation of shortlived and abnormal proteins (Knowles and Ballard, 1976; Seglen and coworkers 1979). The quantitative importance of the latter route in tissues such as muscle with a relatively low lysosomal content has yet to be evaluated.

Proteolysis at neutral pH within the cell and at a location distinct from the lysosomal matrix must be regulated very carefully to prevent excessive protein turnover taking place under steady state conditions. Several neutral proteolytic systems have been characterised in vitro and implicated in protein catabolism in living tissue. Among these are the ATP-dependent proteinases (Goldberg, Strnad and Swamy 1980), the trypsin-like serine proteinase from smooth muscle (Beynon and Kay, 1978; Carney and coworkers 1980) and the Ca^{2+}-activated thiol proteinase (CAF) from skeletal muscle (Dayton and colleagues 1975).

---

[1] Permanent address: Department of Biochemistry, University College, Cardiff, Wales.

The Ca^{2+}-activated proteinase was originally purified from pig skeletal muscle (Dayton and coworkers 1976a) and was shown to have little effect on most of the contractile proteins of muscle.   CAF's major influence seemed to be removal of α-actinin from myofibrils with consequent disappearance of the z-disk (Dayton and colleagues 1976b).

A proteinase having the same polypeptide chains, amino acid composition, and enzymic properties as CAF has recently been purified from bovine platelets (Szpacenko and colleagues 1980), and proteolytic activities similar to CAF have been detected in a variety of tissues including chick oviduct (Waxman, 1979; Vedeckis and coworkers 1980).   All these Ca^{2+}-activated proteases require Ca^{2+}-concentrations of approximately 0.2mM for detectable activity and 1 to 5mM for maximal activity (Dayton and colleagues 1976b).   Free intracellular Ca^{2+}-concentration does not seem to rise above 10μM in normal skeletal muscle (Caputo, 1978; Ebashi, 1976; Fabiato and Fabiato, 1979).   Therefore, it has been unclear how CAF could act under physiological conditions unless unusually high Ca^{2+}-concentrations existed in localised areas or the Ca^{2+} requirement of CAF were altered in some way.   This situation has been exacerbated by the discovery in muscle of a protein inhibitor of CAF (Okitani and coworkers 1976; Waxman and Krebs, 1978).   This CAF inhibitor has recently been purified to homogeneity (Otsuka and Goll, 1980).   Studies with purified CAF and CAF inhibitor have shown that millimolar concentrations of Ca^{2+} are required for CAF inhibitor to bind to CAF and that one mole of CAF inhibitor can bind and inactivate up to ten moles of CAF (Otsuka and Goll, 1980).   Thus, in living tissue, the presence of only small amounts of inhibitor would inactivate any CAF activity that might be switched on in response to localized increases in intracellular Ca^{2+} concentration up to millimolar levels.   It would thus be very difficult for this form of CAF to be active under physiological conditions.

During development of the original procedure for purification of CAF from porcine skeletal muscle (Dayton and coworkers 1976a), it was noticed that a small peak of Ca^{2+}-activated proteolytic activity eluted from DEAE-cellulose columns at a lower ionic strength than that required to elute the CAF peak (see Fig. 2 in Dayton and coworkers 1976a).   This small peak of activity accounted for less than 5% of total Ca^{2+}-activated proteolytic activity in the extracts and therefore was ignored in the further purification of CAF (Dayton and coworkers 1976a).   Recently, however, Mellgren (1980) has shown that dog heart muscle contains a Ca^{2+}-activated protease that is one-half maximally active at a Ca^{2+}-concentration of 40μM.   This low Ca^{2+}-requiring protease was eluted from DEAE-cellulose columns by a concentration of KCl similar to that required for elution of the small peak of activity in our earlier studies.   Consequently this early eluting peak of activity has now been purified from bovine cardiac muscle.

Ground bovine cardiac muscle is suspended in 50mM Tris·acetate, pH 7.5, 4mM EDTA by three 30-sec bursts on a Waring blendor at 11,000 rpm with 30-sec cooling periods between each burst.   The supernatant obtained after centrifugation at 11,000 g for 20 min is subjected to isoelectric precipitation between pH 6.2 and 4.9.   The protein precipitated at pH 4.9 is suspended in 50mM Tris·acetate, pH 7.5, 1mM EDTA, the pH of the solution is adjusted to 7.5 and, after centrifugation at 142,800 g for 1 hr, the clarified solution is salted out between 0 and 65% ammonium sulfate saturation to produce a P_{0-65} crude CAF fraction.   This fraction is then chromatographed on: 1) DEAE-cellulose; 2) phenyl-Sepharose 3) Ultrogel 34.

The initial chromatography on DEAE-cellulose separated two Ca^{2+}- activated proteolytic activities (Fig. 1).   Assay of the two activities indicated that the material eluting first at an average KCl concentration of 120 mM required only micromolar quantities of Ca^{2+} for activation, whereas the protein that eluted later at an average KCl concentration of 235mM required the millimolar concentrations of Ca^{2+} typical of CAF.

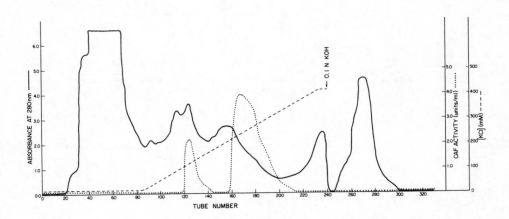

Fig. 1.   Chromatograpny or a P$_{0-65}$ crude CAF fraction (20 g in 900 ml) on a column (5 x 26 cm) of DEAE - cellulose, equilibrated in 20 mM Tris· Acetate buffer, pH 7.5, 1 mM EDTA, 0.1% mercaptoethanol.   A linear KCl gradient (0-400 mM) was used for desorption.   Flow Rate = 85 ml/hr;Fractions = 17 ml.

These two proteolytic activities were designated the "low Ca^{2+}-requiring form" and the "high Ca^{2+}-requiring form", respectively.   Both forms were applied to separate phenyl-Sepharose columns (Fig. 2).   The two proteolytic activities eluted off phenyl-Sepharose columns at identical low concentrations of KCl indicating that both have very hydrophobic areas on their surfaces. Ca^{2+}- requirements of the two proteases remained unchanged after phenyl-Sepharose chromatography.

The proteolytic activities eluted off phenyl-Sepharose were concentrated and chromatographed in succession on the same Ultrogel 34 column.   Both the low and high Ca^{2+}-requiring forms eluted at the same position (Fig. 3), indicating that these two proteases have very similar Stokes' radii.

A. Szpacenko et al.

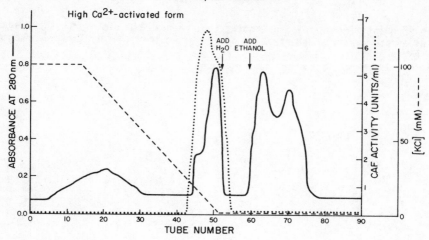

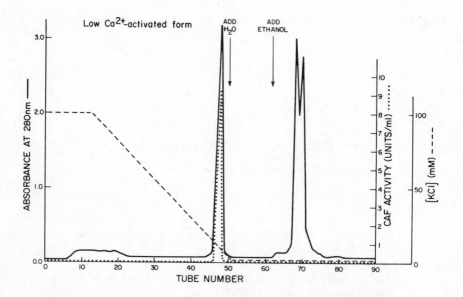

Fig. 2. Chromatography of the "high" (Upper - 89 mg in 90ml)
and "low" (Lower - 17 mg in 56 ml) Ca^{2+}-activated
forms on columns (2.6 x 30 cm) of phenyl-Sepharose,
equilibrated in 100 mM KCl, 20 mM Tris·acetate buffer,
pH 7.5, 1 mM EDTA.   Decreasing linear KCL gradients
(100 mM - 0) were used for desorption. Flow rates =
16 ml/hr.   Fractions = 8 ml。   Water and ethanol
were added at the indicated points to remove tightly
-adsorbed protein.

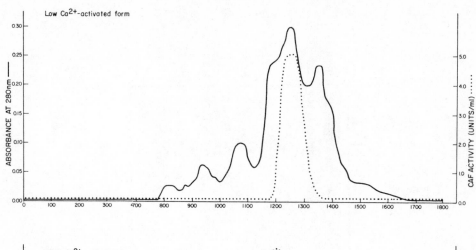

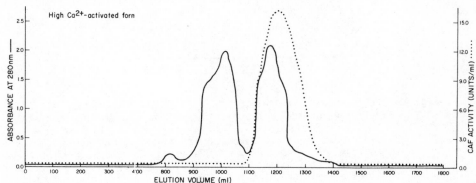

Fig. 3.   Chromatography of the "low" (Upper - 102 mg in
           20 ml) and "high" Ca^{2+}-activated forms (Lower -
           494 mg in 47 ml) on a column (5 x 90 cm) of
           Ultrogel 34, equilibrated in 20 mM Tris· Acetate
           buffer, pH 7.5, 1mM EDTA.   Flow rates = 28 ml/hr.

The Ca^{2+}-activated proteolytic activities from Ultrogel 34 were concentrated by
salting out between 0 and 60% ammonium sulfate saturation and further purified
by preparative polyacrylamide gel electrophoresis.   Each Ca^{2+}-activated protease
(1-5 mg) was loaded onto a 14 x 16 cm polyacrylamide gel slab, 3mM thick, and
electrophoresed at pH 7.5 under non-denaturing conditions until several hours
after the bromophenol blue tracking dye had migrated off the end of the gel.
The gels were removed, a thin slice was trimmed from each side of the gel, and
the slices were stained to locate the major protein bands present.   The high
Ca^{2+}-requiring protease preparation was separated into five bands (Track 1 -
Fig. 4), whereas the low Ca^{2+}-requiring protease electrophoresed as a diffuse,
broad area containing two more concentrated bands (barely discernible in Fig. 4-
Track 2).

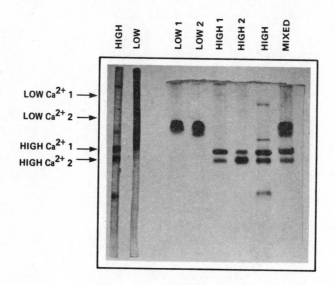

Fig. 4.   Polyacrylamide gel electrophoresis of the high and low Ca^{2+}-
            activated proteolytic activities in a non-denaturing buffer at
            pH 7.5.   From the left, Tracks 1 (=7) and 2 - High and low
            Ca^{2+}-requiring preparations, respectively, as obtained after
            Ultrogel 34 chromatography.   Tracks 3 - 6 - Re-electrophoresis
            of bands obtained from slab gels of tracks 2 and 1, respectively.
            Track 8 - mixture of High Ca^{2+}-1 and low Ca^{2+}-2 from slab gels
            1 and 2.

The areas containing the two major bands (High Ca^{2+}-1 and 2) from the high and
(low Ca^{2+}-1 and 2) from the low Ca^{2+}-requiring forms were sliced from the slab
gels and the protein in these slices was eluted from the gels.   Re-electro-
phoresis of each eluted protein in the same non-denaturing buffer at pH 7.5
showed that both of the proteins eluted from the low Ca^{2+} gel (Tracks 3 and 4)
migrated with less net negative charge than the proteins from the high Ca^{2+} gel
(Tracks 5, 6 and 8 - Fig. 4).

The proteins in the high Ca^{2+}-1 and 2 bands were both active proteolytically
towards casein at concentrations of Ca^{2+} greater than 0.2mM and maximal activity
required 2 to 5 mM Ca^{2+}.   For the low Ca^{2+} proteins, by contrast, only low
Ca^{2+}-2 had proteolytic activity when assayed on casein; this protein was maximally
active at Ca^{2+} concentrations of 50 μM and its activity changed little as the
Ca^{2+} concentration was increased to 5 mM.

When the proteins eluted from bands 1 and 2 of the high Ca^{2+}-requiring preparation
were subjected to electrophoresis on SDS-polyacrylamide gels, high Ca^{2+}-1 had
both the 80,000 and the 30,000-dalton subunits described for the original CAF
preparation (Dayton and colleagues 1976a) whereas high Ca^{2+}-2 had only the 80,000
dalton polypeptide, (Fig. 5).   Both forms were catalytically active, however, so

that the active site of CAF must be located in the 80,000-dalton subunit. Other
studies in our laboratory with a fluorescent sulfhydryl reagent, N-iodoacetyl-N-
(5-sulfonic-1-naphthyl) ethylene diamine have also shown that the essential -SH
group is located in the 80,000 dalton subunit (Hennecke and Goll, unpublished
results).

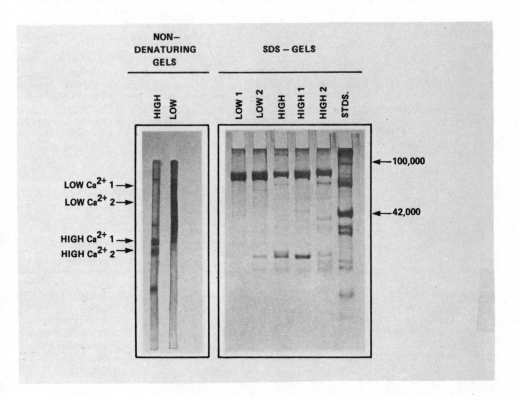

Fig. 5.   Polyacrylamide gel electrophoresis of the high and
           low Ca^{2+}-activated proteolytic activities.
           Left panel (as in Fig. 4) - Non-denaturing gels at
           pH 7.5 of the high and low Ca^{2+}-requiring preparat-
           ions as obtained after Ultrogel 34 chromatography.
           Right panel-SDS gels.   Tracks 1 and 2 - Low Ca^{2+}-1
           and 2 protein bands obtained from slab gels under
           non-denaturing conditions.
           Tracks 3 - 5, Unfractionated High-Ca^{2+}-protein and
           High Ca^{2+}-1 and 2 bands from slab gels.
           Track 6 = Standard myofibrillar proteins from
           rabbit skeletal muscle plus bovine serum albumin.
           α-Actinin = 100,000 and actin = 42,000 daltons
           respectively.

When the low Ca^{2+}-1 and 2 proteins were electrophoresed on SDS-polyacrylamide gels, the low Ca^{2+}-2 form (active) contained the 80,000 and 30,000-dalton subunits whereas the low Ca^{2+}-1 protein (inactive) contained only the 80,000-dalton subunit. Thus the role of the 30,000 dalton subunit remains unclear.   Since it co-migrates with the 80,000-dalton subunit on electrophoresis under non-denaturing conditions of two of the proteolytically active bands (High Ca^{2+}-1 and low Ca^{2+}-2), it seems likely that the 30,000-dalton subunit is part of the CAF molecule, possibly having some as yet undiscovered regulatory function.   Azanza and coworkers (1979) have isolated and characterised a thiol proteinase from rabbit muscle that is activated maximally at 2 mM Ca^{2+} and contains only the 80,000 dalton subunit.   This would appear to resemble the high Ca^{2+}-2 form described above.

Since the low Ca^{2+}-requiring form migrated more slowly on polyacrylamide gel electrophoresis at pH 7.5 (Fig. 4) and eluted at a lower KCl concentration from DEAE-cellulose than the high Ca^{2+}-requiring forms, it seems that the low Ca^{2+}-2 protein is a form of CAF that has less net negative charge and a lower Ca^{2+}-requirement than the forms that have been isolated up to now.   With these exceptions, the low Ca^{2+}-requiring activity is very similar to CAF in many of its properties (summarised in Table 1).

TABLE 1   Comparison of the Properties of the Low Ca^{2+}-
Activated Protease and CAF

Effector or Property	Low Ca^{2+}-Activated Protease	CAF
EDTA (5mM)	Inhibited	Inhibited
Ca^{2+} (0.05mM)	Activated	No effect
Ca^{2+} (5mM)	Activated	Activated
Iodoacetamide (5mM)	Inhibited	Inhibited
Leupeptin	Inhibited	Inhibited
CAF inhibitor	Inhibited	Inhibited
Soybean Trypsin Inhibitor	No effect	No effect
Smooth Muscle Protease Inhibitor (Carney et al. 1980)	No effect	No effect
Polypeptide chains	30,000 & 80,000	30,000 & 80,000
Elution off Ultrogel 34	Identical with CAF	Identical with Low Ca^{2+}-activated Protease
Elution off phenyl-Sepharose	50-0 mM KCl	50-0 mM KCl

Both the low and high Ca^{2+}-activated forms are inhibited at nearly identical rates by incubation with iodoacetamide (Fig. 6).   Hence, both the low and high Ca^{2+}-activated proteases require an unblocked thiol side chain for activity.   The low Ca^{2+}-requiring protease also is completely inhibited by the protein inhibitor of CAF (Table 2).   Thus far, this protein inhibitor has been specific for CAF and has not inhibited any other protease with which it has been tested.   CAF inhibitor inhibits the low-Ca^{2+}-requiring protease at both 50μM and 5mM Ca^{2+} (Table 2). Assuming that CAF inhibitor must bind to the protease to be effective, this result

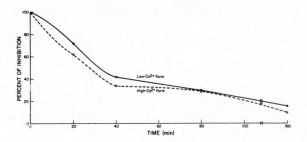

Fig. 6. Inhibition of the high and low Ca^{2+}-activated proteases by iodoacetamide. Each protease was incubated with 5mM iodoacetamide (at reagent to protease
molar ratios of 58 and 116 for the low and high
Ca^{2+} forms respectively) in 50mM Tris·Acetate buffer,
pH 7.5, 1mM EDTA at 25°. Samples were removed at
the times indicated and assayed for proteolytic
activity remaining using casein as substrate.

indicates that CAF inhibitor requires much lower concentrations of Ca^{2+} to bind
to the low Ca^{2+}-requiring protease than the 1mM levels it requires to bind to
the high Ca^{2+}-requiring protease (Otsuka and Goll, 1980).

The high Ca^{2+}-requiring protease purified in this study is identical to the
original CAF and the low Ca^{2+}-requiring protease obtained after preparative
polyacrylamide gel electrophoresis is also similar in many of its properties.

TABLE 2   Inhibition of High and Low Ca^{2+}-Forms of
CAF by CAF Inhibitor[a]

| TYPE OF CAF | 50µM Ca^{2+} | | 5mM Ca^{2+} | |
	NO INHIBITOR	WITH INHIBITOR	NO INHIBITOR	WITH INHIBITOR
HIGH Ca^{2+}	0.0	0.0	35.7	0.8
LOW Ca^{2+}	6.08	0.0	6.66	0.0

[a]Figures are OD$_{278}$ units/mg enzyme protein/30 min. as determined with casein
substrate. Assay conditions: 5.0mg casein/ml, 100mM KCl, 100mM Tris·acetate,
pH 7.5, 10mM 2-mercaptoethanol, 1mM NaN$_3$, CaCl$_2$ as indicated, one mole CAF
inhibitor per 10 moles CAF when added, 20µg CAF/ml, 25.0°C.

It seems likely that the low Ca^{2+}-requiring protease is a new form of CAF that is modified in a way that reduces its Ca^{2+}-requirement to levels near those found physiologically inside cells. We, therefore, tentatively refer to this low Ca^{2+}-requiring protease as "low Ca^{2+}-CAF". The nature of the difference between low Ca^{2+} CAF and high Ca^{2+} CAF that decreases the Ca^{2+}-requirement of the low Ca^{2+} form is unclear but may be related to the higher net negative charge possessed by the originally isolated, high Ca^{2+} CAF.

The discovery of a low Ca^{2+}-CAF suggests that this form is the physiologically active form of CAF and that the originally purified high Ca^{2+}-CAF is an "off" form. Physiological control of CAF activity, however, must be more complex than simple switching between "on" and "off" forms of the protease, because all cells found thus far to contain this protease have also been shown to possess a protein inhibitor of the protease. Our preliminary results with CAF inhibitor and the low Ca^{2+}-CAF indicate that the micromolar levels of Ca^{2+} needed for activity of low Ca^{2+}-CAF are also sufficient for CAF inhibitor to inactivate this form of CAF. Consequently, even low Ca^{2+}-CAF could not be active physiologically unless its interaction with CAF inhibitor was also subject to an "on" and "off" control. Such a complex system of controls is not unexpected if CAF, indeed, acts physiologically at some initial stage of a functionally significant process.

When the ability of high Ca^{2+}-CAF to hydrolyze denatured proteins was compared with the activities of other proteolytic enzymes such as bovine pancreatic trypsin and the trypsin-like smooth muscle protease (Beynon and Kay 1978) it was found that CAF has only about one-tenth of the "general" proteolytic activity exhibited by the other proteases (Kay and colleagues 1980). This lack of general proteolytic activity is likely to be a reflection of the unusual and restricted specificity of CAF and suggests that CAF acts physiologically to accomplish a specific function rather than as a general protease to degrade groups of proteins during their metabolic turnover, such as has been proposed for the function of lysosomal proteases. Since CAF has now been demonstrated to exist in bovine platelets (Szpacenko and colleagues 1980) and may exist in many other cell types as well, CAF probably has a broader role than initiating metabolic turnover of myofibrillar proteins as was originally proposed (Dayton and coworkers 1975). The effects of CAF on platelets and smooth muscle cells indicate that CAF may act to detach actin filaments from their intracellular attachments, whether these attachments are Z-disks in striated muscle cells or plasma membrane attachments in other cells. The mechanism by which CAF accomplishes such detachment is now under study.

ACKNOWLEDGEMENTS

This work was supported by NIH Grants Nos. AM-19864 and HL-20984, by grants from the Muscular Dystrophy Association, and by Arizona Agricultural Experiment Station Project 28, a contributing project to Regional Research Project NC-131. J.K. was a Visiting Professor at the University of Arizona and acknowledges an award from the Wain Fund administered by the Agricultural Research Council. We are very grateful to our colleagues, Gretchen Hennecke and Jane Greweling for their excellent assistance and smiling dispositions and to Marguerite Santucci for assistance with the manuscript.

REFERENCES

Azanza, J-L., J. Raymond, J-M. Robin, P. Cottin,and A. Ducastaing (1979). Biochem. J. 183, 339-347.

Ballard, F.J. (1977). Essays Biochem. 13, 1-37.

Beynon, R.J., and J. Kay (1978) Biochem. J. 173, 291-298.

Caputo, C. (1978). Ann. Rev. Biophys. Bioeng. 7, 63-83.

Carney, I.T., C.G. Curtis, J. Kay,and N. Birket (1980). Biochem. J. 185, 423-433.

Dayton, W.R., D.E. Goll, M.H. Stromer, W.J. Reville, M.G. Zeece, and R.M. Robson (1975). In E. Reich, D.B. Rifkin, and E. Shaw (Eds.). Proteases and Biological Control, Cold Spring Harbor Laboratory. pp 551-577.

Dayton, W.R., D.E. Goll,M.G. Zeece, R.M. Robson, and W.J. Reville (1976a). Biochemistry 15, 2150-2158.

Dayton, W.R., W.J. Reville, D.E. Goll, and M.H. Stromer (1976b). Biochemistry 15, 2159-2167.

Ebashi, S. (1976). Ann. Rev. Physiol. 38, 293-313.

Fabiato, A., and F. Fabiato (1979). Ann. Rev. Physiol. 41, 473-484.

Goldberg, A.L., N.P. Strnad, and K.H.S. Swamy (1980). in D.C. Evered and J. Whelan (Eds). Protein Degradation in Health and Disease Vol. 75 CIBA Foundation Symposia. Excerpta Medica. pp 227-247.

Kay, J., R.F. Siemankowski, L.M. Siemankowski, and D.E. Goll (1980). Manuscript in preparation.

Knowles, S.E., and F.J. Ballard (1976). Biochem. J. 156, 609-617.

Mellgren, R.L. (1980). FEBS Lett. 109, 129-133.

Okitani, A., D.E. Goll, M.H. Stromer, and R.M. Robson (1976). Fedn. Proc. 35, 1746.

Otsuka, Y., and D.E. Goll (1980). Fedn. Proc. 39, 2044.

Seglen, P.O., B. Grinde, and A.E. Solheim (1979). Eur. J. Biochem. 95, 215-225.

Szpacenko, A., Y. Otsuka, D.E. Goll, and M.H. Stromer (1980). Fedn. Proc. 39, 2044.

Vedeckis, W.V., M.R. Freeman, W.T. Schrader, and B.W. O'Malley (1980). Biochemistry 19, 335-343.

Waxman, L. (1979). Fedn. Proc. 38, 479.

Waxman, L., and E.G. Krebs (1978). J. Biol. Chem. 253, 5888-5891.

CHARACTERIZATION OF TWO ALKALINE PROTEINASES FROM RAT SKELETAL MUSCLE

B. Dahlmann, L. Kuehn, M. Rutschmann, I. Block and H. Reinauer

Biochemical Department, Diabetes Forschungsinstitut an der Universität Düsseldorf,
Auf'm Hennekamp 65, D 4000 Düsseldorf, West Germany

ABSTRACT

Two enzymes of the alkaline proteolytic system from rat skeletal muscle were separated and purified. Both enzymes are of the serine-type but are clearly different
from one another with respect to molecular weight, substrate specificity and the
ability of inhibitors to affect their activity. Furthermore, the cleaving pattern
of myofibrillar proteins produced by the enzymes in vitro as well as immunotitration experiments indicate that these alkaline proteinases are distinct enzymes.

KEYWORDS

Alkaline proteinases; isolation and characterization; muscle protein breakdown; rat
skeletal muscle.

INTRODUCTION

Skeletal muscle proteinases can be divided into two major groups according to their
pH-optimum of activity: acidic proteinases of lysosomal origin and alkaline proteinases of extra-lysosomal origin. For some years, we have been interested in the
latter group of proteinases since we found their enzymic activity to show an adaptive behaviour to certain protein catabolic conditions. In particular, the activity of these enzymes is increased in a state of insulin deficiency, e.g. in diabetes
mellitus, induced by injection of streptozotocin in rats (Röthig and coworkers,
1978). In another example, testosterone depletion as a result of castration has
been shown to lead to an enhanced alkaline proteolytic activity in rats (Dahlmann,
Mai and Reinauer, 1980). Restoration of the alkaline proteolytic activity to normal can be achieved when animals are treated daily with insulin (Röthig and co-
workers, 1978) or when castrated animals are administered 0.5 mg testosterone per
day (Dahlmann, Mai and Reinauer, 1980). Importantly, this increase in alkaline proteolytic activity and its return to normal correlates with urinary excretion of
3-methylhistidine (Dahlmann and coworkers, 1979). Since 3-methylhistidine results
exclusively from actomyosin breakdown(Young and Munro, 1978), the phenomena of increased proteinase activity and the enhanced 3-methylhistidine excretion appear to
be causally related.
What is the mechanism underlying this change in activity of the alkaline proteolytic system? As an approach to obtain some insight into this question, a definition
and characterization of the various enzymes involved would be helpful. Towards this
end, a number of alkaline proteolytic enzymes has been isolated in our laboratory,

163

during the past years. This report will focus on two of these enzymes, a description of their purification as well as some of their properties.

PURIFICATION OF THE ENZYMES

The purification scheme developed to isolate the two enzymes in highly purified
form is depicted in Fig. 1. Briefly, skeletal muscle tissue is homogenized in
50 mM Tris/HCl buffer, pH 8.5, containing 1 M KCl. The homogenate is then fractionated with ammonium sulfate to obtain, at 33% saturation, a particulate myofibrillar fraction and a non-myofibrillar, cytosolic fraction. The proteolytic activity
present in the myofibrillar fraction can be solubilized only under conditions leading to the co-solubilization of myofibrillar proteins that is, in buffers with high
concentrations of KCl or, alternatively, by subjecting this fraction to a 30 min.
incubation with bovine trypsin conjugated to Sepharose 4B. After such a treatment,
the activity remains soluble in buffers free of KCl. The enzyme has been purified
further by affinity chromatography on Sepharose-soy bean trypsin-inhibitor, by ion-
exchange chromatography and by gel filtration. With this procedure, an overall
purification of the enzyme, designated myofibrillar proteinase of about 1800 fold
has been achieved (Dahlmann and Reinauer, 1978). The proteolytic activity remaining
in the supernatant after the ammonium sulfate step has an absolute requirement for
KCl to remain soluble. By virtue of this need for salt, a dialysis step against
salt free buffers was used to precipitate the activity and the sediment resuspended
in a small volume of 50 mM Tris/HCl, pH 8.5 containing 1 M KCl. After repeated molecular sieve chromatography to separate from aminopeptidase activity, an alkaline
proteolytic activity is obtained which is about 1500 fold purified (Reinauer and
Dahlmann, 1979). This enzyme has been termed cytosolic proteinase.

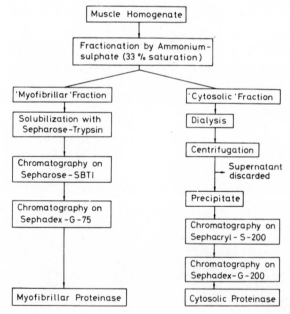

Fig. 1. Purification scheme for the isolation of the myofi-
         brillar and cytosolic proteinase from rat skeletal
         muscle.

CHARACTERIZATION OF THE ENZYMES

Although the two enzymes are quite different from one another, they do share certain properties. As illustrated in Fig. 2, both enzymes have an optimal hydrolytic activity towards azocasein or haemoglobin at pH 9.4-9.6, although an appreciable activity can be detected at neutral pH, as well. In this context it should be stressed again that the cytosolic proteinase is active in the presence of 1 M KCl only. This requirement for the salt results in a decrease of enzymic activity or even its abolishment when the proteinase is assayed or stored in buffers containing KCl concentrations lower than 1 mole/l (Fig. 3).

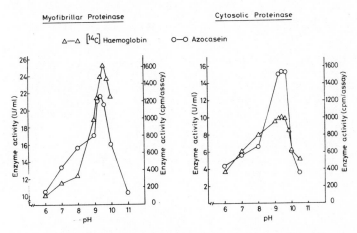

Fig. 2. Effect of pH on the activity of the myofibrillar and the cytosolic proteinase

Experimental conditions were as described by Dahlmann and Reinauer, 1978, except that 1 M KCl was included in buffers for the assay of cytosolic proteinase.

The dependence of cytosolic proteinase activity on high ionic strength precluded the use of electrophoretic techniques such as isoelectric focusing or polyacrylamide gel electrophoresis to further characterize the enzyme. Either technique could, however, be applied to an analysis of the myofibrillar proteinase. When subjected to isoelectric focusing, the enzyme is resolved into several bands with proteolytic activity detectable between pH 5 and 7 (Fig. 4) and, when run on polyacrylamide gels at pH 4.3, the proteolytic activity - as detected by photopaper digestion test (Dahlmann and Reinauer, 1978) - migrates towards the cathode as a single band, and to a position distinct from that of trypsin (Fig. 5).
The question then investigated was to which of the described proteinase groups these two alkaline proteinases belong to. When the effect of various agents was examined, neither SH-blockers nor metal-chelating agents were found to affect the activity of the two enzymes (Table 1). On the other hand, di-isopropylphosphorofluoridate, phenylmethylsulfonylfluoride as well as naturally occuring proteinase inhibitors like soy bean trypsin inhibitor and lima bean trypsin inhibitor completely inhibited the activity of both enzymes, suggesting that their catalytic site is structurally related to serine proteinases. These are the properties which both enzymes have in common. The differences so far elaborated are as follows:
Complete inhibition of the myofibrillar proteinase is observed in the presence of pancreatic trypsin inhibitor, chicken ovomucoid or chicken ovoinhibitor, while the cytosolic proteinase is not affected by these inhibitors. Furthermore, the halo-

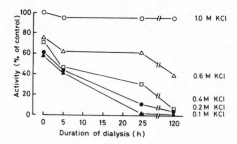

Fig. 3. Effect of KCl on the activity and stability of the
        cytosolic proteinase.

Cytosolic proteinase (50 mM Tris/HCl buffer, pH 8.5, containing 1 M KCl) was diluted with 50 mM Tris/HCl buffer, pH
8.5 to give KCl concentrations of 0.6 M (△—△), 0.4 M
(□—□), 0.2 M (●—●) and 0.1 M (▲—▲). The activity was
measured immediately after dilution or after dialysis for
5, 24 and 120 h at 4°C against buffers with the respective
KCl concentrations. For comparison, the activity of enzyme
kept in the presence of 1 M KCl is given (○—○). The enzymatic activity was measured with azocasein as substrate
(Dahlmann and Reinauer, 1978).

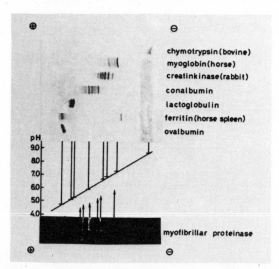

Fig. 4. Isoelectric focusing of purified myofibrillar pro-
        teinase.

Isoelectric focusing was performed with 20 μg protein on LKB
Ampholine PAGE plates, pH 3.5-9.5, according to the instruc-
tions supplied by the manufacturer. For comparison, coomassie stains of some calibration proteins are shown. The proteolytic activity was detected by photopaper test (Dahlmann
and Reinauer, 1978).

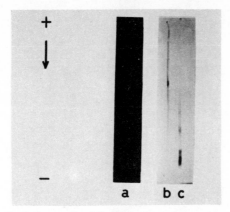

Fig. 5. Polyacrylamide gel electrophoresis of purified myo-
        fibrillar proteinase.

Electrophoretic conditions were as described by Dahlmann
and Reinauer, 1978. Coomassie stains of purified myofibrillar proteinase (b) and bovine trypsin (c). In a parallel
run, the myofibrillar proteinase was detected by photopaper
digestion test (a), according to Dahlmann and Reinauer,
1978.

TABLE 1   Influence of various Effectors and Inhibitors on
           Purified Myofibrillar and Cytosolic Proteinase

Inhibitor	Final concn.	Myofibrillar Proteinase % Inhibition	Cytosolic Proteinase % Inhibition
SH-blocking agents			
Jodoacetamide	1 mM	0	0
pCMB	1 mM	0	0
Metal-chelating agents			
EDTA	1 mM	0	2
o-Phenanthroline	1 mM	0	3
Serine-blocking agents			
Dip-F	1 mM	100	100
Pms-F	1 mM	5	100
	5 mM	81	–
SBTI	0.1 mg/ml	100	100
LBTI	0.1 mg/ml	100	100

Experimental details were as described elsewhere (Dahlmann
and Reinauer, 1978; Reinauer and Dahlmann, 1979).

methylketones 7-amino-1-chloro-3-L-tosylaminoheptan-2-one (TLCK) and 1-chloro-4-
phenyl-3-L-tosylamidobutan-2-one (TPCK) inhibited the myofibrillar proteinase as
did benzamidine, antipain and leupeptin. In contrast, the cytosolic proteinase was
predominantly affected by TPCK, chymostatin and phenylbutylamine (Table 2).
In keeping with the different sensitivity to the action of proteinase inhibitors,
the activity of the two proteinases towards a number of synthetic substrates was
different, too: while the myofibrillar proteinase cleaves substrates containing an
arginine residue, substrates containing an aromatic amino acid like tyrosine are
split by the cytosolic proteinase (Table 3).

B. Dahlmann et al.

TABLE 2   Effect of some Proteinase Inhibitors on Purified
Myofibrillar and Cytosolic Proteinase

Inhibitor	Final concn.	Myofibrillar Proteinase % Inhibition	Cytosolic Proteinase % Inhibition
Bovine pancreatic trypsin-inhibitor	0.5 mg/ml	100	0
Chicken ovomucoid	0.1 mg/ml	100	0
Chicken ovoinhibitor	0.1 mg/ml	100	0
Benzamidine	5 mM	72	0
TLCK	1 mM	61	0
	5 mM	92	0
Antipain	0.01 mg/ml	43	15
TPCK	1 mM	39	30
	3 mM	64	79
Chymostatin	0.01 mg/ml	11	38
Phenylbutylamine	2 mM	0	54
Leupeptin	0.01 mg/ml	27	4
Pepstatin	0.1 mg/ml	-	2

Experimental details were as described by Dahlmann and Reinauer, 1978 as well as by Reinauer and Dahlmann, 1979.

TABLE 3   Activity of Purified Myofibrillar and Cytosolic
Proteinase against Synthetic Substrates of Trypsin
and Chymotrypsin

Substrate	Concn. (mM)	pH	Enzymic activity (mU / mg of enzyme) Myofibrillar Proteinase	Cytosolic Proteinase
Bz–Arg–OEt	1	8	0	0
Tos–Arg–OMe	2	8	0.240	0
Bz–Arg–Nan	2	8	0.027	0
Bz–Tyr–OEt	0.5	8	0	13.76
Ac–Tyr–OEt	0.5	8	0	28.48
Ac–Tyr–Nan	2	8	0	0.088
Glut–Phe–Nan	2	8	0	0.054

For experimental details, see Reinauer and Dahlmann, 1979.

The differences between the two enzymes isolated from rat skeletal muscle also become apparent when testing the ability to be inhibited by purified rat serum proteinase inhibitors, as shown in Table 4. The cytosolic proteinase is fully inhibited
by alpha$_1$-antitrypsin and alpha$_2$-macroglobulin (similar values were found for bovine trypsin, data not shown), whereas the myofibrillar proteinase is only weakly
inhibited by alpha$_1$-antitrypsin and resistent to the action of alpha$_2$-macroglobulin
at saturating concentrations of the inhibitor.
From gel filtration experiments on Sephadex G-75 the apparent molecular weight of
the myofibrillar proteinase was found to be 30 800 (Dahlmann and Reinauer, 1978).
Under identical conditions, the molecular mass of the cytosolic proteinase was determined as 25 000 (Reinauer and Dahlmann, 1979), a further distinction between the
two enzymes.
In another series of experiments, antibodies previously raised in rabbits against
the two enzymes, were assessed by immunotitration (Fig. 6). The results reveal
structural differences between the two enzymes but also point to some possible conformational homology. Thus, the cytosolic proteinase is inhibited by similar IgG
concentrations of sera raised against either the cytosolic proteinase or the myofibrillar proteinase. However, in order to achieve a 50% inhibition of the myofibrillar proteinase, a concentration of anti-cytosolic proteinase IgG is required
which is tenfold in excess of the amount of homologous IgG needed to achieve the
same effect.

TABLE 4   Effect of Rat Serum Proteinase Inhibitors on Proteinase Activity

	Myofibrillar	Cytosolic
	Proteinase	
	Inhibition (% of control)	
α_1-Antitrypsin		
50 µg	N.D.	81
300 µg	33	80
α_2-Macroglobulin		
8 µg	0	77
40 µg	0	78

Enzyme solution (0.1 ml) was preincubated with 0.1 ml of the
indicated inhibitor for 10 min. at 37°C in 50 mM Tris/HCl,
pH 8.5, supplemented with 1 M KCl in samples containing cytosolic proteinase. Proteolytic activity was determined
with azocasein as substrate (Dahlmann and Reinauer, 1978).

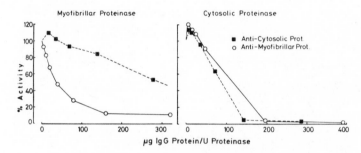

Fig. 6. Immunotitration of purified myofibrillar proteinase
and cytosolic proteinase with anti-cytosolic proteinase -IgG and anti-myofibrillar proteinase-IgG.

Experimental conditions were as given in legend to TABLE 4
except that preincubations lasted for 60 min.

Finally, differences between the proteinases became apparent when their degradative activity towards purified rat skeletal muscle myosin and tropomyosin was examined. As shown in Fig. 7, top panel, the myofibrillar proteinase degrades isolated
myosin heavy chains in a characteristic, time dependent fashion: after 30 min. a
protein with an apparent molecular weight of 72 000 predominates and further cleavage leads to the gradual rise of two polypeptides with 46 000 and 30 000 molecular weight, respectively. In contrast, the cytosolic proteinase produces myosin
fragments primarily in the molecular weight range above 100 000. Only after a prolonged incubation of 2-4 hours, smaller fragments become apparent. Similarily
(Fig. 7, lower panel), tropomyosin is degraded by the myofibrillar proteinase
within 15-30 min., again giving a cleavage pattern which is distinct from that of
an experiment with cytosolic proteinase and tropomyosin.

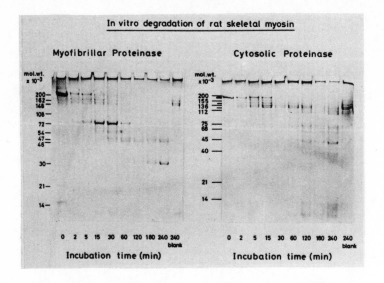

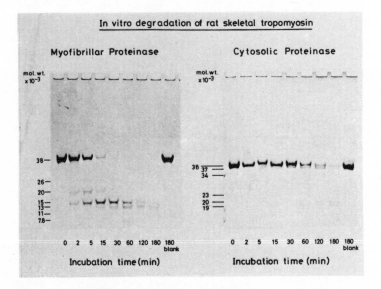

Fig. 7. Degradation of myosin and tropomyosin by purified
myofibrillar and cytosolic proteinase.

Myofibrillar proteins and proteinases were incubated at pH
8.0, 37°C, using the following weight ratios; myosin:protei-
nase 400:1, tropomyosin:proteinase 140:1. At the times indicated, aliquots were removed and subjected to polyacrylamide slab gel electrophoresis as described (Porzio and
Pearson, 1977).

CONCLUSIONS

The data presented quite clearly demonstrate that the cytosolic proteinase and the myofibrillar proteinase are distinct enzyme species. It is known that mast cell proteinases largely contribute to the alkaline proteolytic activity measured in rat skeletal muscle. The cytosolic proteinase described here has properties similar to those of mast cell chymase (Kawiak and coworkers, 1971), such as the requirement for high KCl concentrations, a chymotrypsin-like cleaving pattern and a molecular weight of 25 000. Experiments are in progress to evaluate the significance of this similarity. Independently, however, it must be stressed that the alkaline proteolytic activity present in muscle tissue cannot entirely be attributed to mast cell activity. At least another activity with features distinct from mast cell proteinase, the myofibrillar proteinase, can be isolated from rat skeletal muscle. The data presented in this study suggest that this enzyme is implicated in skeletal muscle protein degradation as observed in catabolic situations like diabetes mellitus.

ACKNOWLEDGEMENTS

These studies were supported by the Deutsche Forschungsgemeinschaft (SFB 113), by the Ministerium für Wissenschaft und Forschung des Landes Nordrhein- Westfalen and by the Ministerium für Jugend, Familie und Gesundheit. We thank Mrs. R. Schwitte and Mrs. M. Koenen for expert technical assistance.

REFERENCES

Dahlmann, B. and H. Reinauer (1978). Purification and some properties of an alkaline proteinase from rat skeletal muscle. Biochem. J., 171, 803-810.

Dahlmann, B., C. Schroeter, L. Herbertz and H. Reinauer (1979). Myofibrillar protein degradation and muscle proteinases in normal and diabetic rats. Biochem. Med., 21, 33-39.

Kawiak, J., W. H. Vensel, J. Komender and E. A. Barnard (1971). Non-pancreatic proteases of the chymotrypsin family. I. A chymotrypsin-like protease from rat mast cells. Biochim. Biophys. Acta, 235, 172-187.

Porzio, M. A. and A.M. Pearson (1977). Improved resolution of myofibrillar proteins with sodium dodecylsulfate-polyacrylamide gel electrophoresis. Biochim. Biophys. Acta, 490, 27-34.

Reinauer, H. and B. Dahlmann (1979). Alkaline proteinases in skeletal muscle. In Holzer, H. and H. Tschesche (Eds.), Biological Functions of Proteinases, Springer Verlag, Heidelberg. pp. 94-101.

Young, V. R. and H. N. Munro (1978). N-methylhistidine and muscle protein turnover: an overview. Fed. Proc., 37, 2291-2300.

HUMAN ERYTHROCYTES PEPTIDASES

Lj. Vitale, M. Zubanović* and M. Abramić

Department of Organic Chemistry and Biochemistry, "Ruđer Bošković" Institute,
*"Pliva", Pharmaceutical and Chemical Works, Zagreb, Yugoslavia

ABSTRACT

Two aminopeptidase type enzymes of human erythrocytes were isolated starting from clarified hemolyzate using DEAE-cellulose chromatography, Sephadex G-200 gel filtration and preparative polyacrylamide gel electrophoresis. One of them was shown to be an aminopeptidase of molecular weight of about 95,000, pI about pH 4.9 and optimal activity in the pH range 6.0 to 7.5. It was activated by Co^{2+} and not by Cl^-, sensitive to EDTA but not to pCMB, DFP and PMSF. The enzyme has a broad specificity towards peptides and amino acid-ß-naphthylamides; Lys-, Phe-, Arg-, Met- and Leu-ßNA being the best substrates. The second enzyme is a dipeptidyl aminopeptidase of molecular weight of about 82,000 and pI about 4.6, with optimal activity in the pH range 7.0 to 9.0 depending on the substrate. It was inhibited by EDTA and pCMB and slightly by DFP. According to its substrate specificity the enzyme is dipeptidyl aminopeptidase III.

KEYWORDS

Human erythrocytes peptidases; protease; aminopeptidase; dipeptidyl aminopeptidase III.

INTRODUCTION

Erythrocytes, like other cells, possess their own set of proteases whose biological role is suggested to be rearrangement of cell constituents during cell maturation, their degradation at the end of the life cycle, and participation in the metabolism of biologically active peptides within the cell or in its environment. The correct understanding of erythrocyte proteolytic system is hindered by the limited knowledge about the participating enzymes. Even though the ery-

173

throcyte is one of the oldest targets of investigation, the number
and characteristics of its proteases are far from being finalized and
known. This is particularily so with respect to peptidases.

As surveyed in Table 1, the data on red blood cell peptidases refer
mainly to the determination of enzymatic activity in hemolyzates,
while their names often come from the used rather than from the best
substrate. Only the peptidase from sheep erythrocytes, rabbit dipeptidyl aminopeptidase and two human arylamidases are purified and
characterized to a greater extent. Erythrocyte peptidases named after
the biologically active peptides used as substrates, like angioten-
sinases (Kokubu and others, 1969; Nast and others,1976; Moore and
others, 1977), are not listed in Table 1, since available data do not
justify their inclusion as a separate category of enzymes.

Having in mind the importance of a better understanding of erythrocyte
metabolic processes, we have undertaken the isolation of the aminopeptidase type enzymes from human erythrocytes and their characterization in the purified form.

ENZYMES ISOLATION

Fresh human erythrocytes were obtained from a blood bank, and then
extensively washed with saline. After each centrifugation the whitish
cell layer and interphase were removed, until smears of the erythrocyte
suspension were free of nucleated cells. Washed cells were frozen
at -20°C, and then lysed by the addition of 2 volumes of distilled
water. The hemolyzate was centrifuged 50 min at 14,500 x g to remove
stroma and then used for enzyme isolation.

Peptidase activity was measured with leucine-ß-naphthylamide (Leu-ßNA)
and Asn^1,Val^5-angiotensin II as substrates. When using Leu-ßNA and
other naphthylamides, the method of Nagatsu and others (1970) was applied, as the degradation of angiotensins was followed by the liberation of amino groups according to Sanger (1945).

Human erythrocyte hemolyzate was shown to possess the ability to hydrolyze both substrates. After the removal of stroma, 83% of the activity towards Leu-ßNA and 95% of the activity towards angiotensin
remained in the supernatant, indicating the solubility of the respective enzymes.
As the first step for enzyme purification, chromatography on DEAE-cellulose was applied (Fig. 1). Hemoglobin passed through the column un-
adsorbed. Enzymes eluted with 0.1 and 0.2 M NaCl were distributed in
several smaller and one major peak, which contained hydrolytic activity towards both substrates. Fractions of the peak with the highest
activity were pooled, concentrated by ultrafiltration and subjected
to gel filtration on Sephadex G-200 (Fig. 2). Besides the additional
2-3 fold purification, gel filtration indicated a slight separation
of activities towards Leu-ßNA and angiotensin.

TABLE 1 Survey of Erythrocyte Peptidases

Name	Source	Purity	Substrates and effectors[a]	References[b]
Tripeptidase	horse, man	part.purified. part.purified.	tripeptides; i.Cd^{2+},cysteine tripeptides; i.pCMB	Adams (1952) Tsuboi (1957), Haschen (1962, 1964, 1965)
	rabbit	& hemolyzate hemolyzate	Gly-Gly-Gly	
Glycyl-glycine dipeptidase	man, rabbit	"	Gly-Gly	Adams (1952a), Haschen (1964, 1965)
Glycyl-leucine dipeptidase	man, rabbit	"	Gly-Leu; i.EDTA; a.Zn^{2+} Gly-Leu	Haschen (1961a, 1964, 1965)
Imidodipeptidase	horse, man	"	X-Pro dipeptides; a. Mn^{2+}	Adams (1952a), Haschen (1964)
Iminodipeptidase	man, rabbit	"	Pro-X dipeptides	Haschen (1964, 1965)
Leucine-aminopeptidase	man, rabbit	"	Leu-amide; i.EDTA, a. Ca^{2+}, Mn^{2+}	Adams (1952a),Haschen(1961, 1964, 1965), Neef (1973)
Alanine-aminopeptidase	man	"	Ala-amide, Ala-ßNA	Haschen (1964), Neef (1973)
Aminopeptidase	man	part.purified.	Leu-ßNA, Ala-ßNA, Gly-ßNA	Behal (1963)
Peptidase A,B,C,D,E	man, horse	electrophero-gram	different peptides	Lewis (1967), Yut (1979)
Peptidase	sheep	purified	N-protected dipeptides, tripeptides; i. DFP	Witheiler (1972)
Dipeptidyl aminopeptidase	man	purified	Arg-Arg-ßNA; i. EDTA, pCMB, DFP	McDonald (1978)
Arylamidase I and II	man	purified	amino acid-ßNA; a. Co^{2+} (I), Cl^- (II)	Mäkinen (1978)

a) i.=inhibited; a.=activated; NA=naphthylamide. b) Only the name of the first author is given.

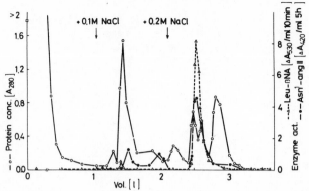

Fig. 1. Chromatography of 300 ml of hemolyzate with 49 g of
protein on DEAE-cellulose. Column size 5 x 20 cm;
10 mM Na-phosphate buffer, pH 6.8, containing 1 mM
ß-mercaptoethanol; flow rate 60 ml/h; temp. 4°C. Substrate conc. 9×10^{-5} M Leu-ßNA and 1.6×10^{-4} M
Asn^1, Val^5-angiotensin II, pH 7.0.

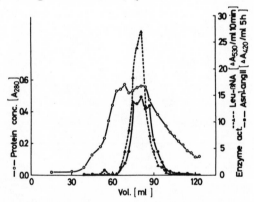

Fig. 2. Gel filtration of enzyme fraction from DEAE-cellulose
step on Sephadex G-200 medium. Sample 38 mg of protein; column size 1.5 x 82 cm; 10 mM Na-phosphate buffer, pH 6.8, containing 1 mM ß-mercaptoethanol; flow
rate 5.3 ml/h; temp. 4°C. Substrates concentrations as
in Fig. 1.

Analysis of the enzyme preparation thus obtained, for proteolytic activity with casein, hemoglobin and azocoll as substrates gave negative results. Polyacrylamide gel electrophoresis and determination of
hydrolytic activities in sliced gels revealed, however, the presence
of two enzymes. One of them degraded only angiotensin and the other
Leu-ßNA as well.

The same technique performed on gel plates was used for the isolation
of the two enzymes. After detection of hydrolytic activities (Fig. 3)

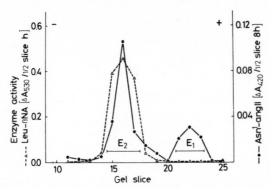

Fig. 3. Polyacrylamide gel electrophoresis of enzyme fraction
from Sephadex G-200 step. Sample 140 µg of protein;
system pH 7.5 - 7%, 40 mA, 3.5 h, 12°C. Enzyme activity was determined in halves of 1.3 x 10 x 2.0 mm gel
slices. Substrates concentration as in Fig. 1.

strips of the gel were cut out and extracted with 10 mM Na-phosphate
buffer, pH 6.8, containing 1 mM ß-mercaptoethanol. Concentrated extracts contained electrophoretically homogeneous enzyme preparations
designated as E_1 and E_2. For Leu-ßNA hydrolyzing activity the overall
purification was around 2000 fold, and the yield about 2.5%.
Asn^1,Val^5-angiotensin II hydrolyzing activity was recovered with the
same yield. The enzymes were kept at -20°C in 40% glycerol.

CHARACTERIZATION OF THE ISOLATED PEPTIDASES

Determination of enzyme type

The lack of proteolytic activity demonstrated that both enzymes isolated from human erythrocytes belong to the peptidases. Hydrolysis of
Leu-ßNA indicated that E_2 is most probably aminopeptidase. The mode
by which the two enzymes degrade substrates was determined by their
action on angiotensin II. Thin-layer chromatography of the degradation products showed that peptides and amino acids were formed in the
reaction with E_1 and E_2, respectively.   The composition of angiotensin II degradation products was determined by comparison with
standards, specific staining, peptide isolation and hydrolysis, and
N-terminal amino acids determination. The results obtained showed
that E_1 and E_2 split angiotensin II as follows:

E_1

             ↓1st   ↓

Asp-Arg-Val-Tyr-Ile-His-Pro-Phe

E_2 ↑1st↑   ↑ ↑ ↑   ↑   ↑ ↑

This led to the conclusion that E_1 is dipeptidyl aminopeptidase and
E_2 aminopeptidase.

Properties of the enzymes

Further characterization of the isolated enzymes included determination of their basic molecular and catalytic properties (Table 2).

TABLE 2   Properties of Human Erythrocytes Peptidases

	Dipeptidyl aminopeptidase	Aminopeptidase
Molecular weight:	$\sim$ 82,000[a,b]	$\sim$ 95,000[a] $\sim$ 40.000[b]
pI:	$\sim$ 4.6	$\sim$ 4.9
pH optimum:	8.6 - 9.0 (Arg_2-ßNA) 7.3 - 8.0 (ang. II)	6.0 - 7.0 (Leu-ßNA) 6.9 - 7.4 (ang. II)
Inhibitors[c]:	EDTA, pCMB	EDTA
Activators:	-	Co^{2+}, Ca^{2+}

[a] Determined by gel filtration. [b] Determined by SDS-polyacrylamide gel electrophoresis. [c] Complete inhibition at concentrations of 10^{-3} M EDTA and 10^{-4} M pCMB. ang. II = Asn^1, Val^5-angiotensin II.

Both enzymes are acidic proteins with molecular weights in the range of 80,000 to 95,000 Daltons. Electrophoretic analysis in the presence of SDS suggested the existence of subunits within the aminopeptidase molecule. Besides the potent inhibitors given in Table 2, aminopeptidase was partially inhibited by 10^{-2} M ϵ-aminocapronic acid and organic solvents. Diisopropylfluorophosphate at a concentration of 10^{-2} M caused 20% inhibition of dipeptidyl aminopeptidase. A parallel experiment with subtilisin demonstrated 100% inhibition. Aminopeptidase was significantly activated by 10^{-3} M Co^{2+} (1.6 fold), and slightly by Ca^{2+} (1.1 fold), while Cl^- did not affect the activity of either enzyme.

The specificity of dipeptidyl aminopeptidase was determined using dipeptidyl-ß-naphthylamides and angiotensins as substrates. Its preferred substrate was Arg-Arg-ßNA with maximal activity in the basic pH region. The data in Table 3 show that the specificity of the enzyme depends on both amino acids of the N-terminal dipeptidyl residue. The lack of hydrolysis of Asp-Arg-ßNA, even after prolonged incubation, and the degradation of octapeptide angiotensin II, which has aspartic acid in position 1, demonstrated the influence of the rest of substrate molecule on the reaction rate. The enzyme did not hydrolyze mono-amino acid naphthylamides, N-protected peptides, dipeptides or tripeptides. According to the described properties the enzyme belongs to a class of dipeptidyl aminopeptidases III. Dipeptidyl aminopeptidase was originally described in the bovine pituitary gland (Ellis and Nuenke, 1967). So far rabbit erythrocytes were reported to possess the same type of peptidase (McDonald and Schwabe, 1978).

TABLE 3 Relative Hydrolysis Rate of Different
         Substrates by Dipeptidyl Aminopeptidase

Substrate	Hydrolysis rate (%)	
	pH 7.0	pH 9.0
Arg-Arg-ßNA	100	220
Ala-Ala-ßNA	14	0
Ala-Phe-ßNA	4	0
His-Ser-ßNA	2	0
Gly-Phe-ßNA	0	0
Gly-Pro-ßNA	0	0
Asp-Arg-ßNA	0	0
Angiotensin II amide	20	−
Angiotensin II	16	−
Angiotensin III	14	−

Dipeptidyl-ßNA: 4.0×10^{-5} M, 0.5% dimethylformamide; angiotensins: 10^{-3} M.

TABLE 4 Relative Hydrolysis Rate of Different
         Substrates by Aminopeptidase

Substrate	Hydrolysis rate (%)
L-Lysine-ßNA	100
L-Phenylalanine-ßNA	88
L-Arginine-ßNA	80
L-Methionine-ßNA	78
L-Leucine-ßNA	48
L-Alanine-ßNA	12
L-Tyrosine-ßNA	3.3*
L-Serine-ßNA	0.7*
L-Valine-ßNA	0.7*
L-Proline-ßNA	0.6*
L-Hydroxyproline-ßNA	0.3*
α,L-Aspartic acid-ßNA	0.2*
α,L-Glutamic acid-ßNA	0.2*
Nα-Benzoyl-L-arginine-ßNA	0.0*
Asn^1,Val^5-angiotensin II	23
Asp^1,Ile^5-angiotensin II	9

Enzyme activity was determined at pH 7.0 with
8.9×10^{-5} M ß-naphthylamides in the presence
of 10^{-3} M Co^{2+} and 1% dimethylformamide, or
with 10^{-3} M angiotensins.
*Calculated from reactions carried out for 14
hours. NA = naphthylamide

For determination of aminopeptidase specificity mono-amino acid-ß-
-naphthylamides were used as substrates (Table 4). The enzyme required a free terminal NH_2-group and preferred the residues of the
basic amino acids lysine and arginine, but also of phenylalanine,
methionine and leucine. Hydrolysis of other tested amino acid-ßNA,
particularily of Asp-ßNA and Glu-ßNA, was detected only after very
long incubation times. Nevertheless, angiotensin II was degraded by
the enzyme. The isolated aminopeptidase hydrolized leucine amide
and peptides of different sizes (di- to deka- peptides were tested)
with aminoacyl preferences similar to those of naphthylamides. This
enzyme most probably corresponds to the arylamidase I described by
Mäkinen and Mäkinen (1978). There are also indications that similar
enzymes are present in the cytosol fraction of the pituitary gland
(Ellis and Perry, 1966) and bovine lens (Swanson and others, 1978).

Demonstration of the existence of an aminopeptidase without strict
specificity in human erythrocytes suggest that some reconsideration
of the data on erythrocyte leucine and alanine aminopeptidase and
certain dipeptidases, may be necessary, particularily with respect
to their names.

REFERENCES

Adams, E., N.C. Davis, and E.L. Smith (1952). Peptidases of erythro-
    cytes. III. Tripeptidase. J. Biol. Chem., 199, 845-856.
Adams, E., M, McFadden, and E.L. Smith (1952a). Peptidases of eryth-
    rocytes. I. Distribution in man and other species. J. Biol. Chem.,
    198, 663-669.
Behal, F.J., R.D. Hamilton, C.B. Kanavage, and E.C. Kelly (1963).
    Studies on aminopeptidases in human blood. Arch. Biochem. Bio-
    phys., 100, 308-312.
Ellis, S., and M. Perry (1966). Pituitary arylamidases and peptidases.
    J. Biol. Chem., 241, 3697-3686.
Ellis, S., and J.M. Nuenke (1967). Dipeptidyl arylamidase III of the
    pituitary. J. Biol. Chem., 242, 4623-4629.
Haschen, R.J. (1961). Characterisierung und Bestimmung der Leucin-
    aminopeptidase menschlicher Erythrocyten. Biochem. Z., 334, 569-
    -575.
Haschen, R.J. (1961a). Characterisierung und Bestimmung der Glycyl-
    -L-leucin-dipeptidase der menschlicher Erythrocyten. Biochem. Z.,
    334, 560-568.
Haschen, R.J. (1962). Characterisierung und Bestimmung der tripep-
    tidase der menschlicher Erythrocyten. Acta Biol. Med. German., 8,
    209-216.
Haschen, R.J., F. Groh, N. Rehfeld, und W. Farr (1964). Das proteoly-
    tische System der menschlichen Erythrozyten. Acta Biol. Med.
    German., 13, 493-503.
Haschen, R.J., W. Farr, und F. Groh (1965). Proteolytische Enzyme und
    Eisweisstoffwechsel der Kaninchen-Erythrozyten während der Zell-
    reifung in vivo. Acta Biol. Med. German., 14, 205-215.

Kokubu, T., H. Akutsu, S. Fujimoto, E. Ueda, K. Hiwada, and Y. Yamamura (1969). Purification and properties of endopeptidase from
rabbit red cells and its process of degradation of angiotensin.
Biochim. Biophys. Acta, 191, 668-676.

Lewis, W.H.P., and H. Harris (1967). Human red cell peptidases.
Nature, 215, 351-355.

Mäkinen, K.K., and P.L. Mäkinen (1978). Purification and characterization of two human erythrocyte arylamidases preferentially hydrolysing N-terminal arginine or lysine residues. Biochem. J., 175,
1051-1067.

McDonald, J.K. and C. Schwabe (1977). Intracellular exopeptidases. In
A.J. Barrett (Ed.), Proteinases of Mammalian Cells and Tissues.
Elsevier/North-Holland Biomedical Press, Amsterdam. pp. 311-391.

Moore, A.F., S. Gurchinoff, W. Brashear, F.M. Bumpus, R. Chang, and
P.A. Khairallah (1977). Angiotensinase activity in red blood cell
membranes and intact adrenal cells. Res. Comm. Chem. Pathol.
Pharmacol., 18, 697-707.

Nagatsu, I., T. Nagatsu,T. Yamamoto, G.G. Glenner, and J.W. Mehl
(1970). Purification of aminopeptidase A in human serum and degradation of angiotensin II by the purified enzyme. Biochim.
Biophys. Acta, 198, 255-270.

Nast, H.P., A. Distler, D. Müller, G. Kantarcioglu, U. Walter, H.P.
Wolff (1976). Untersuchungen über die pH-Abhängigkeit, Hemmbar-
keit und Reaktivierbarkeit von Angiotensin II-amid-spaltenden
Aminopeptidasen menschlicher Erythrozyten. Res. Exp. Med., 168,
89-100.

Neef, L., J.E. Peters und R.J. Haschen (1973). Alanin- und Leuzinami-
nopeptidase in isolierten menschlichen Blutzellen. Z. Inn. Med.,
28, 573-576.

Sanger, F. (1945). The free amino groups of insulin. Biochem. J., 39,
507-515.

Swanson, A.A., B. Albers-Jackson, and J.K. McDonald (1978). Mammalian
lens dipeptidyl aminopeptidase III. Biochem. Biophys. Res. Commun.,
84, 1151-1159.

Tsuboi, K.K., Z.J. Penefsky, and P.B. Hudson (1957). Enzymes of the
human erythrocyte. III. Tripeptidase, purification and specific
properties. Arch. Biochem. Biophys., 68, 54-68.

Witheiler, J.A. and D.B. Wilson (1972). The purification and characterization of a novel peptidase from sheep red cells. J. Biol.
Chem., 247, 2217-2221.

Yut, J., and L.R. Weitkamp (1979). Equine peptidases: Correspondence
with human peptidases and polymorphism for erythrocyte peptidase
A. Biochem. Genet., 17, 987-994.

SERINE PROTEINASES OF MICROORGANISMS

V.M. Stepanov

Institute of Genetics and Selection of Industrial Microorganisms, Moscow, 113545, USSR

ABSTRACT

Individual serine proteinases were isolated from various microbial sources using affinity chromatography, and characterized. It is suggested that microbial serine proteinase might be classified as follows: 1. Subtilisin family, including subfamilies of a) secretory subtilisins (B. subtilis, B. amyloliquefaciens, B. licheniformis, B. thuringiensis), b) intracellular serine proteinases from the same strains, c) SH- containing serine proteinases (Thermoactinomyces vulgaris); 2. Chymotrypsin-like serine proteinases found only in actinomycetes (Str. griseus). Intracellular serine proteinase from B. amyloliquefaciens efficiently splits peptide and denatured protein substrates, but can perform only limited hydrolysis of native proteins thus differing from homologous secretory subtilisins. It is suggested that this feature is of functional importance in the regulation of the enzyme action on proteins within the cell.

KEYWORDS

Microbial serine proteinases; intracellular serine proteinases; proteolysis by intracellular serine proteinases; affinity chromatography of serine proteinases; serine proteinases classification.

This communication is based on results obtained recently in our laboratory and should be considered as a continuation of a report presented during the FEBS Special Meeting on Enzymes in Dubrovnik

183

(Stepanov, 1980). Our goal was to isolate and characterize serine proteinases from various sources including bacteria, actinomycetes and microscopic fungi, with the purpose in mind of enlarging the basis for elucidation of structural, functional and evolutionary relationships within this group of enzymes.

Extracellular serine proteinases of bacilli well known as subtilisins have been thoroughly studied on all levels of their molecular structure as well as their function. Nevertheless, a rather confusing situation arose recently in the evaluation of their evolutionary pattern. It had been generally accepted that subtilisins BPN and Carlsberg presumably produced by different strains of B. subtilis demonstrate exceptionally fast divergence of the primary structures. But then a good deal of taxonomic considerations were revised and it turned out that these enzymes are actually synthesized by different species of bacilli, i.e. by B. amyloliquefaciens and B. licheniformis, respectively. Nevertheless, the extent of structure variations among the extracellular serine proteinases produced by bacilli appears to be quite large and clearly overcomes the rather limited variations found among the intracellular serine proteinases of the same species (Strongin and others, 1978).

The latter assumption was substantiated by the study of the extracellular serine proteinase from B. thuringiensis - a species apparently remote from conventional subtilisin producers (Chestukhina and others, 1979). Pure extracellular serine proteinase has been isolated from the culture filtrate of this bacteria by ammonium sulfate fractionation and affinity chromatography on the sorbents that contain covalently bound phenylboronate (Akparov, Stepanov, 1978) or cyclopeptide bacillichin (Table 1). The enzyme is completely inactivated by phenylmethylsulphonylfluoride (PMSF), a specific inhibitor of serine proteinases. As shown by gel filtration and PAG electrophoresis in the presence of SDS, the enzyme has a molecular weight of 29,000, pI 8.4, has maximal activity and stability at pH 8.5, but rapidly loses its activity at pH lower than 4 or higher than 10, as well as above 60 $^{\circ}$C (Table 4). B. thuringiensis proteinase hydrolyses azocasein, bovine serum albumin, synthetic chromogenic substrate - benzyloxycarbonyl-L-Ala-L-Ala-L-Leu-p-nitroanilide (ZAALpNA) and possesses esterolytic activity.

TABLE 1   Purification of B. thuringiensis extracellular   serine
proteinase

Step	Total protein mg	Activity (ZAALpNA) Total units	Specific units/mg	Purification factor	Yield %
Cultural filtrate	19,800	2256	0.114	1	100
(NH$_4$)$_2$SO$_4$ precipitation (55-85%)	507	1365	2.69	24	60
Chromatography on CHPB (phenylboronate)-Sepharose	12	873	73	638	38
Chromatography on bacillichin-Ultrogel	5.4	745	120	1052	33

According to its physico-chemical characteristics and interactions
with substrates and inhibitors the enzyme ought to be related
to the subtilisins. Its amino acid composition (Table 5) also resembles that typical for subtilisins, e.g., subtilisin BPN, but at
the same time it reveals some traits in common with the intracellular serine proteinases of bacilli - rather a high content of
lysine, glutamic acid and phenylalanine, and a low content of valine. It might be supposed that the intracellular enzyme served as a
parent structure for evolution of the extracellular forms, which,
therefore, retain its features to a certain degree.

It is to be noted that an extracellular serine proteinase isolated
from the culture filtrate of Bacillus subtilis, strain 797, and capable of clearing a suspension of E. coli cells, is also enriched
in dicarboxylic amino acid residues (Asp), but has a remarkably low
glycine content. Hydrophobic chromatography on DNP-hexamethylene
diamine-Sepharose or Ultrogel was of decisive importance for successful purification of this rather unstable enzyme (Borovikova
and others, 1980). Along with such rather different enzymes, one
encounters subtilisin variants that are very close to other members
of the same subfamily. Thus, subtilisin produced by B. subtilis,
strain 72, that has been isolated by affinity chromatography on

phenylboronate-liganded Sepharose, turned to be a structural variant of subtilisin Carlsberg. N-terminal sequences of these two enzymes traced up to 35th residue differ by only two residues (Akparov and others, 1979).

The application of affinity sorbents allowed us to purify serine proteinases from crude preparations of cellulolytic enzymes, e. g., from the surface culture of the fungus Trichoderma lignorum (A.V. Gaida, G.N. Rudenskaya). The main obstacle encountered during the purification of these enzymes is the presence of very active cellulases, as well as other carbohydrases, in Trichoderma cultures, thus excluding the use of conventional ion exchangers or affinity sorbents based on cellulose, dextrans and agarose. To overcome this difficulty, a specifically designed ion exchanger on a silica base, aminosilochrom, has been used in the initial steps of the procedure including the separation of carboxylic proteinase present in the starting material (Rudenskaya, Osterman, Stepanov, 1980). Bacitracin- or phenylboronate-containing affinity sorbents were also successfully used (Table 2).

TABLE 2   Purification of Trichoderma lignorum serine proteinase

Step	Total protein mg	Activity (ZAALpNA) total units	Activity (ZAALpNA) specific units/mg	Purification	Yield %
$(NH_4)_2SO_4$ precipitation	41,500	430,000[+]	10.4[+]		100
Aminosilochrom pH 5.6	16,000	370,000[+]	22.5[+]		86
Aminosilochrom pH 8.0	25,200	10,300	0.41	1	100
Bacitracin-sepharose	1350	7550	56	140	73
Sephadex G-25	130	6420	61	150	62

[+]Milk-clotting activity that corresponds to both serine and carboxylic proteinases.

Serine proteinase was isolated in our laboratory (Stepanov and others, 1980) from the cultural filtrate of thermophilic actinomycetes - Thermoactinomyces vulgaris. Purification of this enzyme was achieved by affinity chromatography on bacitracin or phenylboronate-containing sorbents (Table 3). A comparable procedure has been applied to the purification of the closely related enzyme "termitase" produced by another strain of Thermoactinomyces vulgaris, isolated in the GDR (cf. Frömmel and others, 1978).

TABLE 3 Purification of Thermoactinomyces vulgaris serine proteinase (from the strain INMI-4a).

Step	Total protein A_{280}	Activity (ZAALpNA) total units	Activity (ZAALpNA) specific units/A_{280}	Purification	Yield %
Cultural filtrate	8000	756	0.09	1	100
Bacitracin-Sepharose chromatography	96	894	9.3	103	118
Aminosilochrom chromatography	33	588	17.9	198	77
Sephadex G-25	24	422	17.5	194	56

The amino acid composition of this proteinase (Table 5), as well as many of its physico-chemical characteristics (Table 4), resemble those of the subtilisins, but the N-terminal sequence of the enzyme reveals no convincing homology with a bacterial analog:

Tyr-Thr-Pro-Asn-Asp-Pro-Tyr-Phe-Ser-Ser-Arg-Glu-Tyr

The most remarkable feature of this enzyme is the presence of a free SH group in its structure, that can be blocked by organo-mercurials with concomitant loss of enzyme activity. Comparable observations have been made on "termitase" or an enzyme from Str. rectus in other laboratories. The functional role of the SH group remains obscure. There are indications that it might be redundant for the catalytic step, but its constant presence in the enzymes produced by apparently distant species - Th. vulgaris and Str. rectus - compels one to be cautious in rejecting any functional importance

of this group. Moreover, the presence of a cysteine residue with
comparable properties in carboxypeptidase Y and its analogs, enzymes that also belong to serine proteinases, increases the suspicion
that the thiol group in question might carry out a yet unknown
function, not being involved directly in the catalytic mechanism.

Hence, microbial serine proteinases might be preliminarily classified as follows:

Family of evolutio-narily related en-zymes	Subfamily	Taxon	Examples of enzymes that belong to subfamily
	Secretory subtilisins	Bacilli	B.licheniformis B. subtilis B.thuringiensis
		Fungi	T.lignorum Aspergillus flavus
		Streptomycetes	Str. griseus
Subtilisins	Intracellular serine proteinases	Bacilli	B.amyloliquefaciens B. subtilis, 168 B.licheniformis
		Enterobacteria	E. coli
	SH-containing serine proteinases	Streptomycetes	Thermoactinomyces vulgaris, Str. rectus
Carboxypeptidase Y-like enzymes		Yeasts	Carboxypeptidase Y
Chymotrypsin-like serine proteinases		Streptomycetes	Streptomyces griseus SGPA and SGPB
		Myxobacter	-lytic protease from Myxobacter

This scheme is full of uncertainities and gaps that at least partially reflect our lack of knowledge of microbial proteolytic enzymes.
Nevertheless, one can immediately see the absence of chymotrypsin-
like enzymes in bacteria, and the puzzling presence of enzymes that
belong to two families of streptomycetes.

Intracellular serine proteinases have been isolated earlier in our
laboratory (Stepanov, 1980; Strongin and others, 1978, 1979). Their
chemical and functional characteristics might be summarized in the
following way:

1. These enzymes and the corresponding secretory serine proteinases
reveal homology in their primary structures to the extent that leaves no doubt as to the similarity of their tertiary structures.

2. The intracellular enzymes are more conservative in primary structures than the extracellular ones, the former being apparently involved in functional contacts with other components within the cell.

3. Intracellular serine proteinases exist as dimer molecules,
whereas the secretory subtilisins are completely devoid of quarter-
nary structure. This difference might be functionally important.

4. The intracellular serine proteinases, being more active than
subtilisins against short peptides, show almost no activity against
proteins, e.g. haemoglobin, a mixture of B. subtilis intracellular
proteins.

This rather puzzling contradiction induced us to look more closely
into the action of the intracellular serine proteinase (ISP) from
B. amyloliquefaciens on protein substrates. As shown by V.I. Osto-
slavskaya in our laboratory, ISP preferably splits the Leu-Tyr bond
in the oxidized B-chain of insulin, which is also cleaved by secretory subtilisins, along with a few others. Hence, the specificity
of ISP appears to be more restricted when compared with that of subtilisin. Nevertheless, this difference, important as it might be,
is not strong enough to explain the failure of ISP to attack protein substrates. A.N. Markaryan and A.Ya. Strongin in our laboratory followed the action of ISP on pancreatic ribonuclease A (RNAse A).
ISP splits off the N-terminal S-peptide from RNAse A forming RNAse
S by a reaction typical of subtilisins. Whereas the first step of
this reaction can be performed equally well by both the intra- and
extracellular enzyme, an important difference was observed during
the subsequent stage. Nicked RNAse S turned to be by far more stable to further ISP action than to digestion with subtilisin. In the
latter case, it was necessary to interrupt the reaction in time to
save RNAse S from extended degradation. On the contrary, only mar-
ginal action of ISP on RNAse A was observed.

These experiments lead us to the assumption that ISP is not capable
of digesting native proteins that retain compact tertiary structure.
On the contrary, subtilisins can hydrolyse non-denatured proteins,
although slower than unfolded substrates. Therefore we compared
qualitatively the action of ISP and subtilisin BPN on native and
denatured (reduced and carboxymethylated) RNAse A. As was mentioned above, RNAse A is converted by both proteinases into RNAse S
with subsequent extended degradation of the latter by subtilisin
but not by ISP. On the other hand, reduced and carboxymethylated
RNAse A was efficiently degraded by ISP as well as by subtilisin.
Hence, the limitation of ISP action on RNAse (and apparently on
other proteins) depends on the compactness of their tertiary structure. Evidently, the action of subtilisin on the native proteins is
also hampered by their tertiary structure, but this handicap is by
far less critical than in the case of ISP.

This observation allows us to solve the above-mentioned contradiction between high ISP activity against peptide substrates and its
failure to digest proteins. Obviously, the activity of this enzyme
is directed toward unfolded polypeptide chains, e.g., against the
extended stretches of large proteins (thus leading to limited proteolysis of B. thuringiensis crystal-forming protein, immunoglobu-
lins, RNA polymerase of E. coli, RNAse A) or against unfolded proteins (reduced, carboxymethylated RNAse, B-chain of insulin). Hence,
the intracellular serine proteinases of bacilli (and related ISP of
other microorganisms) might participate in the intracellular degradation of proteins, provided the latter are made vulnerable to the
attack of these enzymes by their partial or complete unfolding.

So far as the unfolding of the protein is a prerequisite for its
degradation by ISP, complete or partial denaturation is to be considered as a likely candidate for a selectivity controlling stage
of intracellular proteolysis. It would be premature to decide in
favour of any mechanism leading to protein denaturation within the
cell, but the recently described (Hershko and others, 1980) energy-
dependent attachment of a proteinous factor to virtual substrate
protein might be one of them.

It is tempting to suppose that the same proteolytic enzyme, e.g.,
ISP of bacilli, might perform the degradation of abnormal protein
tertiary structure which is intrinsically unstable.

Within the pattern of this hypothesis, the incapacity of ISP to degrade native proteins receives a plausible explanation, since it
prevents deleterious non-selective hydrolysis of nicking of
functioning native proteins within the cell. The formation of di-
meric tertiary structure characteristic of ISP might by itself
serve as a device to restrict its activity against compact globulae.
If the active sites of two identical subunits are facing one another, the space between them might be too narrow to accomodate a
voluminous molecule, thus preventing its hydrolysis. On the other
hand, even a rather narrow cleft would accept the unfolded chain
or loosely packed loop of the protein substrate that would explain
ISP efficiency on unfolded substrates. These two hypotheses ,
being far from proved, allow the design of new experiments in the
field of protein turnover and the mode of action of the enzymes
involved in this process.

TABLE 4   Properties of microbial   serine proteinases

| Microorganism | Molec. weight | pI | pH optimun | Loss of activity after treatment with | |
				PMSF	PCMB
B.amyloliquefaciens (subtilisin BPN)	28000	8.1	8.5	100	0 (%)
B.amyloliquefaciens (ISP)	30000x2	4.3	7-10	100	20
B.subtilis 797	28000	8.2	8-8.5	100	-
B.thuringiensis (extracellular)	29000	8.4	8.5	100	-
Thermoactinomyces vulgaris	28000	8-9	8.2	100	97
Trichoderma lignorum	21000	6.8	8.4-8.8	100	0

TABLE 5 <u>Amino acid composition of serine proteinases</u>
(in residues per molecule)

The names of the enzymes are abbreviated as follows: intracellular serine proteinase of <u>B.amyloliquefaciens</u> - ISP. Extracellular proteinases from: <u>B.subtilis</u> 797 - BSU; <u>Bacillus thuringiensis</u> - BTU; <u>Trichoderma lignorum</u> - TLI; <u>Thermoactinomyces vulgaris</u> - TVU; subtilisin BPN - BPN; subtilisin Carlsberg - CAR.

Amino acid	BPN	CAR	ISP	BSU	BTU	TLI	TVU
Asp	28	28	36	37	28	23	35
Thr	13	19	13	17	16	16	21
Ser	37	32	25	34	18	27	29
Glu	15	12	33	15	29	11	23
Pro	14	9	13	14	12	10	16
Gly	33	35	36	17	32	27	33
Ala	37	41	32	36	31	24	43
Val	30	31	19	28	19	12	21
1/2 Cys	0	0	0	0	0	2-3	1
Met	5	5	5	4	5	2	-
Ile	13	10	13	11	12	8	13
Leu	15	16	25	17	18	10	9
Tyr	10	13	7	9	11	8	13
Phe	3	4	7	4	10	7	4
Trp	3	1	-	-	4	-	-
Lys	11	9	20	18	16	3	8
His	6	5	4	5	4	3	3
Arg	2	4	4	4	8	9	4
Sum	273	274	296	271	273	203	276

REFERENCES

Akparov, V.Kh., and Stepanov, V.M. (1978) Phenylboronic acid as a ligand for biospecific chromatography of serine proteinases. J.Chromatography, 155, 329-336.

Akparov, V.Kh., Belyanova, L.P., Baratova, L.A., and Stepanov, V.M. (1979) Subtilisin 72-serine proteinase of B.subtilis, strain 72, resembling subtilisin Carlsberg. Biokhimiya, 44, 886-891.

Borovikova, V.P., Aksenovskaya, V.E., Lavrenova, G.I., Kislukhina, O.V., Kalunyantz, K.A., and Stepanov, V.M. (1980). Isolation and characterization of lytic enzyme from B.subtilis 797. Biokhimiya, 45, 1524-1533.

Chestukhina, G.G., Epremyan, A.S., Akparov, V.Kh., Azizbekyan, R.R., Netyksa, E.M., and Kotova, T.S. (1979). Extracellular serine proteinase of B.thuringiensis. Khimiya Proteolyticheskikh Fermentov, Uglich, p. 67.

Frömmel, C., Hausdorf, G., Höhne, W.E., Behnke, U., Ruttloff, H. (1978) Characterisierung einer Protease aus Thermoactinomyces vulgaris (Thermitase). 2. Acta Biol.Med.Germ., 37, 1193-1204.

Hershko, A., Ciechanover, A., Heller, H., Haas, A.L., Rose, I.A., (1980) Proposed role of ATP in protein breakdown: Conjugation of proteins with multiple chains of polypeptide of ATP-dependent proteolysis. Proc.Nat.Acad.Sci.USA, 77, 1783-1786.

Rudenskaya, G.N., Osterman, A.L., and Stepanov, V.M. (1980) Carboxylic proteinase from Trichoderma lignorum. Biokhimiya, 45, 710-716.

Stepanov, V.M. (1980) Intracellular serine proteinases of bacteria-isolation, chemistry, evolutionary aspects. In P. Mildner and B. Ries, Eds., Enzyme regulation and mechanism of action, Pergamon Press, Oxford, N.Y.

Stepanov, V.M., Rudenskaya, G.N., Nesterova, N.G., Kupriyanova,T.I., Khokhlova, Ju.M., Usaite, I.A., Loginova, L.G., and Timokhina, E.A. (1980) Serine proteinase from Thermoactinomyces vulgaris, strain INMI-4a. Biokhimiya, 45, 1871-1880.

Strongin, A.Ya., Isotova, L.S., Abramov, Z.T., Gorodetsky, D.I.,and Stepanov, V.M. (1978) Two related structural genes coding two homologous serine proteinases in the Bacillus subtilis genome. Molec.Gen.Genet., 159, 337-339.

Strongin, A.Ya., Gorodetsky, D.I., Kuznetzova, I.A., Yanonis, V.V., Abramov, Z.T., Belyanova, L.P., Baratova, L.A. and Stepanov, V.M. (1979) Intracellular serine proteinase of Bacillus subtilis strain Marburg 168. Biochem.J., 179, 333-339.

STREPTOMYCES RIMOSUS ALKALINE AND TRYPSIN—LIKE SERINE PROTEINASES

M. Renko, M. Longer, M. Pokorny*, V. Turk and Lj. Vitale**

*Department of Biochemistry, J. Stefan Institute, University E. Kardelj, 61000 Ljubljana; *Research and Development Institute, Krka, Pharmaceutical and Chemical Works, 68000 Novo mesto; **Institute Ruđer Bošković, 41000 Zagreb, Yugoslavia*

ABSTRACT

An alkaline and a trypsin-like serine proteinase were purified from the culture filtrate of Streptomyces rimosus by ultrafiltration, acetone fractionation, ion exchange chromatography, gel filtration and affinity chromatography on Kunitz soybean trypsin inhibitor - Sepharose. Both enzymes were isolated in electrophoretically pure form and characterized.

KEYWORDS

Serine proteinases; alkaline serine proteinase; trypsin-like serine proteinase; Streptomyces rimosus proteinases; microbial proteinases; proteinases.

INTRODUCTION

It has been shown that Streptomyces rimosus produces a variety of hydrolases (Pokorny and Vitale, 1980). The presence among them of serine and metallo-proteinases was confirmed, while carboxyl and thiol proteinases were not present (Pokorny and co-workers, 1979). In the course of our further investigations we isolated and characterized serine alkaline proteinase (Renko and co-workers, in press).

This report describes the modified purification procedure for the alkaline proteinase and its properties, as well as the isolation and characterization of the new enzyme, serine trypsin-like proteinase.

PURIFICATION AND PROPERTIES

The enzyme activity of the serine alkaline proteinase was detected towards hemoglobin (Anson, 1939) or casein (Kunitz, 1947). Trypsin-like proteinase activity was determined using N-α-benzoyl-DL-arginine-p-nitroanilide (BAPNA) as substrate

(Erlanger and co-workers, 1961).

All operations, unless specified otherwise, were performed at 4 ^{o}C. The purification scheme is presented in Fig. 1. The culture filtrate (8 l) was centrifuged to remove insoluble material, concentrated by ultrafiltration to a final volume of about 500 ml, and then fractionated with acetone (30-65%). The precipitate obtained was dissolved in 0.01 M sodium acetate buffer containing 5 mM $CaCl_2$, pH 5.5, and dialyzed overnight against the same buffer. The dialyzed sample was then applied to a column of CM-Sephadex C-50 (4 x 10 cm) equilibrated with the same buffer.

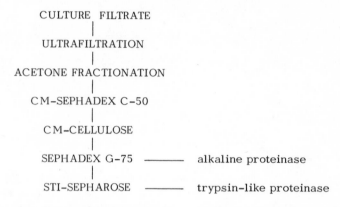

CULTURE FILTRATE
|
ULTRAFILTRATION
|
ACETONE FRACTIONATION
|
CM-SEPHADEX C-50
|
CM-CELLULOSE
|
SEPHADEX G-75 ———— alkaline proteinase
|
STI-SEPHAROSE ———— trypsin-like proteinase

Fig. 1. Purification scheme of <u>S. rimosus</u> serine proteinases.

The break-through fractions were pooled, acidified to pH 4.2 with acetic acid and then applied to a CM-cellulose column (3 x 20 cm) equilibrated with 0.01 M sodium acetate buffer, pH 4.2. The column was washed with the starting buffer and the fractions containing proteinase activity (toward hemoglobin and BAPNA as substrates) were eluted with a linear gradient of NaCl (0.0-0.25 M). The active fractions were pooled, concentrated by ultrafiltration and applied to a Sephadex G-75 superfine column (3 x 100 cm), equilibrated with 0.01 M sodium acetate buffer, pH 5.5, containing 0.1 M NaCl. A typical elution profile is shown in Fig. 2. The second and the major protein peak is active toward hemoglobin and represents pure serine alkaline proteinase.

The first protein peak shows a high activity toward BAPNA and is further purified by affinity chromatography on Kunitz soybean trypsin inhibitor (STI) linked to Sepharose 4B resin. STI inhibitor was coupled to CNBr-activated Sepharose 4B by the procedure recommended by the manufacturer (Pharmacia, 1979). The BAPNA active fractions were dialyzed overnight against 0.1 M Tris HCl, pH 8.5, containing 0.5 M NaCl and 5 mM $CaCl_2$, and applied to an affinity column equilibrated with the same buffer. Under these conditions BAPNA activity remained bound to the column. The enzyme was eluted with 1 mM HCl containing 0.5 M NaCl and 5 mM $CaCl_2$, pH 3.0, and collected in 0.1 M Tris HCl buffer, pH 9.0. The BAPNA active fractions were free of alkaline proteinase. The isolated enzyme is a trypsin-like serine proteinase.

In typical purification from the culture filtrate we obtained for the alkaline proteinase 150 mg of the pure enzyme with a 100-fold purification and a 25% yield, and for the trypsin-like proteinase 2 mg of the enzyme with a 480-fold purification and only about 2% recovery.

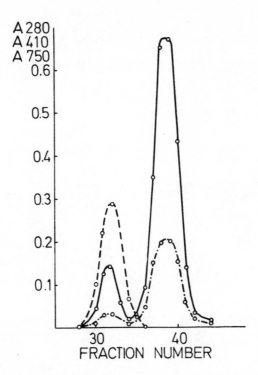

Fig. 2. Sephadex G-75 superfine elution profile of serine
proteinases: ——— the absorbance at 280 nm; ---- the trypsin-
like enzyme activity at A_{410}; —·—· alkaline enzyme activity at
A_{750}.

The purity of alkaline and trypsin-like serine proteinases was checked by polyacrylamide gel electrophoresis (pH 8.4), and a single protein band was obtained in each
case (Fig. 3).

Serine alkaline proteinase: The molecular weight was estimated by gel filtration
through a Sephadex G-75 column and by SDS polyacrylamide gel electrophoresis, and
was calculated to be 22,000. Isoelectric focusing shows one active band with a pI value of 4.90 and one small inactive band (Fig. 4). The optimal pH for casein hydrolysis is between pH 7.0 - 10.5, for azocoll pH 8.5 (determined according to Yoshida
and Noda, 1965), whereas hemoglobin hydrolysis shows two pH optima at pH 4.5 and
11.0. The enzyme also hydrolyzes synthetic substrates (Table 1) as well as proteins.
It can be seen that the enzyme splits p-nitrophenyl, ethyl and methyl esters, hexapeptide and octapeptide. The enzyme is stable in the pH range 4.0-9.0. Also, it was
completely stable for at least 6 months at -25 °C. The enzyme is very sensitive to
1 mM DFP and PMSF, but not to TPCK. Its amino acid composition (Table 2) shows
about 210 amino acid residues with one methionine and two cysteines.

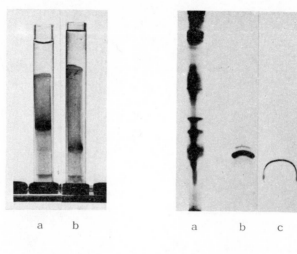

a   b   a   b   c

Fig. 3.   Fig. 4.

Fig. 3. Polyacrylamide gel electrophoresis: a) trypsin-like
serine proteinase; b) alkaline serine proteinase.

Fig. 4. Isoelectric focusing of serine proteinases: a) protein
standards; b) alkaline serine proteinase; c) trypsin-like serine proteinase.

TABLE 1 Substrate Specificity of S.rimosus Alkaline Proteinase

Synthetic substrate	Hydrolysis
N-α-benzoyl-D,L-Arg-p-nitroanilide	–
Glutaryl-L-Phe-p-nitroanilide	+
t-Boc-L-Ala-p-nitrophenyl ester	+
N-CBZ-L-Ala-p-nitrophenyl ester	+
N-CBZ-L-Tyr-p-nitrophenyl ester	+
N-α-Tosyl-L-Arg-methyl ester	+
N-benzoyl-L-Tyr-ethyl ester	+
N-CBZ-L-Phe-β-naphthylamide	–
L-Tyr-β-naphthylamide	–
Ala-Ala-Ala	–
Leu-Trp-Met-Arg-Phe-Ala	+
Asp-Arg-Val-Tyr-Ile-His-Pro-Phe (Angiotensin II)	+

Trypsin-like serine proteinase: The molecular weight is about 28,000, as determined by gel chromatography and SDS polyacrylamide gel electrophoresis. The isoelectric point, pI, is 4.5 (Fig. 4). The optimal activity is at pH 8.0-9.0 on BAPNA as
substrate. The enzyme is stable between pH 6.0-8.5 and very unstable below pH 5.0
and above pH 9.0. The inhibitors PMSF and TLCK (at the concentration of 1.6 mM)

and Kunitz soybean trypsin inhibitor (at the concentration of 1 mg/ml) completely inhibit the enzyme activity. Calcium ions (5×10^{-3} M) have no effect on the enzyme stability. Preliminary amino acid composition shows two methionines, two cysteines and a high content of aspartic and glutamic acid.

TABLE 2 Amino Acid Composition of S.rimosus Serine Alkaline Proteinase.

Asp	16	Met	1
Thr	16	Ile	9
Ser	17	Leu	11
Glu	21	Tyr	10
Pro	7–8	Phe	5–6
Gly	30	His	6
Ala	21	Lys	16
Cys	2	Trp	2–3
Val	14	Arg	3

DISCUSSION

The data presented demonstrate concomitant isolation of two serine S. rimosus proteinases and their characterization. The purification scheme outlined for the alkaline and trypsin–like proteinases is especially convenient for the isolation of the alkaline proteinase, whereas the second enzyme was obtained in rather a low yield. This is probably the result of instability of the trypsin–like enzyme in the acid pH region, and also of rather low content of this enzyme in the culture filtrate. Therefore the purification procedure needs further modification. The molecular weights of both enzymes are in the same range as other known alkaline and trypsin–like serine proteinases (Morihara, 1974). The isoelectric point of the alkaline proteinase is different from other Streptomyces species, whereas pI for the trypsin–like enzyme is similar to S. erythreus trypsin–like enzyme (Yoshida and co–workers, 1971). Another trypsin–like enzyme was isolated from S. rimosus with an indication that the value of the isoelectric point is around pH 8 (Chauvet and co–workers, 1976). Both isolated enzymes were classified in the group of serine proteinases on the basis of the effect of specific inhibitors.

REFERENCES

Anson, M.L. (1939). The estimation of pepsin, trypsin, papain and cathepsin with hemoglobin. J.Gen.Physiol., 22, 79–89.

Chauvet, J., J.P. Dostal, and R. Archer (1976). Isolation of a trypsin–like enzyme from Streptomyces paromomycinus (paromotrypsin) by affinity adsorption through Kunitz inhibitor Sepharose. Int.J.Peptide Prot.Res., 8, 45–55.

Erlanger, B.F., N. Kokowsky, and W. Collen (1961). The preparation and properties of two new chromogenic substrates of trypsin. Arch.Biochem., 95, 271-278.

Kunitz, M.J. (1947. Crystalline soybean trypsin inhibitor. II. General properties. J.Gen.Physiol., 30, 291-310.

Morihara, K. (1974). Comparative specificity of microbial proteinases. In A. Meister (Ed.), Advances in Enzymology, Vol. 41, John Wiley, New York, pp. 179-243.

Pokorny, M., Lj. Vitale, V. Turk, M. Renko, and J. Žuvanić (1979). Streptomyces rimosus extracellular proteases. 1. Characterization and evaluation of various crude preparations. Europ.J.Appl.Microbiol.Biotechnol., 8, 81-90.

Pokorny, M., and Lj. Vitale (1980). Enzymes as by-products during biosynthesis of antibiotics. In Lj. Vitale and V. Simeon (Eds.), Industrial and clinical enzymology, Trends in Enzymology, Vol. 61, Pergamon Press, Oxford, pp. 13-25.

Rassulin, Yu.A., L.R. Radžabov, E.D. Kaverzneva, V.A. Shibnev, G.S. Erkomaish-vili (1974). Isolation and properties of an elastolytic enzyme of Actinomyces rimosus. Izv.Akad.Nauk SSSR, Ser.Khim., 3, 687-693.

Renko, M., M. Pokorny, Lj. Vitale, and V. Turk. Streptomyces rimosus extracellu-lar proteases. 2. Isolation and characterization of serine alkaline proteinase. Europ.J.Appl.Microbiol.Biotechnol. (in press).

Yoshida, E., and H. Noda (1965). Isolation and characterization of collagenases I and II from Clostridium hystolyticum. Biochim.Biophys.Acta, 105, 562-574.

Yoshida, N., A. Sasaki, and H. Inoue (1971). An anionic trypsin-like enzyme from Streptomyces erythreus. FEBS Letters, 15, 129-132.

PURIFICATION, PROPERTIES AND APPLICATION OF THERMITASE, A MICROBIAL SERINE PROTEASE

R. Kleine, U. Rothe, U. Kettmann and H. Schelle

Physiologisch–chemisches Institut der Universität Halle/S. und Ingenieurhochschule Köthen, DDR

ABSTRACT

Thermitase, an extracellular serine protease of <u>Thermoactinomyces vulgaris</u> was purified by two different single-step procedures, either by "reversed" polyacrylamide gel electrophoresis or by chromatography on porous Glass (CPG-10) on a large scale.

Thermitase was found to be a thermostable (optimum temperature at 80 $^\circ$C against protein substrates), cationic (isoelectric point at pH 9), alkaline (optimum pH between pH 8 and 9) single chain proteinase containing and binding calcium ions.

Besides casein, gelatin and other soluble proteins thermitase hydrolyzes the insoluble proteins elastin and collagen significantly. In its substrate specificity thermitase resembles elastase and subtilisin.

The purified enzyme can be stabilized at pH 8 and 55 $^\circ$C by additives of both low molecular weight (e.g. Ca^{++} ions, Tween 20, basic amino acid derivatives, organic solvents) and high molecular weight (e.g. gelatin and other basic polypeptides).

These properties are the basis for the application of the enzyme concerning which a condensed review is given.

KEYWORDS

Purification; stabilization; thermostability; application.

INTRODUCTION

The growing use of enzymes as highly specific and very efficient catalysts for a variety of commercial and biological processes and analytical applications in numerous fields depends on the ready availability of reasonably pure, stable and low cost enzymes. One precondition to attaining this goal is the development of simple and rapid purification procedures. Furthermore, the use of thermostable enzymes is another advantage since elevated temperatures during enzymatic processes allow increased turnover-rates.

In the present paper we report on the single-step isolation of a thermostable serine

protease from the culture filtrate of <u>Thermoactinomyces vulgaris</u>, termed <u>thermita-</u>
<u>se</u> (Kleine and Rothe, 1977; Behnke and co-workers, 1978a, 1978b; Frömmel and
co-workers, 1978), and on some properties and fields of application of the purified
enzyme.

MATERIALS AND METHODS

<u>Culture medium.</u> The tenfold concentrated and dialyzed (against distilled water)
culture filtrate from <u>Thermoactinomyces vulgaris</u> (Behnke and co-workers, 1978a)
served as starting material for thermitase isolation.

<u>Polyacrylamide gel electrophoresis (PAGE)</u> was carried out according to Maurer
(1971) using 7.5% gels and 10 mM Tris-glycine buffer, pH 8.3, with or without added Na-dodecyl sulphate. "Reversed" preparative PAGE for one-step isolation of
thermitase from the culture medium of <u>Th. vulgaris</u> was performed as previously
described (Rothe and Kleine, 1979).

<u>Adsorption chromatography</u> using controlled pore glass CPG-10 was done according
to Kleine and co-workers (1978). The preferred CPG-Lots used had mean pore diameters of 160 to 300 Å with surface areas of between 60 and 150 m^2/g and particle
size of 40 to 80 μm (Electro-Nucleonics, USA, obtained by Serva Feinbiochemica,
Heidelberg, FRG).

<u>Protein concentration</u> was determined spectrophotometrically using the following
specific absortivities ($A_{1cm}^{1\%}$ at 280 nm and pH 6 to 7): 16.0 for thermitase, 22.2
for porcine elastase, 15.4 for trypsin, 20.2 for chymotrypsin, 11.7 for subtilisin
BPN′, 17.6 for thermolysin and 14.2 for proteinase K. The concentration of pronase
and collagenase (<u>Clostridium histolyticum</u>) was measured by the method of Lowry
and co-workers (1951), with bovine serum albumin as a standard. All these enzymes , with the exception of thermitase, were obtained from Serva Feinbiochemica,
Heidelberg, FRG.

Enzyme Assays

<u>Unspecific proteinase activity</u> was assayed by measuring the liberation of either the
amino groups from gelatin (1%) with the trinitrobenzene sulfonic acid (TNBS) method,
or the coloured and trichloroacetic acid-soluble peptides from azocasein (0.5%)
according to Langner and co-workers (1971) and Langner (1973), with slight modifications.

<u>Elastolytic activity</u> was determined by measuring the solubilized coloured peptides
from Congo red (CR)-elastin (10 mg of powdered CR-elastin per ml of incubation
mixture) at 480 nm (Shotton, 1970).

<u>Collagenase-like activity</u> was measured with the artificial substrate phenylazoben-
zyloxycarbonyl- (PZ)-L-Pro-L-Leu-Gly-L-Pro-D-Arg as described by Wünsch and
Heidrich (1963), with the exception that solutions were assayed in 0.1 M Tris-HCl,
pH 8.0.

<u>Esterolytic activity</u> was measured potentiometrically in a pH-stat (Radiometer) at
pH 8 and 55 °C using various N-acylated (Z-, Suc-, Ac or Boc- groups) peptide
methyl esters, composed of 1 to 5 amino acid residues, as substrates. Their synthesis and properties are described elsewhere (Fittkau and co-workers, Könnecke
and co-workers, both in preparation).

<u>Amidolytic activity</u>   was determined photometrically at pH 8.0 and 55 $^{\circ}$C in 25 mM
Tris HCl buffer using N-acetylated (Ac)- and N-succinylated (Suc)-L-alanine$_n$-p-
-nitroanilides (n = 1 to 5) as substrates. The increase in optical density was recorded at 405 nm (Eppendorf Photometer with M 1100 recorder and tempered cuvettes).

RESULTS AND DISCUSSION

SINGLE-STEP ISOLATION PROCEDURE FOR THERMITASE

<u>Isolation of thermitase by "reversed" preparative polyacrylamide gel electropho-
resis (PAGE)</u>

The isolation of proteins by PAGE consists either in the extraction of the separated
protein bands from the corresponding cut and minced gel pieces, or in the elution
of the single protein zones during electrophoresis by means of an elution chamber
(Maurer, 1971). Avoiding the obvious disadvantages of these procedures encountered during large-scale isolation of enzymes, we applied a new separation technique.
The starting point was the disc-electrophoretic behaviour of the proteins of the culture filtrate from <u>Th. vulgaris</u>. Of all proteins only thermitase migrates toward the
cathode during PAGE at pH 8.3 (Fig. 1, track 1 to 4). During SDS-PAGE, however,
the enzyme moves anodically together with all the other protein components of the
starting material (Fig. 1, track 5 and 6).

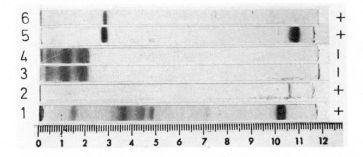

Fig. 1. Polyacrylamide gel electrophoresis of culture filtrate
and thermitase in the presence (gel strip 5 and 6) and absence
(gel 1 to 4) of 0.1% SDS. The gels were loaded either with 50ul
of culture filtrate (gel 1,3,5) or 50 ug of thermitase (gel 2,4,
6) and subjected to electrophoresis (8 mA/gel) at pH 8.3 for
two (gel 1,2,5,6) or four (gel 3,4) hours. Proteins were visualized by staining the gels with 0.1% Coomassie Brilliant Blue
G-250 in 12.5% trichloroacetic acid, followed by washing with
7% acetic acid.

The reasonable approach of isolating thermitase by the gel cutting and extraction
technique proved to be less attractive for routine application. To overcome these
difficulties, we developed the new technique of "reversed" electrophoresis, that is,
removal of the anionic material by its migration into the gel lying between the separation chamber and the anode. The cationic thermitase, however, moves in the

opposite direction toward the cathode.

The separation principle consists in pumping the dialyzed and concentrated culture filtrate between the gel plug and the buffer solution of the "crystallization" chamber which contains 15 to 20% polyethylene glycol to increase the specific gravity of the buffer. Due to the different electric charges all undesirable components including the brownish coloured material move into the polyacrylamide gel layer whereas thermitase concentrates near the cathode. Since this protease has the tendency to crystallize in concentrated solutions, especially in the presence of neutral salts, buffers or polyols (Kleine and Rothe, 1977, unpublished results), the addition of polyethylene glycol promotes spontaneous crystallization of the separated enzyme. Further details are and will be given elsewhere (Rothe and Kleine, 1979 and in preparation). With this equipment in continuous operation by the recycling technique it is possible to isolate over a three-day period approximately 70 mg of thermitase crystals from 150 to 200 ml of culture medium, corresponding a yield of about 40%. The recovery can be enhanced by quantitative precipitation with ammonium sulphate of the enzyme in solution in the mother liquor of the crystallization chamber. Concerning purity and activity, the isolated enzyme did not differ from that purified by ion exchange chromatography (Kleine and Rothe, 1977) or by adsorption chromatography on CPG-10 (see below).

Isolation of thermitase by adsorption chromatography on CPG-10 glass

CPG was originally used as a suitable matrix for immobilizing enzymes (Messing, 1969). Its use for the separation of proteins by adsorption chromatography has been reported independently by Mizutani and Mizutani (1976), and by Bock and co-workers (1976). The Mizutanis (1977) further found that besides hydrophobic interactions aminosilanol bonding between the cationic groups in the protein and the silanol groups of the hydrated glass beads was one major force for the adsorption of proteins on glass surfaces. Since thermitase proved to be a cationic protein (Fig. 1), CPG seemed suitable for a simple and effective isolation of this enzyme by adsorption chromatography on a large scale. Corresponding tests with various CPG-lots confirmed this assumption. It was found that glass beads with mean pore diameters of 160 to 300 Å were best suited to selective adsorption of thermitase in aqueous medium.

The purification procedure starts by loading the CPG-column (bed volume 500 to 1000 ml, equilibrated with distilled water) with the concentrated and dialyzed culture filtrate in a ratio of about 1:1 (v/v). After removal of practically all contaminating materials by exhaustive washing of the column with water, the only adsorbed compounds, thermitase and small amounts of an anionic protease component, were eluted separately by a gradual rise in ionic strength with ammonium acetate, pH 6.0. The elution profile of a large-scale experiment is given in Fig 2. The initial breakthrough peak with water as eluent comprises dyestuffs and practically all the anionic proteins of the crude concentrate. The gradual increase of the eluent concentration yields a good separation of the anionic protease from thermitase which elutes with 0.1 M ammonium acetate (Fig. 2, peak 2). The yield amounts about 430 mg of thermitase or almost 90% of the total proteolytic activity of the eluates. The post-thermitase fraction (peak 3) contains approximately 55 mg of protein, equal to 10% of the total activity. Concerning its electrophoretic and proteolytic behaviour this protein is identical with thermitase, only its gelatinase activity is significantly reduced.

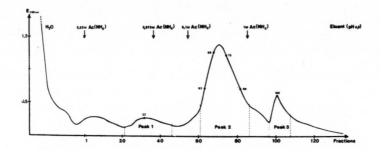

Fig. 2. Elution profile of the large-scale isolation of thermitase from the culture filtrate of <u>Thermoactinomyces vulgaris</u>
by adsorption chromatography on CPG-10. 0.5 L of starting
material, dialyzed and concentrated tenfold, was applied to
a 580 ml/column of CPG-10. The elution was carried out as
indicated with a constant flow of 100 ml/hr. Fractions of
20 ml were collected, pooled as indicated by vertical dotted
lines and assayed for proteolytic activity and protein content
(see Fig. 3).

For experiments which require extremely pure protein material, only peak-2 fractions were pooled and subjected to an additional gel filtration step on Sephadex G-50
with water as eluent to remove some lower molecular weight material. Finally, the
pooled fractions with thermitase activity from the CPG- or Sephadex column were
either liophylized or mixed with solid ammonium sulphate up to 1.5 M to obtain the
enzyme in a crystallized state. The liophylized product, though possessing a somewhat (about 15%) lower specific activity than the solute, is better to handle for reasons of increased stability (compared to the solute) and solubility (thermitase crystals are sparingly soluble in water). The mean values for the overall yields of several experiments (liophylized enzyme) were near 85%. Using two 750 ml CPG columns it is possible to isolate up to two grams of thermitase weekly, using 4 runs,
i.e., 2 experiments per column per week.

Distribution patterns of the proteolytic activity in the eluates

As can be seen from Fig. 3, the main part of the total proteolytic activity, from 84%
(against azo-casein) to 94% (against gelatin), is enriched in peak 2 (see Fig. 2) ,
whereas peak 1 has practically no elastinolytic (against Congo red-elastin and Suc-
cinyl-trialanyl-p-nitroanilide) and only 2% and 4% of the caseinolytic and gelatinoly-
tic activity, respectively. The remaining activity, eluted with 1 M ammonium acetate (peak 3), resembles that of peak 2 with the exception of its weak gelatin-splitting activity.

A somewhat different picture is obtained if the activity per mg of protein is compared (not shown). As expected, the best activities toward all five substrates tested
are localized in the main peak 2. In contrast, the isolated anionic protease component, centered around fraction 32 of Fig. 2, is characterized by an extremely weak

elastinolytic and a high collagenase–like activity per mg of protein. The lack of elastinolytic activity may be explained by the acidic character of this protease. According to Gertler and Hayashi (1971), unspecific electrostatic binding of the positively
charged groups of the enzyme to the negatively charged groups of the elastin molecule is essential for activity. This condition is fullfilled by thermitase (pI at pH 9)
but not by the anionic protease (pI near pH 9).

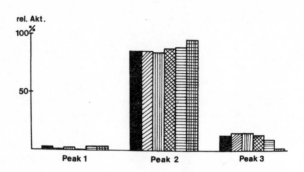

Fig. 3. Distribution of protein and proteolytic activity of the
pooled fractions 1, 2 and 3 of Fig. 2. The columns mean
(from left to right) protein content, total activity against
Suc–Ala$_3$–p–nitroanilide, CR–elastin, PZ–Pentapeptide and
gelatin. Both the protein content and overall activity of peak
1, 2 and 3 equal 100. Relative data are given.

PROPERTIES OF THE PURIFIED ENZYME

General Characteristics

Table 1 summarizes some physico–chemical and kinetic properties of thermitase.
It is a single–chain cationic, thermostable serine protease with an essential sulphy-
dril group,necessary for full activity and stability (Behnke,Ruttloff and Kleine, 1980).

TABLE 1 Some Properties of Purified Thermitase

Molecular weight and structure	35,000; 305 residues; no subunits detectable
A–280/A–260	2.7 to 2.8 (in H$_2$O)
Isoelectric point	9.0
Potential inhibitors	PMSF, HgCl$_2$, DIFP
Optimum temperature	60 °C (esterolysis)
	70 °C (amidolysis)
	80 °C (proteinolysis)
Optimum pH	8 to 9 (all substrates)
pH stability at 55°/1 hr	100% between pH 5.7 and 7.2
	50% at pH 4.8 and 7.5
	20% at pH 4.5 and 8.0
	5% at pH 4.0 and 8.5

Substrate Specificity – Comparison with other Proteases

In order to get a real assessment relative to other proteases, the enzymatic properties of thermitase were compared with those of porcine elastase, subtilisin BPN´, and
Cl. histolyticum collagenase, and additionally, with the apparently unrelated proteases thermolysin, proteinase K, pronase, as well as trypsin and chymotrypsin.

Polypeptides. As can be seen from Fig. 4, thermitase has a strong proteinase action
compared with subtilisin, elastase and collagenase on substrates such as CR–elastin
and gelatin. Only thermolysin (against azo–casein, CR–elastin and PZ–peptide), proteinase K (on azo–casein and gelatin) and pronase (on gelatin and PZ–peptide) were
more active than thermitase. The elastin–splitting activity of thermitase could be
further demonstrated by disc–electrophoretic analysis of the elastin–protease incubation mixtures and by the clarification of an elastin suspension, stabilized by agarose, either by measuring the decrease of absorbance at 578 nm or qualitatively on
slides covered with elastin–containing agarose by the spot technique (not shown) .
The significant hydrolysis by thermitase of PZ–Pro–Leu–Gly–Pro–D–Arg, a substrate
for microbial collagenases (Wünsch and Heidrich, 1963), could be substantiated by
its pronounced ability to solubilize reconstituted collagen from rat tail tendons. This
property was not shared with the other proteases tested except collagenase. In accordance with its very weak hydrolytic action on synthetic trypsin substrates (Kleine
and Rothe, 1977) thermitase is unable to split poly–arginine or poly–lysine whereas
poly–alanine is strongly hydrolyzed to small alanylpeptides with dialanine as a major
product (Kleine, Brömme and Fittkau, 1980). Since the same behaviour is found
using N–acylated oligopeptides, it is reasonable to assume that thermitase will preferentially attack polypeptides near their termini and to a lesser extent in their middle part.

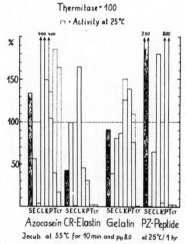

Fig. 4. Comparison of nine proteases with respect to their acti-
on on four different substrates. Activity of thermitase = 100 .
S, subtilisin; E, elastase; C, collagenase; L, thermolysin;
K, proteinase K; P, pronase; T, trypsin; CT, chymotrypsin.

Synthetic substrates. The known high hydrolytic affinity of thermitase for the specific elastase substrate acetyl-trialanine methyl ester (Kleine and Rothe, 1977) was the starting point for elucidating its structural requirements for catalysis in more detail. To compare the various substrates, the respective K_M and k_{cat} values were determined and the resulting k_{cat}/K_M ratio was used as a reliable parameter for catalytic efficiency. In analogy with the best investigated serine proteases such as the related elastase (Bieth and Wermuth, 1973), k_{cat}/K_M may be a second order acylation rate constant, whatever the substrate (ester or anilide), as catalysis with thermitase also preceeds via an acylenzyme intermediate (Hausdorf, 1979). As can be seen from Tables 2 and 3, thermitase is highly active towards N-protected alanine or other hydrophobic amino acid residues containing esters with a preference for Z(benzoyloxycarbonyl)- and Ac(acetyl)-trialanine esters. Among the numerous Z-dipeptide esters tested, the best fitted residues in position P_1 proved to be leucyl, methionyl, phenylalanyl and alanyl, provided that leucyl occupies the P_2 position. The residue in P_2 (Table 3) also seems of great importance for maximum second order rate constants. Considering both binding sites in thermitase (S_1, S_2), optimal Z-dipeptide ester substrates will be Z-Val-Phe OMe, Z-Ala-LeuOMe and Z-Phe-Phe OMe. Finally, the size and hydrophobicity of the protecting group, located in P_3 in the case of the dipeptide ester influences both K_M and k_{cat}, as the following values assayed with Ac-, Boc- and Z-Leu-Phe OMe demonstrate. The values for K_M were 50, 18 and 4.2 mM; the respective k_{cat} 7500, 427 and 2500.sec^{-1} (pH 8 and 55°C).

TABLE 2 Thermitase-Catalyzed Hydrolysis of N-acyl-L-Alanine Methyl Esters and –p-Nitroanilides at pH 8.0 and 55 $^{\circ}$C (^{x}at 25°C)

Substrate (without org. solv.)	Esterolysis			Amidolysis		
	K_M [mM]	k_{cat} [s^{-1}]	k_{cat}/K_M [M^{-1}·s^{-1}]	K_M [mM]	k_{cat} [s^{-1}]	k_{cat}/K_M [M^{-1}·s^{-1}]
Z Ala₁	15	370	$2.5 \cdot 10^4$		n.d.	
Ac Ala₁	*10	16	$1.6 \cdot 10^3$	3.3	0.007	2.1
Suc Ala₁	*2	15	$7.5 \cdot 10^3$	4.8	0.074	15.5
Z Ala₂	0.6	1157	$1.7 \cdot 10^6$		n.d.	
Ac Ala₂	*1.3	1000	$7.5 \cdot 10^5$	1.6	0.33	200
Boc Ala₂	0.9	2850	$3.2 \cdot 10^6$		n.d.	
Suc Ala₂		n.d.		3.3	0.20	60
Z Ala₃	1.0	2363	$2.4 \cdot 10^6$		n.d.	
Ac Ala₃	*0.9	3217	$3.6 \cdot 10^6$	1.25	1.60	$1.3 \cdot 10^3$
Suc Ala₃	*2.0	560	$2.8 \cdot 10^5$	0.36	0.90	$2.5 \cdot 10^3$
Z Ala₄	0.15	600	$4.0 \cdot 10^6$		n.d.	
Ac Ala₄	*0.16	1518	$9.4 \cdot 10^6$	0.12	11.8	$9.8 \cdot 10^4$
Suc Ala₄	*0.23	737	$3.2 \cdot 10^6$	0.02	19.5	$9.7 \cdot 10^5$
Suc Ala₅	*0.08	142	$1.8 \cdot 10^6$	0.07	5.0	$7.4 \cdot 10^3$ˣ

All nitroanilides and Z-derivatives were determined in presence of 2.5 and 10 per cent of DMF, respectively.

TABLE 3 Comparison of k_{cat}/K_M Values (Q) of Z-Dipeptide
Esters Hydrolyzed by Thermitase

Variation in P_1		Variation in P_2	
Substrate	$Q.10^5$	Substrate	$Q.10^5$
Z-Leu-Leu OMe	9.0	Z-Val-Phe OMe	29.0
Z-Leu-Phe OMe	6.0	Z-Phe-Phe OMe	10.0
Z-Leu-Met OMe	6.9	Z-Pro-Phe OMe	7.4
Z-Leu-Tyr OMe	1.9	Z-Ala-Phe OMe	6.9
Z-Leu-Ala OMe	1.3	Z-Gly-Phe OMe	1.8
Z-Leu-Val OMe	0.1	Z-Ala-Ala OMe	16.0
Z-Leu-Pro OMe	0.01	Z-Val-Ala OMe	7.7
Z-Leu-Ile OMe	0.008	Z-Gly-Ala OMe	0.8

The tests (pH-stat technique) were performed at pH 8 and 55°C
in the presence of 30% (all Z-Leu-peptides, Z-Val- and Z-Ala-
-Phe OMe) or 20% dimethyl formamide.

Here already, these results point out that the active centre of thermitase may be
composed of several subsites as in the case of elastase. As expected for nucleophilic covalent catalysis, the efficiency of anilide substrates might be significant lower
than that of their corresponding methyl ester derivatives. Therefore, the high este-
rase/anilidase ratio for the k_{cat} values (e.g. 2000 for $AcAla_3$ OMe/$AcAla_3$-pNA;
Table 2) is similar to the corresponding values of other serine proteases such as
elastase (1540) and subtilisin (10^5) (Table 4) or chymotrypsin (650) (Gertler,
Weiss and Burstein, 1977). The absolute k_{cat} values, however, are all higher, including those for N-acetylated di- and tetra esters of subtilisin and elastase (Table 4
and unpublished results).

TABLE 4 Kinetic Constants of Three Related Proteases at pH 8.0
and 55 $^{\circ}$C

Substrate	Thermitase			Subtilisin			Elastase I		
	K_M	k_{cat}	V_s	K_M	k_{cat}	V_s	K_M	k_{cat}	V_s
ATEE	16	314	520	31	233	508	not hydrolyzed		
TAME	40	37	62	30	7	15			
Ac Ala OMe	10	16	27	120	72	157	153	6.7	16*
Ac Ala$_3$ OMe	0.9	3220	5320	1.6	2300	5000	1.6	860	2000
Ac Ala$_3$ pNA	1.25	1.6	2.6	0.03	0.023	0.05	0.07	0.56	1.34
Suc Ala$_3$ pNA	0.4	0.9	1.4	0.14	0.02	0.04	0.6	106	24.6

K_M : mM/L k_{cat} : sec^{-1} V_s : $\mu M \cdot min^{-1} \cdot mg$ Protein^{-1} * at 25°C

Relating to the kinetic constants of the poor anilide substrates, it should be mentioned that with lengthening of the peptide chain from dito tri- and oligo-peptides, thermitase displays a growing peptidolytic property, discernible in the liberation of

amino acid p-nitro-anilides from the respective N-acyl-tri- or -oligopeptide-p-
-nitroanilide. The same holds true for subtilisin and elastase (Kleine, Bröme and
Fittkau, 1980).

Stabilization of the Enzyme

As can be seen from Table 5, the purified enzyme can be stabilized at elevated temperatures in alkaline medium by various lower and higher molecular weight effectors. Among the numerous additives tested, Ca^{++} ions, non-ionic detergents, organic solvents, derivatives of basic amino acids and polypeptides, especially gelatin,
proved to be good stabilizers.

TABLE 5 Stabilization of Thermitase by Low and High Molecular
Weight Additives at pH 8.0 and 55 ^{o}C.

Additive	Concentration	1 hr	2 hrs	24 hrs
Control	–	20	10	0
Calcium ($CaCl_2$)	1 mM	150	120	50^+
Tris-HCl, pH 8	75 mM	$80^•$	$70^•$	–
Triton X-100	0.5-1%	100	72^+	25^+
Tween 20	1%	118^+	110^+	85^+
Arginine βNA	0.045%	110	–	48^+
Lysine- βNA	0.037%	190	–	28^+
Alanine- βNA	0.028%	53	–	22^+
Suc-Ala$_3$-p NA	0.012%	78	–	19^+
Protamine	0.022%	93	–	25
Trasylol	0.011%	110	85	50
Cytochrome C	0.034%	115	71	27
Collagen	0.5%	98	–	–
Gelatin	1.0%	310	–	295
Poly-arginine	0.03%	175	–	37
Concanavalin A	0.008%	95	80	58
Dimethyl formamide	37.0%	$160/200^+$	–	–

Remaining activity (untreated control without effector = 100)
is given.
After the indicated preincubation time (1, 2, 24 hrs) 50 or
100 µl aliquots of heated thermitase (0.1 mg/ml) were with-
drawn, mixed with substrate and residual activity was determined at pH 8/55 ^{o}C. β,NA β-naphthylamide; p NA, p.nitroani-
lide.

Substrates: $^•$ = Suc-Ala$_3$-p NA, 2.5×10^{-4} M;
            $^+$ = Azo-casein, 0.5%; all other activities were
determined with acetyl-tyrosine ethyl ester, 0.023 M.

POSSIBLE FIELDS OF APPLICATION FOR THERMITASE

It is well known that economic and financial factors influence the fields of practical use of an enzyme. Therefore it is understandable that for many purposes the culture broth containing the enzyme is simply concentrated and used without further purification. This may be true also for thermitase. The bulk enzyme, in crude or partially purified state, has been applied in the food industry, e.g. for production of wafers, bread crumbs for dredging or for preparing of partially hydrolyzed and easily digestible food proteins. Some other fields of application may be derived from the properties mentioned above. Owing to its conspicuous long-term stability in the presence of gelatin and its strong gelatin-splitting activity, which is shared with the anionic protease component (Fig. 2), thermitase should be used in due time for the recovery of silver salts from used films. The excellent elastinolytic and collagenolytic properties should be enable thermitase to find uses in tenderizing meat prior to the cooking process or for the disintegration of waste products of meats and bones, especially from poultry.

On the other hand, its use in medicine and biochemistry requires the highly purified enzyme. For example, the careful disintegration of bony and tooth tissues or other elastic and collagenolytic fibres also containing organs and cell cultures is feasible with thermitase. By virtue of its high esterolytic activity, even in the presence of high concentrations of organic solvents, e.g. 50 per cent dimethyl sulfoxide, thermitase might be suitable in the field of peptide synthesis for efficient but gentle esterification of peptide ester derivatives. Finally, thermitase will also be useable for the direct synthesis of new peptide bonds in a two-solvent-phase system.

In summary, the single-step procedure presented for the large-scale isolation of thermitase offer possibilities for its broad application in both theoretical and practical fields. The reported properties characterize thermitase as a thermostable serine protease endowed with high elastinolytic and collagenolytic and esterolytic activities. Finally, the purified enzyme can be used at elevated temperatures for long-term experiments at alkaline pH if one of the stabilizers, listed in Table 5, is added.

REFERENCES

Behnke, U., E. Schalinatus, H. Ruttloff, W.E. Höhne, and C. Frömmel (1978). Acta Biol.Med.Germ., 37, 1185-1192.

Behnke, U., R. Kleine, M. Ludewig, and H. Ruttloff (1978). Acta Biol.Med.Germ., 37, 1205-1214.

Behnke, U., H. Ruttloff, and R. Kleine (1980). J.Appl.Biochem.2 (in press).

Bieth, J., and C.G. Wermuth (1973). Biochem.Biophys.Res.Commun. 53, 383-390.

Bock, H.G., P. Skene, S. Fleischer, P. Cassidy, and S. Harshman (1976). Science (Wash.), 191, 380-383.

Frömmel, C., G. Hausdorf, W.E. Höhne, U. Behnke, and H. Ruttloff (1978). Acta Biol.Med.Germ., 37, 1193-1204.

Gertler, A., and K. Hayashi (1971). Biochim.Biophys.Acta, 235, 378-380.

Gertler, A., Y. Weiss, and Y. Burstein (1977). Biochemistry, 16, 2709-2715.

Hausdorf, G. (1979). Proteinchemische Charakterisierung der Thermitase, einer thermostabilen Protease aus T. vulgaris. Diss., Berlin.

Kleine, R., and U. Rothe (1977). Acta Biol.Med.Germ., 36, K27-K33.

Kleine, R., U. Rothe, H. Ruttloff, and U. Behnke (1978). DDR Wirtschaftspatent,

Nr. WP C 07 G/210135.

Kleine, R., D. Brömme, and S. Fittkau (1980). First joint Biochem.Symposium GDR-CSSR, Reinhardsbrunn, May 1980, Poster Nr. 19.

Langner, J., S. Ansorge, P. Bohley, H. Kirschke, and H. Hanson (1971). Acta Biol. Med.Germ., 26, 935-951.

Langner, J.(1973). Acta Biol.Med.Germ., 31, 1-18.

Lowry, O.H., N.J. Rosebrough, A.L. Farr, and R.J. Randall (1951). J.Biol.Chem., 193, 265-275.

Maurer, H.R. (1971). Disc electrophoresis. W. de Gruiter, Berlin.

Messing, R.A. (1969). J.Amer.Chem.Soc., 91, 2370-2375.

Mizutani, T., and A. Mizutani (1976). J. Chromatogr., 120, 206-210.

Mizutani, T., and A. Mizutani (1977). Analyt.Biochem., 83, 216-221.

Shotton, D.M. (1970). In S.P. Colowick, and N.O. Kaplan (eds.), Methods in Enzymology, Vol. 19. Academic Press, New York. pp. 113-140.

Wünsch, E., and H.G. Heidrich (1963). Z.Physiol.Chem.,333, 149-151.

COMPARATIVE SPECIFICITY OF MICROBIAL ACID PROTEINASES

K. Morihara

Shionogi Research Laboratories, Shionogi & Co., Ltd.,
Fukushima-ku, Osaka 553, Japan

ABSTRACT

The specificities of various acid proteinases from fungi or yeasts were comparatively determined using synthetic peptides as substrates. Swine pepsin was used in a comparative study. The pepstatin-sensitive proteinases exhibited their specificity against aromatic or hydrophobic amino acid residues on both sides of the splitting point in peptide substrates, and some differences were observed depending upon their source. Some of the enzymes, possessing trypsinogen-activating ability, also showed their specificity against L-lysine residue on the carboxyl side of the splitting point. In comparison, a pepstatin-insensitive enzyme A from <u>Scytalidium</u> showed considerably broad specificity. The hydrolysis of peptides by pepstatin-sensitive enzymes was markedly accelerated by elongating the peptide chain by three amino acid residues on both sides of the splitting point, while the elongation of two residues produced the same effect with the pepstatin-insensitive enzyme. Thus, $Z-(Ala)_2-Lys_{\uparrow}-(Ala)_3$ was most susceptible ($\uparrow$ shows the bond split) to the former enzymes possessing trypsinogen-activating ability, and $Z-Phe-Leu_{\uparrow}-Ala-Ala$ for the latter one. Another pepstatin-insensitive enzyme from <u>Scytalidium</u> showed negligible activity on the peptides used in this study.

KEYWORDS

Swine pepsin; pepstatin; diazoacetyl-<u>DL</u>-norleucine methyl ester; 1,2-epoxy-3-(p-nitrophenoxy)propane; trypsinogen-activating ability; synthetic peptides for microbial acid proteinases; primary specificity; effect of secondary interaction; Z-Phe-Leu-Ala-Ala; $Z-(Ala)_2-Lys-(Ala)_3$.

INTRODUCTION

Acid proteinases, which are most active in acidic pH ranges, are widely distributed in fungi and yeasts (Matsubara and Feder, 1971) but are seldom found in bacteria. Most enzymes are sensitive to pepstatin, diazoacetyl-<u>DL</u>-norleucine methyl ester (DAN) and 1,2-epoxy-3-(p-nitrophenoxy)propane (EPNP) (Tang, 1976), similar to the case with pepsin. It has been established (Hsu and others, 1977;

Subramanian, 1978) that two aspartic acid residues constitute their active
sites, as in pepsin. Therefore, they are also called carboxyl proteinases. The
pepstatin-sensitive acid proteinases are usually divided into two groups such as
pepsin-like and chymosin-like ones according to their activity of clotting milk
casein (Matsubara and Feder, 1971). Some of the microbial enzymes have been
found (Hofmann and Shaw, 1964) to show very high trypsinogen-activating ability.
Murao, Oda and Matsushita (1972) have isolated Scytalidium lignicolum which can
produce pepstatin-insensitive acid proteinases. The enzymes show different
behavior against DAN and/or EPNP from those of pepstatin-sensitive ones.

N-Acylated dipeptides had been used as substrates for pepsin until 1970, but the
activity was very low. Since then, Medzihradszky and co-workers (1970) and Oka
and Morihara (1970) have independently found that the molecular size of the
peptide substrate is significant for its susceptibility and have synthesized
some efficient peptide substrates of large molecular size for pepsin (Fruton,
1976). The present study was undertaken to compare the specificities of various
acid proteinases from microorganisms having different enzymatic characteristics,
as mentioned above, using peptides synthesized in our laboratory. Pepsin was
used for a comparative study.

MATERIALS AND METHODS

Enzymes. Crystalline swine pepsin (2X cryst) and crystalline acid proteinase
from Rhizopus chinensis (3X cryst) were supplied by Worthington Biochem. Corp.,
New Jersey, and Seikagaku Kogyo Co., Tokyo, respectively. Crystalline acid
proteinase from Rhodotorula glutinis and Cladosporium sp. were kindly donated by
Prof. S. Murao of Osaka Prefectural University. Crystalline chymosin-like
enzyme from Mucor pusillus was supplied by Prof. K. Arima of Tokyo University.
Acid proteinases from Aspergillus niger and Aspergillus saitoi were purified
according to the method described previously (Oka and Morihara, 1973a) using
Proctase B (Meiji Seika Co., Tokyo) and Molsin (supplied by Dr. F. Yoshida of
the Noda Institute for Scientific Research) as starting materials, respectively.
Crude extract of a chymosin-like enzyme from Mucor miehei were a kind gift from
Novo Ind., Copenhagen. The acid proteinase was purified according to the method
of Ottesen and Rickert (1970).

All of the enzyme preparations used were homogeneous by disc gel electrophoresis
at pH 6. Some enzymatic properties of the respective enzymes have been studied
by many workers (Morihara and Oka, 1973; Morihara and others, 1979; Oda, Funakoshi
and Murao, 1973; Oda, Kamada and Murao, 1972; Oda and Murao, 1974; Oda and
others, 1975; Oda and others, 1976; Oka and Morihara, 1973a; Satoi and Murao,
1973; Takahashi and Chang, 1976); they are summarized in Table 1.

Enzymes obtained from both A. niger and A. saitoi or M. miehei and M. pusillus
showed almost identical kinetic parameters against peptide substrates used in
this study. Accordingly, the data of the enzymes from A. niger and M. miehei
were adopted as those for the enzymes from Aspergillus and Mucor, respectively,
in the following experiments.

Peptide substrates. Most of the peptides were synthesized in this laboratory
according to the method described in the previous papers (Oka and Morihara,
1973a, 1973b, 1974; Morihara and Oka, 1973). Abbreviated designations of peptides
or their derivatives conform to the tentative rules of the IUPAC-IUB Commission
on Biochemical Nomenclature. Except where otherwise specified, the constituent
amino acids were all of L-configuration.

TABLE 1 Enzymatic Properties of Microbial Acid
Proteinases

Acid proteinase from	Opt. pH	Proteolytic Act. ([PU])	Trypsinogen activation [a] (nM/min/mgE)	Inhibitor[b] Pepstatin[c]	DAN	EPNP
Pepsin (swine)	1.5	13.8	0	$+(10^{-9})$	+	+
Aspergillus niger (Proctase B)	3.0	10.0	800	$+(10^{-8})$	+	+
Aspergillus saitoi	3.0	8.2	820	$+(10^{-8})$	+	+
Rhizopus chinensis	3.0	8.7	180	$+(10^{-8})$	+	+
Cladosporium sp.	3.0	7.7	150	$+(10^{-6})$	+	+
Rhodotorula glutinis	3.0	11.1	24	$+(10^{-7})$	+	+
Mucor miehei	3.0	2.9	0	+	+	+
Mucor pusillus	3.0	1.4	0	$+(10^{-8})$	+	+
Scytalidium lignicolum A	3.0	3.0	<10	$-(10^{-3})$	-	-
Scytalidium lignicolum B	2.0	1.7	0	-	-	+

a, [Trypsinogen] = 2.5×10^{-6} M, pH 3.4, 40°C, 10 min. b, +, inhibition; -, no
inhibition. c, The inhibition was determined at [E]/[I] = 1/100. The K_i
values are shown in the parentheses.

Determination of hydrolysis of various synthetic peptides. The determination
was performed by the usual ninhydrin method (Oka and Morihara, 1973a). Except
where otherwise specified, the pH of the reaction mixture was adjusted at pH 3.2
for microbial enzymes with 0.05 M acetic acid and at 1.5 for pepsin with 0.05 M
acetic acid-HCl. The initial velocity was determined at 40°C with the enzyme
concentration suitably adjusted.

The sites of action of the enzymes on the substrates were determined by paper or
thin-layer chromatography of the hydrolyzates, using authentic compounds for
comparison, or by high-voltage paper electrophoresis at pH 1.9. The ninhydrin
color yields of various compounds including the product fragments have been described in the previous papers (Oka and Morihara, 1973a, 1973b; Morihara and
Oka, 1973), in which the percentage yields were calculated using L-leucine as a
standard.

In cases where the substrates were split at more than one peptide bond, the
ratios of the ninhydrin color intensities of the two or three product fragments
were determined by paper chromatography or paper electrophoresis, as shown in
the previous paper (Oka and Morihara, 1974). The initial velocities of hydrolysis of the respective peptide bonds in a substrate were calculated from the
total rate of ninhydrin color production using hydrolysis factors for the corresponding peptide bonds. These factors are the ratios of the amounts of respective
product fragments formed to the total amount of fragments on the paper chromato-
gram. Calculation of these factors was based on the assumption that the ratios
of the rates of production of the various product fragments are constant during
the initial stage of reaction (up to 30% hydrolysis).

Kinetic study. In all cases the data were in accord with Michaelis-Menten
kinetics over the range of substrate concentrations employed (1.5-6.0 mM for
peptide substrates containing lysine, and 0.15 to 1.0 mM for the other peptides);

five to six runs were performed for each determination of K_m and k_{cat}; the enzyme concentration (0.005-0.1 mg/ml) was chosen to give reliable data for the initial rate of hydrolysis. The reaction mixture usually contained 10% of dimethylformamide owing to the small solubility of peptide substrates. For the calculation of k_{cat}, the molecular weights of these enzymes were assumed to be; pepsin, 33,000 (Williams and Rajagopalan, 1966); Aspergillus niger enzyme, 34,000 (Horiuchi and others, 1969); Rhizopus enzyme, 35,000 (Tsuru, Hattori and Fukumoto, 1970); Mucor miehei enzyme, 38,000 (Ottesen and Rickert, 1970); Rhodotorula enzyme, 30,000 (Oda, Kamada and Murao, 1972); Cladosporium enzyme, 38,000 (Murao, S., private communication); Scytalidium enzyme A, 40,000 (Oda and others, 1976); Scytalidium enzyme B, 20,000 (Oda and others, 1975).

RESULTS

Effect of Molecular Size of Peptide Substrates

Table 2 indicates that the acid proteinases from Aspergillus, Rhodotorula and Scytalidium required N-acylated tetrapeptides or peptides of larger molecular size to become susceptible to the enzymes, while the other enzymes were active against N-acylated dipeptides as seen with pepsin.

TABLE 2   Effect of Molecular Size of Peptide Substrates[a]

Substrate[b]	Susceptible (Acid proteinase from)	Not susceptible
Z-Phe↑Tyr Ac-Phe↑Tyr(I$_2$)	Pepsin, Rhizopus, Mucor, Cladosporium	Aspergillus, Rhodotorula, Scytalidium (A, B)
Z-Phe↑Leu-Ala Z-Phe↑Tyr-Ala	Pepsin, Rhizopus, Mucor, Cladosporium	Aspergillus, Rhodotorula, Scytalidium (A, B)
Z-Phe↑Leu△Ala▲Ala Z-Phe↑Tyr△Ala▲Ala	Pepsin, Rhizopus, Mucor, Cladosporium, Aspergillus, Rhodotorula, Scytalidium A (△), Scytalidium B (▲)	

a, Reaction mixture: [S] = 0.5 mM, [E] = 0.02 mg/ml, [Dimethylformamide] = 10%, pH 3.2 (pH 1.5 for pepsin), 37°C, 20 h. b, ↑ show the bonds split by all the enzymes except Scytalidium enzymes (A and B).

Primary Specificity

The primary specificity of the microbial enzymes was determined by comparison using Z-tetrapeptides as substrates. The P_1- and P_1'-specificity of pepstatin-sensitive proteinases was determined using Z-AA↑Leu-Ala-Ala or Z-Phe↑AA-Ala-Ala (AA = various amino acid residues; ↑ shows the bond split), respectively, as the substrate. Amino acid residues in peptide substrates are numbered by the nomenclature of Schechter and Berger (1967). The results are summarized in Table 3, which shows that the enzymes were specific for aromatic or bulky amino acid residues at P_1-position, and some difference could be seen in their specificity at the position depending upon their source. Thus, like pepsin, the enzymes from Aspergillus, Mucor and Cladosporium were more specific for a phenylalanine residue than for a leucine one, whereas the enzymes from Rhizopus and Rhodotorula showed almost equal preference of hydrolysis for phenylalanine (or tyrosine) and leucine. A striking difference can be seen in their specificity for a lysine residue; the specificity roughly corresponded to their trypsinogen-activating ability.

TABLE 3 Primary Specificity of Pepstatin-Sensitive Acid Proteinases

Peptide $P_1-P_1{}'P_2{}'P_3{}'$	Acid proteinase from					
	Aspergillus	Rhizopus	Mucor	Rhodotorula	Cladosporium	Pepsin
	k_{cat}/K_m (sec^{-1} M^{-1})$^{\underline{a}}$					
ZGly$^{\downarrow}$-LeuAlaAla	0	0	0	0	0	0
ZAla-LeuAlaAla	0	0	0	0	$5^{\underline{b}}$	0
ZGlu-LeuAlaAla	0	0	$2^{\underline{b}}$	0	$1^{\underline{b}}$	$1^{\underline{b}}$
ZLeu-LeuAlaAla	18(1.7)	60(1.2)	6(2.2)	14(5.9)	63(2.3)	57(3.4)
ZPhe-LeuAlaAla	50(0.6)	33(1.0)	67(0.7)	18(1.0)	223(0.6)	213(0.3)
Z-DPhe-LeuAlaAla	0	0	0	0	0	0
ZTyr-LeuAlaAla	21(0.8)	50(1.5)	36(26)	17(1.2)	126(3.1)	10(0.9)
ZLys-LeuAlaAla	35(8.5)	5(19)	0	$5^{\underline{b}}$	44(4.6)	0
ZPhe$^{\downarrow}$-GlyAlaAla	0	0	0	0	0	0
ZPhe-AlaAlaAla	$1^{\underline{b}}$	$4^{\underline{b}}$	$6^{\underline{b}}$	$4^{\underline{b}}$	26(3.1)	19(5.3)
ZPhe-GluAlaAla	$3^{\underline{b}}$	0	$9^{\underline{b}}$	0	$12^{\underline{b}}$	$5^{\underline{b}}$
ZPhe-LeuAlaAla	50(0.6)	33(1.0)	67(0.7)	18(1.0)	223(0.6)	213(0.3)
ZPhe-DLeuAlaAla	0	0	0	0	0	0
ZPhe-TyrAlaAla	11(1.8)	7(7.3)	41(25)	21(1.9)	198(5.8)	1030(1.1)
ZPhe-LysAlaAla	0	0	$4^{\underline{b}}$	0	$7^{\underline{b}}$	$9^{\underline{b}}$

$\underline{a}$, The K_m values (mM) are shown in the parentheses. $\underline{b}$, V/[S] at [S] = 0.5 mM.

Table 3 further indicates that the pepstatin-sensitive enzymes were also specific for aromatic or bulky amino acid residues at $P_1{}'$-position, although there was

some difference between the microbial enzymes and pepsin. That is, pepsin was most specific for a tyrosine residue at the position, while the microbial enzymes were either more specific for a leucine residue or equally specific for both tyrosine and leucine residues.

The primary specificity of the pepstatin-insensitive proteinase A from Scytalidium was determined using Z-Phe-AA$_{\uparrow}$Ala-Ala, Z-Ala-Phe$_{\uparrow}$AA-Ala and others as substrates. Table 4 indicates that the enzyme A was not only specific against hydrophobic or bulky residues such as L-leucine at both the P_1 and $P_1{}'$-positions, but also showed rather broad specificity in comparison with the pepstatin-sensitive ones. For example, the Glu-Ala bond was hydrolyzed considerably by the former enzyme, but the Glu-Leu bond was hydrolyzed negligibly by the latter ones (refer to Table 3). Also the Phe-Gly bond was susceptible to the enzyme A, but

TABLE 4   Primary-Specificity of Scytalidium
Enzyme A

Peptide P_2 P_1 P_1' P_2'	K_m (mM)	k_{cat}/K_m (sec^{-1} M^{-1})
Z Phe Gly↓Ala Ala	-	<10[a]
Z Phe Ala-Ala Ala	-	420[a]
Z Phe Glu-Ala Ala	3.7	2,540
Z Phe Leu-Ala Ala	1.0	10,700
Z Phe DLeu-Ala Ala		0
Z Phe Tyr-Ala Ala	1.6	1,540
Z Phe Lys-Ala Ala	0.8	1,770
Z Ala Phe↓Gly Ala	1.5	310
Z Ala Phe-Leu Ala	0.6	4,600
Z Gly Leu-Gly Phe		220[a]
Z Gly Phe-Leu Ala	1.3	1,900

a, Calculated on the assumption that the K_m is
1 mM.

never to the pepstatin-sensitive enzymes (refer to Table 3).

Effect of Secondary Interaction

The effect of elongating the peptide chain on both sides of the splitting point
(i.e., secondary interaction) was studied by hydrolysis using the pepstatin-sensitive enzymes. Table 5 indicates that elongation of the peptide chain on the
N-terminal side from the splitting point resulted in an increase of hydrolysis
with all the microbial enzymes as well as pepsin. Some difference can be seen
among them in the rates of promotion depending upon their source; i.e., elongation of the peptide chain from P_1 or P_2 to P_3 with L-alanine resulted in marked
increase of hydrolysis with the enzymes from Aspergillus, Rhizopus, and
Rhodotorula and with pepsin, and a considerable increase with the enzymes from
Cladosporium, but not with Mucor enzyme. On elongation of the peptide chain
from P_1' or P_2' to P_3' with L-alanine, the peptides also became very susceptible

TABLE 5   Effect of Elongation of the Peptide Chain on
the N-Terminal Side on Hydrolysis by Acid Proteinases[a]

Peptides P_4 P_3 P_2 P_1 P_1'	Acid proteinase from					
	Aspergillus	Rhizopus	Mucor	Rhodotorula	Cladosporium	Pepsin
			(nM/min/mg enzyme)			
ZPhe↓Tyr	0	0.3	0.6	0	0	2.7
ZGlyPhe-Tyr	0	4.2	6.1	0	2.9	6.7
ZAlaPhe-Tyr	0	1.0	16.7	≦0.1	2.2	22.0
Z-DAlaPhe-Tyr	0	3.1	2.0	0	1.5	1.0
ZGlyGlyPhe-Tyr	0	3.6	1.0	0	2.0	59.1
ZAlaGlyPhe-Tyr	2.5	73.1	9.7	1.0	9.0	584.0
Z-DAlaGlyPhe-Tyr	0	2.0	≦0.1	0	1.5	3.5

a, [S] = 0.5 mM, [E] = 0.02-0.5 mg/ml.

TABLE 6 Effect of Elongation of the Peptide Chain on the C-Terminal Side on Hydrolysis by Acid Proteinases[a]

Peptides $P_1 P_1'P_2'P_3'$	Acid proteinase from					
	Aspergillus	Rhizopus	Mucor	Rhodotorula	Cladosporium	Pepsin
			(nM/min/mg enzyme)			
ZPhe-↓Leu	0	0.6	0	0	0.1	0
ZPhe-LeuNH$_2$	0	≦0.1	0	0	0	0
ZPhe-LeuGly	0	3.0	2.1	0	6.1	1.7
ZPhe-LeuAla	0	6.3	3.8	≦0.1	25.5	2.9
ZPhe-Leu-DAla	0	≦0.1	0.8	0	≦0.1	0
ZPhe-LeuAlaAla	17	35	27	10	91	50
ZPhe-LeuAla-DAla	10	13	11	2	45	10

a, [S] = 0.5 mM, [E] = 0.02-0.5 mg/ml.

to all the enzymes tested, although there were some differences in the rates of promotion depending upon their sources, as seen in Table 6.

Z-Phe-Leu-Ala-Ala was hydrolyzed much or little by pepstatin-sensitive enzymes at positions other than the main splitting point (↑): Aspergillus enzyme and pepsin (Z-Phe-Leu-Ala-Ala), Rhizopus enzyme (Z-Phe-Leu-Ala-Ala), Mucor enzyme (Z-Phe-Leu-Ala-Ala) Rhodotorula enzyme (Z-Phe-Leu-Ala-Ala) and Cladosporium enzyme (Z-Phe-Leu-Ala-Ala). This may be ascribed to the difference not only in primary specificity but also in the effect of secondary interaction.

To investigate the effect of elongation of the peptide chain at the P_3 and P_3' positions, further study was undertaken with the results given in Table 7. Since these peptides were all susceptible to the microbial enzymes possessing trypsinogen activating ability at the peptide bond containing the carboxyl group of L-lysine, we could examine the effect of elongating the peptide chain at the P_3 and P_3' positions. Of the substrates used, Z-(Ala)$_2$-Lys-(Ala)$_3$ was the most susceptible to the enzymes from Aspergillus, Rhizopus, Rhodotorula and Cladosporium. While, Mucor enzyme and pepsin showed very little peptidase activity on the peptide. The substrate was also significant because of its high

TABLE 7 Effect of Molecular Size of Peptide Substrates on Hydrolysis by Microbial Acid Proteinases

Peptides $P_3 P_2 P_1 P_1'P_2'P_3'$	Acid proteinase from			
	Aspergillus	Rhizopus	Rhodotorula	Cladosporium
		k_{cat}/K_m (sec^{-1} M^{-1})[a]		
ZLys-↓AlaAla	<1	<1	0	20(10)
ZLys-AlaAlaAla	16(11)	10(10)	8(23)	40(7.6)
ZLys-LeuAlaAla	156(4.3)	111(2.8)	20(26)	738(0.8)
ZAlaLys-AlaAlaAla	200(11)	63(5.1)	26(22)	34(13)
ZAlaAlaLys-AlaAlaAla	10,500(2.9)	4,750(6.1)	341(34)	3,110(2.8)

a, The K_m values (mM) are shown in the parentheses.

solubility. Recently, Hofmann, Cunningham, and Hodges (1979) synthesized Ac-(Ala)$_2$-Lys-Phe(NO$_2$)-(Ala)$_2$-NH$_2$ as a substrate for penicillopepsin. The k$_{cat}$/K$_m$ (420,000 sec^{-1} M^{-1}) was comparable to that of the most efficient substrates for pepsin (Fruton, 1976).

The influence of molecular size in peptide substrates with the pepstatin-insensitive proteinases (A and B) from Scytalidium was completely different from that of the pepstatin-sensitive ones mentioned above. The results summarized in Table 8 indicate that almost maximum activity was attained with N-acylated tetrapeptides for hydrolysis of the enzyme A, in which the presence of amino acid residues at the P$_2$ and P$_2$' positions was essential. The enzyme A may have a more compact active site than usual pepstatin-sensitive acid proteinases.

TABLE 8   Effect of Molecular Size of Peptide
Substrates on Hydrolysis by Scytalidium
Enzyme A and B

Peptide						A	B
P$_3$	P$_2$	P$_1$	P$_1$'	P$_2$'	P$_3$'	k$_{cat}$/K$_m$ (sec^{-1} M^{-1})[a]	
	Z	Phe↓Leu				0	0
	Z	Phe-Leu	Ala			<10[b]	0
Z	Phe	Leu-Ala	Ala			10,700 (4)	<10[b,c]
	Z	Lys-Ala	Ala			0	0
	Z	Lys	Ala-Ala	Ala		140[b]	<10[b,c]
	Z	Lys	Leu-Ala	Ala		2,090 (2.1)	30[b,c]
	Z	Lys	Ala-Ala	Ala	Ala	2,190 (2.5)	-
	Z	Ala	Lys-Ala	Ala	Ala	7,970 (3.1)	15[b]
Z	Ala	Ala	Lys-Ala	Ala	Ala	2,600 (1.0)	45[b]

a, The K$_m$ values (mM) are shown in the parentheses.

b, Calculated on the assumption that the K$_m$ is 1 mM.

c, The C-terminal peptide bond was split.

On the other hand, the enzyme B showed very low activity towards the peptides presented here, although its activity increased with an increase in the molecular size of the peptide substrates. It is not clear whether the enzyme B has a larger active site than usual pepstatin-sensitive acid proteinases, because the peptide substrates used here may not have been adequate substrates for the enzyme. Iio and Yamasaki (1976) have shown that an acid proteinase from Aspergillus niger (Proctase A; commercially available from Meiji Seika, Co., Tokyo) shows similar characteristics against the inhibitors (pepstatin, DAN and EPNP) to those of the enzyme B. They have also shown that the specificity of the former enzyme against the oxidized insulin B chain differed considerably from that of the usual pepstatin-sensitive enzymes. The latter enzymes are known (Morihara, 1974) to show very similar specificitiy against the B chain irrespective of their source. Oda and Murao (1976) have shown that the specificity of the Scytalidium enzyme A and B is rather like to that of pepstatin-sensitive enzymes.

Currently available data (Subramanian, 1978) support the generalization that all acid proteinases which are sensitive against pepstatin, irrespective of their source, have homologous amino acid sequences and essentially similar three-dimensional structures. The present study indicates some differences exist between the pepstatin-sensitive enzymes from microbial origin and pepsin not only in their primary specificity but also in the effect of secondary inter-action. It should be interesting if the differences can be deduced from their structures. Also, the interesting question remains as to whether the structures of the Scytalidium enzymes are like those of usual pepstatin-sensitive acid proteinases.

ACKNOWLEDGEMENT

The author is greatly indebted to Dr. T. Oka of his wholehearted cooperation in this work as well as to Drs. S. Murao and K. Oda for their interest and en-couragement.

REFERENCES

Fruton, J. S. (1976). The mechanism of the catalytic action of pepsin and related acid proteinases. In A. Meister (Ed.), Advances in Enzymology, Vol. 44. An Interscience Publication, New York. pp. 1-36.

Hofmann, T., and R. Shaw (1964). Proteolytic enzymes of Penicillium janthinel-lum. I. Purification and properties of a trypsinogen-activating enzyme (peptidase A). Biochim. Biophys. Acta, 92, 543-557.

Hofmann, T., A. Cunningham, and R. S. Hodges (1979). Kinetic studies on penicil-lopepsin with a new substrate, AC-ALA-ALA-LYS-NPH-ALA-ALA-NH$_2$. In Abstracts of XIth International Congress of Biochemistry (Toronto), p. 229.

Horiuchi, A., M. Honjo, M. Yamasaki, and Y. Yamada (1969). Studies on acid proteases. I. Homogeneity and molecular weight of acid proteases from Aspergillus niger var. Macrosporus (Proctase A and Proctase B). Sci. Pap. Coll. Gen. Educ. Univ. Tokyo, 19, 127-139.

Hsu, L. N., L. T. J. Delbaere, M. N. G. James, and T. Hofmann (1977). Penicil-lopepsin from Penicillium janthinellum crystal structure at 2.8 Å and sequence homology with porcine pepsin. Nature, 266, 140-145.

Iio, K., and M. Yamasaki (1976). Specificity of acid proteinase A from Aspergillus niger var. macrosporus towards B-chain of performic acid oxidized bovine insulin. Biochim. Biophys. Acta, 429, 912-924.

Matsubara, H., and J. Feder (1971). Other bacterial, mold, and yeast proteases. In P. D. Boyer (Ed.), The Enzymes, Vol. 3, 3rd ed. Academic Press, New York. pp. 723-744.

Medzihradszky, I. M., K. Voynick, H. Medzihradszky-Schweiger, and J. S. Fruton (1970). Effect of secondary enzyme-substrate interactions on the cleavage of synthetic peptides by pepsin. Biochemistry, 9, 1154-1162.

Morihara, K., and T. Oka (1973). Comparative specificity of microbial acid proteinases for synthetic peptides. III. Relationship with their trypsinogen activating ability. Arch. Biochem. Biophys., 157, 561-572.

Morihara, K. (1974). Comparative specificiy of microbial proteinases. In A. Meister (Ed.), Advances in Enzymology, Vol. 41. An Interscience Publica-tion, New York. pp. 179-243.

Morihara, K., H. Tsuzuki, S. Murao, and K. Oda (1979). Pepstatin-insensitive acid proteases from Scytalidium lignicolumn. Kinetic study with synthetic peptides. J. Biochem. (Tokyo), 85, 661-668.

Murao, S., K. Oda, and Y. Matsushita (1972). New acid proteases from Scytalidium lignicolum M-133. Agr. Biol. Chem. (Tokyo), 36, 1647-1650.

Oda, K., M. Kamada, and S. Murao (1972). Some physicochemical properties and substrate specificity of acid protease of Rhodotorula glutinis K-24. Agr. Biol. Chem. (Tokyo), 36, 1103-1108.

Oda, K., S. Funakoshi, and S. Murao (1973). Some physicochemical properties and substrate specificity of acid protease isolated from Cladosporium sp. No. 45-2. Agr. Biol. Chem. (Tokyo), 37, 1723-1729.

Oda, K., and S. Murao (1974). Purification and some enzymatic properties of acid protease A and B of Scytalidium lignicolum ATCC 24568. Agr. Biol. Chem. (Tokyo), 38, 2435-2444.

Oda, K., S. Murao, T. Oka, and K. Morihara (1975). Some physicochemical properties and substrate specificity of acid protease B of Scytalidium lignicolum ATCC 24568. Agr. Biol. Chem. (Tokyo), 39, 477-484.

Oda, K., and S. Murao (1976). Action of Scytalidium lignicolum acid proteases on insulin B-chain. Agr. Biol. Chem. (Tokyo), 40, 1221-1225.

Oda, K., S. Murao, T. Oka, and K. Morihara (1976). Some physicochemical properties and substrate specificities of acid proteases A-1 and A-2 of Scytalidium lignicolum ATCC 24568. Agr. Biol. Chem. (Tokyo), 40, 859-866.

Oka, T., and K. Morihara (1970). Specificity of pepsin. Size and property of the active site. FEBS Lett., 10, 222-224.

Oka, T., and K. Morihara (1973a). Comparative specificity of microbial acid proteinases for synthetic peptides. I. Primary specificity. Arch. Biochem. Biophys., 156, 543-551.

Oka, T., and K. Morihara (1973b). Comparative specificity of microbial acid proteinases for synthetic peptides. II. Effect of secondary interaction. Arch. Biochem. Biophys., 156, 552-559.

Oka, T., and K. Morihara (1974). Comparative specificity of microbial acid proteinases for synthetic peptides. IV. Primary specificity with Z-tetra-peptides. Arch. Biochem. Biophys., 165, 65-71.

Ottesen, M., and W. Rickert (1970). The isolation and partial characterization of an acid protease produced by Mucor miehei. Compt. Rend. Trav. Lab. Carlsberg, 37, 301-325.

Rajagopalan, T. G., W. H. Stein, and S. Moore (1966). The inactivation of pepsin by diazoacetyl-norleucine methyl ester. J. Biol. Chem., 241, 4295-4297.

Satoi, S., and S. Murao (1973). Inhibition of acid proteases by a pepsin in-hibitor (S-PI). Agr. Biol. Chem. (Tokyo), 37, 2579-2587.

Schechter, I., and A. Berger (1967). On the size of the active site in proteases. I. Papain. Biochem. Biophys. Res. Commun., 27, 157-162.

Subramanian, E. (1978). Molecular structure of acid proteases. Trends Biochem. Sci., 3, 1-3.

Takahashi, K., and W. J. Chang (1976). The structure and function of acid proteases. V. Comparative studies on the specific inhibition of acid proteases by diazoacetyl-DL-norleucine methyl ester, 1,2-epoxy-3-(p-nitro-phenoxy)propane and pepstatin. J. Biochem. (Tokyo), 80, 497-506.

Tang, J. (1976). Pepsin and pepsinogen: Models for carboxyl (acid) proteases and their zymogens. Trends Biochem. Sci., 1, 205-208.

Tsuru, D., A. Hattori, and J. Fukumoto (1970). Studies on mold proteases. III. Some physicochemical properties and amino acid composition of Rhizopus chinesis acid protease. J. Biochem. (Tokyo), 67, 415-420.

Williams, R. C., and T. G. Rajagopalan (1966). Ultracentrifugal characterization of pepsin and pepsinogen. J. Biol. Chem., 241, 4951-4954.

PROTEASES IN CULTURE FILTRATES OF ASPERGILLUS NIGER AND CLAVICEPS PURPUREA

I. Kregar, I. Maljevac**, V. Puizdar, M. Derenčin**, A. Puc*,
and V. Turk

*Department of Biochemistry, J. Stefan Institute, 61000 Ljubljana; *LEK, Pharm.
and Chemical Works, 61000 Ljubljana and **TOK, Factory for Production of
Organic Acids, 66250 Ilirska Bistrica, Yugoslavia*

ABSTRACT

Aspartic proteinases were isolated from culture filtrates of Aspergillus niger and Claviceps purpurea using ultrafiltration, affinity chromatography on immobilized pentapeptide inhibitor pepstatin and gel filtration. Some biochemical characteristics of both enzymes were determined.

KEYWORDS

Aspergillus niger proteinases; Claviceps purpurea proteinases; acid microbial proteinases; aspartic proteinases; carboxyl proteinases.

INTRODUCTION

Considerable interest in proteolytic enzymes of microbial origin has been noted in the past decade. These enzymes can also be obtained as by-products when an antibi-otic, alkaloid or citric acid is the main product of a fermentation process. Culture filtrates are therefore sources of various enzymes, including proteinases, which can have wide industrial applications. It is known that the genus Aspergillus is a source of various aspartic, DFP-sensitive and metallo-proteinases, some of which have already been isolated and characterized (Matsubara and Feder, 1971). There are, however, no data on proteases which are secreted in the medium during submerged growth of C. purpurea. It was noted only by Rehaček and coworkers (1971) that extracellular proteinases were continuously excreted and reached a peak on the 10th day of culture growth. The purpose of our work was to investigate the proteinases secreted by A. niger and C. purpurea.

MATERIALS AND METHODS

Aspergillus niger ATCC 10577 was grown as a surface culture for industrial citric acid production. At the end of fermentation (170-190 hrs) the biomass was filtered off and suspended in tap water. After 24 hrs it was filtered and the culture filtrate obtained was used as the source of enzymes. Claviceps purpurea was grown in sub-merged culture under conditions for ergot alkaloid biosynthesis. After 240 hrs the culture filtrate was obtained by centrifugation.

Ultrafiltration

Amicon UM-10 membranes or Diaflo hollow fibers H1P10 (Amicon, USA) were used for concentration of the culture filtrate and of the pooled fractions containing proteolytic enzymes.

Gel Chromatography

Sephadex G-75 was used for the separation of proteins and for the molecular weight determination (Whitaker, 1963).

Affinity Chromatography

Immobilized pentapeptide inhibitor pepstatin on AH-Sepharose 4B (Pharmacia, Sweden) was prepared by carbodiimide coupling (Murakami and Inagami, 1975 Kregar and co-workers, 1977).

Other Determinations

Proteolytic activity was determined by the modified method of Anson (1939).

RESULTS AND DISCUSSION

Separation of concentrated culture filtrates on Sephadex G-75, as shown in Fig. 1 for A. niger , revealed the presence of two proteolytic peaks active toward Hb, pH 3.5

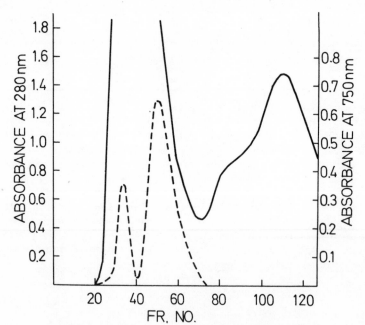

Fig. 1. Gel chromatography of A.niger culture filtrate on Sephadex G-75. ——— elution profile of protein, A_{280}; ---- proteolytic activity toward haemoglobin pH 3.5 A750.

<u>C. purpurea</u> culture filtrates (Fig. 2) also showed proteolytic activity toward casein in neutral pH. Preliminary characterization showed that the low molecular active fractions were inhibited by pepstatin and can therefore be classified as carboxyl or

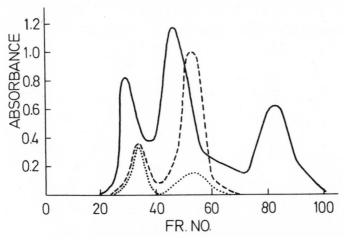

Fig. 2. Gel chromatography of <u>C. purpurea</u> culture filtrate on Sephadex G-75. —— elution profile of protein, A_{280} ---- proteolytic activity toward haemoglobin pH 3.5, A_{750}; •••• proteolytic activity toward casein, pH 7.0

TABLE 1 Purification of aspartic proteinase of A. niger and
C. purpurea

Step	Total activity Anson units (A.U.)		Specific activity A.U./mg protein		Yield %	
	A.niger	C.purp.	A. niger	C. purp.	A. niger	C.purp.
Culture filtrate	45,640	14,400	0.11	0.38	100	100
Concentrate	17,040	5,760	0.83	1.60	37	40
50-75% methanol fraction	8,960		1.09		19.7	
Pepstatin–Sepharose 4B	6,540	2,950	39.50	52.00	14.3	20.5
Sephadex G-75	5,735	1,970	82.50	61.25	12.5	13.7

aspartic proteinases (Hartley, 1960). In order to isolate this group of enzymes we devised a method based upon affinity chromatography on immobilized pepstatin, a strong inhibitor of aspartic proteinases. Purification as shown in Table 1, can start either with the culture filtrate or with preliminarily separated fractions on Sephadex G-75. In either case concentration by ultrafiltration is necessary. Methanol fractionation enriched the aspartic proteinase content in <u>A. niger</u> samples, but this step was not used in <u>C. purpurea</u> samples. Prior application to pepstatin-Sepharose

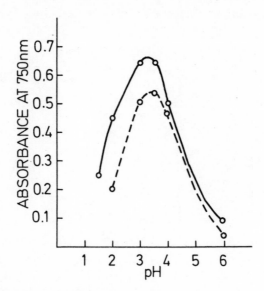

Fig. 3. pH dependence of proteolytic activity toward haemoglobin for A. niger (———) and C. purpurea (----)
proteinase.

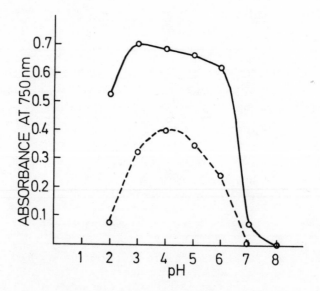

Fig. 4. pH stability of A. niger (———) and C. purpurea (----)
proteinase.

pH of the sample was brought to 3.5 and NaCl concentration to 1 M. Because microbial acid proteinases are known to be unstable above pH 6, the enzyme was eluted from the column at pH 6.1 with 0.2 M universal buffer in 1.5 M NaCl and immediately after desorption acidified to pH 5. Below pH 6 the enzyme was not released from the column. The final step in the purification procedure was gel chromatography on Sephadex G-75. Proteolytic activity was eluted as a single peak, with some additional UV absorbing material in A. niger samples. Electrophoresis on polyacrylamide gel showed that both enzymes were essentially pure. A. niger proteinase consisted of two multiple forms having isoelectric points of 4.0 and 4.5 as determined by isoelectric focusing. The molecular weight of C. purpurea proteinase was 43,000 and 38,000 for the A. niger enzyme. pH optimum for haemoglobin hydrolysis was close to pH 3.5 (Fig. 3). The enzymes were stable between pH 3 - 6 (Fig. 4). Pepstatin completely inhibited their activities.

Many acid proteinases from various Aspergillus genus have been already described (Matsubara and Feder, 1971; Lobareva and Stepanov, 1978). Molecular weights reported vary from 30,000 to 60,000; even a value as high as 150,000 has been reported for a proteinase from Aspergillus orizae (Tsujita and Endo, 1978). All have pH optima for the hydrolysis of protein substrates between pH 3 to 4. Isoelectric points are within pH 3 to 4 and pH stability within the pH 3 to 6 range. Bosmann (1973) isolated two acid proteinases from A. niger having molecular weights of 49,000 and 56,000 and isoelectric points of 4.1 and 4.5 which are similar with those of our enzyme. According to these data, the aspartic proteinase that we have isolated appears to be quite similar in its properties to other Aspergillus acid proteinases.

The properties of C. purpurea aspartic proteinase are also similar to other microbial acid proteinases. Additional experiments regarding specificity will be necessary for detailed classification of the enzymes. The method described for the isolation of acid proteinases was so far not used for microbial proteinases, although often applied to animal tissue proteinases.

ACKNOWLEDGEMENT

We thank Mrs. M. Pregelj, Mrs. S. Turk and Mr. K. Lindič for their excellent technical assistance. This work was supported by the grant from the Research Community of Slovenia.

REFERENCES

Anson, M.L. (1939). The estimation of pepsin, trypsin, papain and cathepsin with haemoglobin. J.Gen.Physiol., 22, 79-89.

Bosmann, H.B. (1973). Protein catabolism II. Identification of neutral and acidic proteolytic enzymes in Aspergillus niger. Biochim.Biophys.Acta , 293, 476-489.

Hartley, B.S. (1960). Proteolytic enzymes. Ann.Rev.Biochem. , 29, 45-72.

Kregar, I., I. Urh, H. Umezawa, and V. Turk (1977). Purification of cathepsin D by affinity chromatography on pepstatin Sepharose column. Croat.Chem.Acta, 49, 587-592.

Lobareva, L.S., and V.M. Stepanov (1978). Microbial carboxyl proteinases. Uspehi Biol.Khimii (USSR), 19, 83-105.

Matsubara, H., and J. Feder (1971). Other bacterial, mold and yeast proteases. In
    P.D. Boyer (Ed.), The Enzymes , Vol. 3 , 3rd ed. Academic Press, New York.
    pp. 723-744.
Murakami, K., and T. Inagami (1975). Isolation of pure and stable renin from hog
    kidney. Biochem.Biophys.Res. Comm., 62, 757-763.
Rehaček, Z., P. Sajdl, J. Kozova, K.A. Malik, and A. Ričicova (1971). Correla-
    tion of certain alterations in metabolic activity with alkaloid production by sub-
    merged Claviceps. Appl.Microbiol., 22, 949-956.
Tsujita, Y., and A. Endo (1978). Purification and characterization of the two mole-
    cular forms of membrane acid protease from Aspergillus oryzae. Europ.J.
    Biochem., 84, 347-353.
Whitaker, J. (1963). Determination of molecular weight of proteins by gel filtration
    on Sephadex. Analyt.Chem.,35, 1950-1953.

Industrial Application

THE VARIETY OF SERINE PROTEASES AND THEIR INDUSTRIAL SIGNIFICANCE

C. Dambmann and K. Aunstrup

Novo Research Institute, 2880 Bagsvaerd, Denmark

ABSTRACT

Industrial use of the serine proteases Subtilisin Carlsberg and trypsin is well established and important to the industry. A large number
of other serine proteases from animal and microbial sources are known
but their industrial use was insignificant, until recently a new group
of highly alkaline serine proteases from alkolophilic <u>Bacillus</u> species
were introduced.

The reasons for this will be discussed with basis in the properties
of the enzymes. Furthermore problems in connection with production of
serine proteases from various sources will be described, especially in
relation to contaminating non-serine proteases. Finally the variation
in microbial proteases will be correlated with the classification of
the enzyme producing microorganism.

KEYWORDS

Proteases, serine proteases, bacillus proteases, medical proteases,
detergent proteases.

INTRODUCTION

Proteolytic enzymes are useful tools in the laboratory, valuable terapeutic agents and they also find large scale application in industry.

Proteases are the most important enzymes, several thousand tons are
produced annually and the commercial value is well over 100 Million
US$.

A list of the most important commercially available proteases is given
in Table 1. It reveals that all types of proteases are of commercial
importance, both sulphydryl enzymes, acid proteases, metalloproteases
and serine proteases, but it is also seen that only one or a few sources of each type is of significance.

TABLE 1   Commercial Protease Preparations of Industrial Use

```
SH-proteases:                                               G-1082
        Papain
Acid proteases:
        Rennin
        Pepsin (bovine, porcine, chicken)
        Mucor miehei, M. pusillus
        Endothia parasitica
        Aspergillus oryzae, A. niger
Metallo proteases:
        Bacillus subtilis
        Bacillus thermoproteolyticus
        Serratia marcescens
Serine proteases:
        Trypsin, chymotrypsin
        Plasmin
        Subtilisin Carlsberg (B. licheniformis)
        Subtilisin Novo (B. amyloliquefaciens)
        Proteases from alkalophilic Bacillus spp.    © NOVO INDUSTRI A/S
```

The economically most important serine protease is Subtilisin Carlsberg derived from Bacillus licheniformis, which is used in detergents. Trypsin is still an important enzyme, indespensable in both medical and technical applications. New highly alkaline proteases from alkalophilic Bacillus species are advancing, while traditional products like Subtilisin Novo from Bacillus amyloliquefaciens and Chymotrypsin are of dwindling importance.

Many other serine proteases, especially of microbial origin have been described in the scientific literature and some have been developed by industry, as revealed by the number of patents, yet they have found no practical application. The reasons for this will be illustrated in the following with basis in the properties of the enzymes.

        TRYPSIN

Crude trypsin preparations, such as Pancreatin, have been produced for many years. They have been used in e.g. the leather industry, the silk manufacturing, the rubber industry, and as digestive agents. From an industrial point of view the most interesting of these applications is that in the leather industry. It is the result of the work of Dr. Otto Röhm, founder and partner of Röhm and Haas, Darmstadt, Germany.

Dr. Röhm was one of the pioneers in modern enzyme technology, and as early as 1905 he made his first important contribution. The bating process, which is used in the production of leather from hides and skins, included a treatment with dog faeces, Dr. Röhm found that the components of the dog faeces, which were active during the bating process were proteolytic enzymes, and that an enzyme preparation made from pancreas could be used in place of dog faeces. This was an industrial improvement, as it made work in the tanneries more agreeable for the workers, and it also made the bating process more reliable, because the pancreatic preparations could be standardized to a certain activity. When this discovery was made, it was only natural to produce the enzymes from pancreas and use them in the bating process.

It is an interesting fact that the first enzymatic detergent also contained Pancreatin, and again Dr. Röhm was the inventor. In 1913 Dr. Röhm was granted a German patent in which he protected the application of tryptic enzymes for washing purposes.

The reason trypsin was selected for these purposes was probably its ready availability. Useful enzyme preparations could be prepared simply by collecting and drying pancreatic glands. Furthermore pH optimum of trypsin was approximately right for the applications.

In 1932, Northrop and Kunitz began their classic investigations of enzymes in the pancreas, during which they crystallized trypsin from ox pancreas. For a long period of time, it was always understood that bovine trypsin was referred to when discussing this enzyme. Only many years later attention was given to trypsin prepared from other animal species, especially porcine trypsin.

From an industrial point of view, the most interesting difference between porcine and bovine trypsin is exhibited by their stability in aqueous solution. At neutral or basic pH, porcine trypsin is considerably more stable than bovine trypsin.

In the 1960s we developed a crystallization process that could be used in large scale and it was therefore possible to introduce crystalline trypsin as a commercial product sold in kilogramme lots.

This preparation has found application in treatment of burns and necrotic wounds, where it will hydrolyse dead tissue and coagulated blood, whereas the living tissue is protected by its content of trypsin inhibitors.

The good stability at neutral pH is here an advantage since it increases the intervals between often painful bandage shifts.

The commercial preparations are often stabilized by the addition of Ca^{++}.

In recent years the pure trypsin preparation has found an important application in preparation of cells for tissue culture. The function is similar to that in would treatment, dissolution of dead material whereas the living is left untouched.

CHYMOTRYPSIN

The other pancreatic serine protease, chymotrypsin has been of minor industrial interest, but it has found an interesting application in cataract operations. Here it is used to weaken the zonular threads in which the lens is suspended, so that the lens can be readily removed.

Chymotrypsin was initially the only enzyme used for this purpose, but it has recently been found that trypsin can be used, too.

As the name indicates, chymotrypsin has good milk coagulating properties, but it cannot be used for cheese manufacture, because the hydrolysis of casein goes too far.

PLASMIN

This serum protease should also be mentioned although it is only produced on a rather limited scale.

The application of Plasmin is based on its strong and specific binding to fibrin, i.e. injected into the blood stream it will bind to any thrombi present.

Due to its proteolytic activity it will dissolve the thrombi, thus being a terapeutic agent.

It can also be used as a diagnostic agent. Thrombi, e.g. in the extremities, can be difficult to localize, but if Plasmin is marked with a shortlife radioactive isotope such as Technetium and injected, it will bind to the thrombus, which can be detected and localized by its emission.

SUBTILISIN CARLSBERG

The original idea of Dr. Röhm to add enzymes to detergents, was many years later, around 1940, taken up by Dr. Jaag in Switzerland. At first his products were based on pancreatic enzymes too, but later he tried microbial proteases in an attempt to improve the effect.

A detergent solution is a complex mixture, rather unfriendly to enzymes It has a high pH, it contains sequestering agents, surfactants and in Europe also sodium perborate as a bleaching agent. So, the requirements to the ideal detergent protease are many and difficult to meet.

In 1959 a new detergent product, BIO 40, based on a protease from B. subtilis, was marketed. This was a considerable improvement, but still the ideal detergent protease had not been found.

Then a year later, a much better protease was introduced, namely Subtilisin Carlsberg, produced by B. licheniformis. It was introduced as a commercial enzyme, ALCALASE®, by NOVO in 1960.

The product achieved commercial success, and during the following years enzyme was added to more and more detergents so that, in 1969, almost 50% of all detergents manufactured in Europe and in the USA contained proteolytic enzymes. Thus, protease from B. licheniformis had become by far the most important microbial enzyme from an economical point of view (Dambmann et al, 1971).

It is estimated that bacillus proteases amount to 35% of the global enzyme sales, and although a number of different bacillus proteases are manufactured, Subtilisin Carlsberg is still the dominating enzyme.

Subtilisin Carlsberg meet most of the requirements to a good detergent protease. It consists of a single peptide chain, and since cystine and cysteine residues are absent, no disulphide bridges are formed, which make it stable against oxidation.

It has a broad specificity and will hydrolyse most types of peptide
bonds, and some ester bonds. Activators are not required, and Ca ions
are not necessary for stability as is the case for many of the serine
proteases. This means that sequestering agent like EDTA or tripoly-
phosphate do not inactivate the enzyme. The enzyme is stable over a
broad pH range. The optimum pH-value depends on the substrate and reaction conditions. A typical pH-stability curve is shown in Fig. 2.
The temperature-stability is shown in Fig. 3, and it is seen that the
enzyme is stable up to a temperature of 50-60° C.

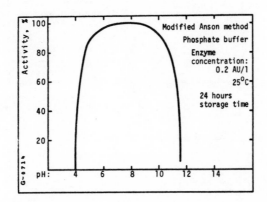

Fig. 2   Influence of pH on the stability of Alca-
         lase

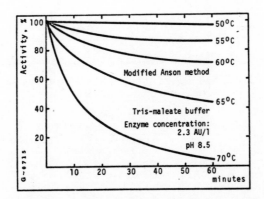

Fig. 3   Influence of temperature on the stability
         of Alcalase

Subtilisin Carlsberg has good activity for hydrolysis of urea denatured haemoglobin between pH 6 and 11 as shown in Fig. 4 and at temperatures as high as 60° C as seen in Fig. 5.

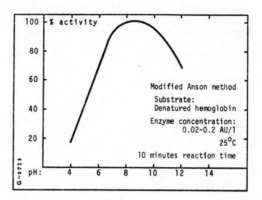

Fig. 4   Influence of pH on the activity of Alcala-
          se

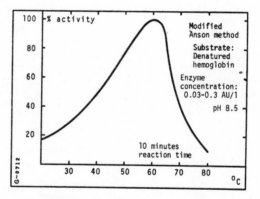

Fig. 5   Influence of temperature on the activity
          of Alcalase

Verbruggen (1975) made a very thorough study of pure and industrial
grade Subtilisin Carlsberg using crossed immunoelectrophoresis. He
concluded that the organism produces two primary protease components,
a major component, which is Subtilisin Carlsberg and a minor unidenti-
fied component, which constitutes only a small fraction of the enzyme
preparations.

This minor component has been isolated and characterized at NOVO (results to be published). It is a non-serine protease with a molecular
weight close to that of Subtilisin Carlsberg, but with different enzyme chemical properties. It is considerably less stable and from an
application point of view it seems to be of no interest.

The interesting thing about this protease is its immunological properties. Serologically it is different from Subtilisin Carlsberg and
it has a higher allergenicity than Subtilisin Carlsberg. Being present
in the commercial B. licheniformis derived protease products, it contributes to their allergenicity.

To enzyme industry it is important to minimize all safety risks. The
allergenicity of many proteins presents a potential safety risk to
those handling enzymes in the factories, and much have been done to
avoid this risk factor, especially by reducing dust levels.

Recently we have succeeded in making a novel B. licheniformis protease product, exhibiting substantially attenuated allergenic proper-
ties.This was done by using B. licheniformis strains, which have been
mutated to block their synthesis of proteases other than Subtilisin
Carlsberg.

The broad specificity of Subtilisin Carlsberg and the limited extent
of hydrolysis, it can exert, make it suitable for another interesting
application, namely the modification of food proteins.

In today's world, with its increasing demand for food proteins, the
interest has naturally been directed against plant proteins, which
can be produced faster and in higher yields than animal proteins. Of
particular interest is soy protein, because of its nutritional value
and favourable amino acid composition, but to make it acceptable in
various foods, it is necessary to change its functional properties.

Application of a protease is an attractive means of obtaining better
functional properties of food proteins without deteriorating their
nutritional value. The degradation of the protein into peptides of
varying chain lengths generally renders the product more soluble,
especially at the isoelectric point, and other functional properties
such as viscosity, emulsification capacity and water absorption capacity are also influenced.

The degree of hydrolysis (DH) is defined as the ratio of the number
of peptide bonds cleaved to the total number of peptide bonds. DH is
a relevant figure, because it can be shown that for a given protein
and enzyme system, the properties of the hydrolyzate are largely determined by its DH-value.

Figure 6 shows hydrolysis curves for ALCALASE$^{®}$ (Subtilisin Carlsberg)
and various proteins.

SUBTILISIN NOVO

This alkaline serine protease is usually present as a side activity
in commercial preparations of Bacillus α-amylase. It was first purified in 1954 by Hagihara, who used the commercial preparation Bacterial Protease Nagarse (B.P.N.) as a source of the protease. He therefore named the enzyme Subtilisin BPN. Later a similar enzyme was isolated from Bacterial Proteinase NOVO, and called Subtilisin NOVO.
Further studies revealed that the two enzymes were identical.

Subtilisin NOVO is produced by the B. subtilis variety now called

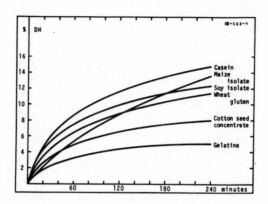

Fig. 6 Hydrolysis of proteins with Alcalase - pH-
stat method

B. amyloliquefaciens. This is primarily an α-amylase producing microorganism, but mutants are known that produces alkaline protease in
good yields. However, the productivity of the available strains is √
still only about 10% of that reported for B. licheniformis.

It is interesting to note that the properties of Subtilisin NOVO in
√ many respects are very similar to those of Subtilisin Carlsberg.

They have the same molecular weight and there is an extensive homology
between them. Activity and stability of different pH-values and temperatures are very similar, and only the specificities are somewhat different.

But this also means that the application of the two enzymes are more
or less the same. This, combined with the higher costs of production
for Subtilisin NOVO caused by the lower yield, makes it understandable
that the market for Subtilisin NOVO is insignificant compared to that
of Subtilisin Carlsberg.

HIGHLY ALKALINE PROTEASES

The commercial success of the detergent proteases in the 1960s initi-
ated a search for alkaline proteases that would act as even better detergent proteases than Subtilisin Carlsberg. The properties that were
looked for were primarily good activity and stability under washing
conditions, i.e. pH 9-10, temperatures above 50⁰ C, and the presence
of various surfactants and sequestering agents.

The search was also extended to microorganisms living under extreme
conditions, and in 1967 it was found that especially stable proteases
were formed by some members of the genus Bacillus, which could grow
at high pH-values (pH 7-11).

Some of these proteases are characterized by having maximum activity
up to pH 12. Fig. 7 shows the activity at different pH values for one

of the highly alkaline proteases, ESPERASE® . They are serine proteases like the subtilisins and consists of single peptide chains free
from disulphide bridges, which means that they are oxidation stable.
Also, the specificity for peptide bond hydrolysis is broad, and they
have esterase activity. The molecular weight is approx. 27,000 and
the iso-electric point about 11.

Of special interest is their compatibility with non-phosphate sequestering agent, such as citrate and nitrilotriacetic acid (NTA).

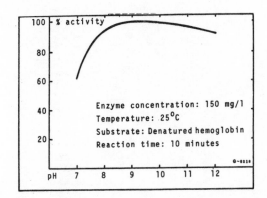

Fig. 7   Activity of Esperase at different pH-val-
         ues

From the above mentioned properties it is obvious that these proteases
are valuable enzymes for incorporating into detergents. But it should
also be mentioned that because of their good activity and stability
at high pH values they have made a new process for enzymic dehairing
possible.

NEW DEVELOPMENTS

For some years the traditional washing habits have gradually changed
towards the use of lower washing temperatures, which is due partly to
the rising energy costs and partly to the increasing use of synthetic
fibres, which can only tolerate moderate temperatures. The consequence
of this is a decreased washing efficiency. This could be compensated
for by the incorporation of enzymes, but due to the lower enzyme activity at lower temperatures the amount of enzyme to be added will
normally economically be prohibitive.

Therefore, the enzyme industry has allocated many resources in order
to find new proteases with a better detergent efficacy at low temperatures than the existing ones.

Recently we have developed a new alkaline protease, which is produced
by a strain of the Bacillus subtilis group. It is still in a development phase, and is temporarily called SP 226.

This new alkaline serine protease is serologically different from Subtilisin Carlsberg and Subtilisin NOVO, which again are serologicall different from each other. These differences can be demonstrated by an Ouchterlony immunodiffusion test as shown in Fig. 8. A serological screening among strains of different species in the Bacillus Subtilis group has shown that the new protease is produced by several strains, and it also showed that none of the Bacilli in this screening produced more than one of the three serotypes of alkaline serine protease.

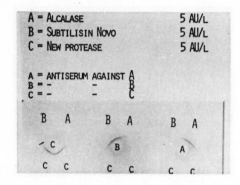

Fig. 8   Ouchterlony immunodiffusion test

It is interesting to compare SP 226 with a Subtilisin Carlsberg based product, like ALCALASE® . They have the same molecular weight and specific activity, and the variation of activity and stability with pH is very similar. The variation of activity with temperature, on the other hand, is clearly different as shown in Fig. 9. Also the highest temperatures at which they are stable, are different. SP 226 is stable up to 50° C, while ALCALASE® is stable up to 60° C.

Another interesting difference is seen in the degree of inhibition by certain stain components. It has been shown that SP 226 is only partly inhibited by egg white and ovomucoid inhibitor (sigma ovomucoid inhibitor 11-0), while Subtilisin Carlsberg is almost 100% inhibited. The results of an experiment, where the inhibition of SP 226 is compared with that of Subtilisin Carlsberg and Subtilisin NOVO, is shown in Table 2.

These results indicate that SP 226 will be a good low-temperature detergent protease and especially when the protein stains contain native egg. Preliminary model washing experiments have shown that SP 226 has a better detergent efficiency at moderate temperatures towards most test swatches, when compared with ALCALASE® . A particular good effect was observed upon soiling with native egg, and washing temperatures below 40° C gave the best results. These results are interesting enoug to justify further investigations of SP 226 as a possible low-temperature detergent protease.

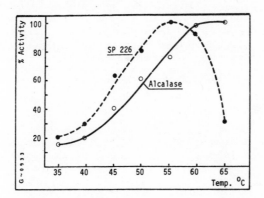

Fig. 9   Effect of temperature on activity of SP
         226 and Alcalase

TABLE 2   Inhibition of SP 226, Subtilisin Carls-
          berg, and Subtilisin Novo by egg white
          and ovomucoid inhibitor

	Residual activity in % after incubation for 30 min.	
G~0901	Egg white diluted in buffer	Ovomucoid inhibitor
SP 226	54	60
Subtilisin Carlsberg	2	4
Subtilisin Novo	18	22
Method of analysis: Hemoglobin method		
pH: 7.5		

PRODUCTION OF SERINE PROTEASE

Pancreas Enzymes

In its simplest form, pancreas proteases are prepared by mincing and
drying defatted pancreas. In this way other pancreas enzymes, primarily amylase will contaminate the preparations and part of the protease will be present in the form of proenzymes, which must be activated before or during use.

Today most preparations are purified by removing insoluble materials
and the protease is activated by autodigestion before drying.

Pancreas is a very rich source of enzymes. The trypsinogen content is about 23% and the chymotrypsinogen content 10-14% of the weight of the protein in the gland.

The production of pancreatic enzymes is a well established process, which now presents few problems. The most important feature of the process is the procurement of pancreatic glands of satisfactory quality.

Bacillus Proteases

The bacillus proteases are produced by submerged cultivation of a suitable strain. The process is now well established and presents no special problems. It is worth noting that the bacteria produce large amounts of enzyme, concentrations of over 5 g pure enzyme protein per liter broth is not unusual.

Recovery of the enzymes is also relatively simple, solids and colloids are removed from the broth, which subsequently is spray dried or the enzyme is recovered by solvent or saltprecipitation.

The major problems in the process are to avoid microbial contamination of the final product and to prepare this in a non-dusting form. The reason for this is the aggressive and irritating effect the proteases have on mucous tissue, and most important their allergenic properties.

Various methods have been developed to make dustfree products and the methods are constantly improved. The dust level at present is so low that it becomes a problem to find analytical methods sensitive enough to detect the dust level.

The methods used in making dustfree preparations are granulation and covering of the grannules with wax, e.g. as is shown in Fig. 10.

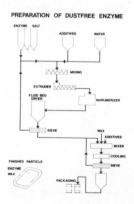

Fig. 10

SERINE PROTEASES FROM OTHER MICROBIAL SOURCES

Serine porteases are known from many microbial genera and species.

Following the success of the detergent proteases in the early nineteen-sixties, a number of these were investigated more thoroughly in order to find an enzyme, which would be superior to Subtilisin Carlsberg.

The most obvious genus to search is the genus Bacillus. In this more than 15 species produce serine protease, but apart from those already mentioned, none proved better than Subtilisin Carlsberg. A very thermostable protease has been found in Bacillus caldolyticus, it has a temperature optimum at 70° C but due to extremely low yields and difficulties in cultivating the organism, the enzyme was never commercially developed.

A complex of proteases can be derived from many streptomyces species. This complex also includes serine proteases and some of these preparations have been commercialized to a limited extent.

One example is the enzyme complex from Streptomyces griseus, marketed under the name Pronase.

A somewhat similar enzyme complex from Streptomyces fradiae was marketed by Merck as a dehairing agent for use in the leather industry. The preparation possesses keratinolytic activity. The preparation was not commercially successful, probably because of a combination of too high cost, too low quality of the leather produced, and difficulties in control of the dehairing process.

In Japan the Ajinomoto Company developed an alkaline serine protease from the thermophilic Streptomyces rectus. The enzyme showed promising properties with regard to stability and activity in detergents, but it was never commercialized.

Serine proteases are also found in many fungi but none are used commercially apart from the activity found in the protease complexes produced by Aspergillus oryzae.

Enzyme preparations of this type are primarily used because of their content of acid protease and in most cases they are used at low pH, where the serine protease has low activity.

SERINE PROTEASES AND TAXONOMY

Initially the classification of the protease producing bacillus species was unclear and little attention was paid to the correct name of the enzyme source. Concequently many publications leave one with the impression that one enzyme, e.g. Subtilisin Carlsberg is produced by several species. This is not in agreement with the studies on trypsin from animal species, where it has been observed that trypsin is species specific and that the difference in trypsins from various species reflect the relatedness of the species.

At NOVO we have started studies of the Bacillus species, which produce protease. So far we have found that all strains of B. licheniformis, which we have been able to obtain, produce Subtilisin Carlsberg, if they produce extracellular protease. We have obtained similar results with limited numbers of strains belonging to other species of the genus.

The method used to identify the protease was crossed immunoelectro- phoresis in monospecific antiserum towards the protease in question.

When the antiserum has been prepared, the method is quick and gives a clear cut answer.

It is expected that this method will become a useful tool in the clas- sification of <u>Bacillus</u> sp.

CONCLUSION

Originally the serine proteases for industrial use, were obtained from ready available sources, such as pancreas glands, accepting that the properties were not always optimal.

During the 1950s it became obvious that serine proteases with special properties were needed. To the enzyme producers it was then natural to look at the microbial proteases, both because of the large variety of microbial enzymes and because of the new fermentation techniques de- veloping at that time.

Today it seems that no further serine proteases for industrial use are needed. But if new proteases with special properties become inter- esting we feel confident that enzyme industry will be able to develop such products.

REFERENCES

Dambmann, C., P. Holm, V. Jensen, and M.H. Nielsen (1971). <u>Developments in Industrial Microbiology</u>, <u>12</u>, 11.
Verbruggen, R. (1975). <u>Biochem.J.</u>, <u>151</u>, 149-155.

EXOGENOUS PROTEINASES IN DAIRY TECHNOLOGY

P.F. Fox

Department of Food Chemistry,
University College, Cork, Ireland

SUMMARY

The principal applications of proteinases in dairy technology are in cheese manufacture. The enzymatic primary phase and non-enzymatic secondary phases of rennet coagulation of milk are reviewed. Aspects of veal rennet substitutes are briefly discussed and developments in immobilized rennets considered in detail. The possibility of accelerating cheese ripening <u>via</u> added proteinases is also considered.

Minor applications of proteinases including production of protein hydrolyzates, protein modification and baby food manufacture are reviewed.

KEYWORDS

Rennet coagulation, rennet substitutes, immobilized rennets, accelerated cheese ripening, protein hydrolysis, protein modification.

The Proteins of Milk

Bovine milk contains $\sim$ 3.5% protein of which $\sim$ 80% is casein. Casein is a heterogenous group of four principal phosphoproteins and phosphoglycoproteins ($\alpha_{s1} \sim$ 40%; $\alpha_{s2} \sim$ 10%; $\beta \sim$ 40%; $\kappa \sim$ 10%) which have remarkably high heat stability but which are insoluble in their isoelectric region (i.e. $\sim$ pH 4.6). In addition there are normally small amounts of γ-caseins, proteose-peptone and λ-caseins derived from α_{s1}- and β-caseins by postsecretion proteolysis by indigenous milk proteinase(s). The remaining 20% of milk protein constitutes the whey protein or non-casein-nitrogen fraction which is heat-labile [complete denaturation in 5 min at 90°C (Lyster, 1970)]. It consists of four principal proteins: β-lactoglobulin ($\sim$ 50%), α-lactalbumin ($\sim$ 20%), blood serum albumin ($\sim$ 10%) and lactoglobulin ($\sim$ 10%) and several minor proteins.

In addition to differences in heat stability and solubility at pH 4.6, the most significant difference between the caseins and the whey proteins is their coagulability by rennets and their natural colloidal form. The whey proteins occur naturally as monomers or low degree oligomers whereas the caseins exist as micelles of average molecular weight $\sim 10^8$, average diameters $\sim$ 100 nm and containing 500-1,000 monomers. In the micelle, monomers associate first <u>via</u> hydrogen and/or hydrophobic bonds to form sub-micelles of average molecular weight $\sim 2 \times 10^6$ and average diameters of 10-15 nm. Sub-micelles aggregate to form micelles which are held together principally by calcium and phosphate ions or molecules [commonly referred to as colloidal calcium phosphate (CCP), which represents $\sim$ 6% of the dry weight of the micelles] with contributions from hydrogen and/or hydrophobic bonds. The micelles are highly solvated ($\sim$ 2 g H_2O/g protein) and are normally thermodynamically very stable. The principal stabilizing factors are CCP and κ-casein. Removal of CCP by acidification or calcium chelators disrupts the micelles and reduces stability to Ca^{2+}.

Although κ-casein is quantitatively a minor constituent it is the principal micelle stabilizer. κ-Casein is an amphophilic molecule, the amino terminal $\frac{2}{3}$ of which is hydrophobic while the C-terminal segment, rich in carbohydrate and polar amino acids, is strongly hydrophilic. It is probably concentrated on the micelle surface (although views on this are by no means unanimous) with its hydrophilic region exposed to the surrounding aqueous medium, thereby promoting stability. Many proteinases cleave off the C-terminal region of κ-casein rendering the residual micelle unstable and coagulable in the presence of Ca^{2+} at temperatures > 20°C. This is the first characteristic step in the manufacture of most cheese varieties and will be discussed in greater detail later.

Major reviews and texts on milk proteins include McKenzie (1971), Lyster (1972), Swaisgood (1973), Farrell and Thompson (1974), Whitney and co-workers (1976), Schmidt and Payens (1976), Slattery (1976), Davies and Holt (1979).

Proteinases in Dairy Technology

Because of its protein content, the physical state of the protein, and the general environmental conditions, proteinases, more than any other enzyme group, with the possible exception of lipases, play a major role in dairy technology. Proteinases in milk and dairy products originate from three distinctly different sources:

Indigenous proteinase(s): a constituent part of the milk as excreted, is responsible for post-secretion modification of milk proteins and possibly processing characteristics. The literature has been reviewed by Humbert and Alais (1979).

Endogenous proteinase: proteinases secreted *in situ* by microorganisms in post-secretion milk or dairy products. The most important of these are the proteinases secreted by psychrotrophic bacteria which cause spoilage in milk and dairy products (cf. Cogan, 1977; Law, 1979) and the proteinases secreted by starter and non-starter microorganisms in ripening cheese and which are primarily responsible for the formation of small peptides, amino acids and flavour precursors (O'Keeffe, Fox and Daly, 1976, 1978; Gripon and co-workers, 1977; Visser and de Groot-Mostert, 1977).

Exogenous proteinases: proteinases from external sources added to milk or dairy products to induce specific changes.

This review will be concerned exclusively with the use of exogenous proteinases in dairy technology.

EXOGENOUS PROTEINASES

In dairy technology the use of exogenous proteinases is dominated by rennets which is financially the most important group of enzymes used in the entire food industry (Skinner, 1975). In addition there are currently some minor applications and some potentially large applications still at the experimental or developmental stages.

RENNETS

Cheese is manufactured essentially as summarized in Fig. 1.

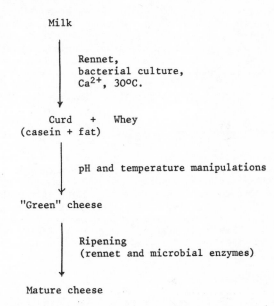

Fig. 1. Summary flowsheet of cheese manufacture

Rennet might be defined as "a crude preparation of any proteinase suitable for cheese manufacture". Traditional rennets are generally prepared by salt (NaCl) extraction of stomachs of milk-fed calves. Lamb and kid rennets are also used as coagulants in several countries but are not commercially available (Anifantakis and Green, 1980). Chymosin is the principal proteinase in rennets from very young animals but the proportion of pepsin increases as the animal matures. Changes in the slaughter pattern of young calves and increasing world-wide production of cheese have led to a shortage of veal rennet and given rise to an active search for substitutes (Nelson, 1975). At present four substitutes (bovine pepsin, porcine pepsin and acid proteinases from *Mucor meihei* and *Mucor pusillus*) are fairly widely used with acceptable results under certain circumstances (Nelson, 1975; Green, 1977).

All the commonly used milk-clotting enzymes belong to the acid proteinase family (Foltmann, 1978), all members of which possess fundamentally the same catalytic centre with two essential aspartate residues. In the tertiary structure, these two residues are buried in a cleft that may serve as a binding site for a peptide containing at least six amino acids. The chemistry, structure and function of the acid proteinases has been well studied and reviewed (e.g. Hofmann, 1974; Tang, 1977). In addition, the literature on chymosin has been reviewed by Foltmann (1966, 1971), pepsin (pig) by Fruton (1971) and milk coagulants generally by Ernstrom (1974) and Green (1977).

Considerable progress has been made on the primary structures of the principal milk clotting enzymes. The complete sequences of porcine pepsin (Sepulveda and co-workers, 1975) and calf chymosin (Foltmann and co-workers, 1979) have been established. The sequence of the first 65 residues of bovine pepsin is known (Harboe and Foltmann, 1975) as well as some additional shorter segments (Foltmann, 1978). The sequence of substantial segments of *M. meihei* acid proteinase has been established (Foltmann, 1978) but no information appears to be available on the sequence of *M. pusillus* proteinase. Penicillopepsin, which is not used as a milk coagulant, has also been fully sequenced (Hsu and co-workers, 1977). The sequences of these acid proteinases show a high degree of homology (Foltmann and Pedersen, 1977). Tertiary structures for porcine pepsin, calf chymosin and proteinase from *Endothia parasitica* have been proposed (cf. Tang, 1977).

Enzymatic Coagulation of Milk

The rennet coagulation of milk occurs in two distinct phases: a primary enzymatic phase in which the protective κ-casein is specifically hydrolyzed thereby loosing its stabilizing ability and a secondary non-enzymatic phase during which the enzyme-altered micelles aggregate and gel under suitable conditions. A tertiary phase may be designated during which a slow general proteolysis of the casein occurs, in practice, during cheese ripening. General reviews on the topic include Lindqvist (1963), Garnier and co-workers (1968), Mackinlay and Wake (1971) and Ernstrom (1974).

Enzymatic Phase

Linderstrom-Lang postulated in 1929 that rennet coagulation of milk involved enzymatic hydrolysis of a protective colloid; this view was substantiated when Waugh and von Hippel (1956) demonstrated that κ-casein was the primary target of chymosin action and was confirmed by Wake (1959). Due mainly to heterogeneity of κ-casein with respect to carbohydrate (Mackinlay and Wake, 1965) a heterogenous mixture of peptides is split from milk or κ-casein (Alais, 1956, 1963). Following many unsuccessful attempts, the peptide bond of κ-casein hydrolyzed during the enzymatic stage was identified by Delfour and co-workers (1965) as that between the phenylalanine and methionine residues 105-106; this was confirmed by Jolles, Alais and Jolles (1968). This period in the study of rennet coagulation was comprehensively reviewed by Mackinlay and Wake (1971). Other acid proteinases used as rennets probably also hydrolyse this bond (cf. Ernstrom, 1974; Green, 1977).

Hydrolysis of κ-casein between residues 105 and 106 produces a hydrophobic N-terminal peptide referred to as para-κ-casein and representing $\sim \frac{2}{3}$ of the molecule and a hydrophilic "macropeptide" from the C-terminal region which contains most of the carbohydrate of κ-casein. The reaction may be conveniently followed by the formation of 2 or 12% TCA-soluble N (Nitschmann and Bohren, 1955) or, more specifically, by the release of TCA-soluble N-acetyl neuraminic acid (Nitschmann, Wissmann and Henzi, 1957; de Koning, Jenness and Wijnand, 1963).

The most sensitive indicator of rennet action on κ-casein is its loss of solubility at pH 5.2 and 5°C (Bingham, 1975).

The pH optimum for the enzymatic stage of rennet coagulation is 5.1-5.5, being somewhat dependent on the reaction conditions (Humme, 1972). This value is considerably higher than the pH optimum exhibited by chymosin on other protein substrates: 3.5 on acid-denatured haemoglobin (Foltmann, 1964), 4.0 on the B-chain of insulin (Fish, 1957) or 3.7-4.0 on phenyl sulphite esters (Hill, 1969), but is close to the value of 4.7 for the hydrolysis of H.Ser-Leu-Phe-Met-Ala-OMe (Hill, 1969), H.Leu-Ser-Phe(NO$_2$)-Leu-Ala-OMe (Raymond and co-workers, 1972), H.Leu-Ser-Phe-Nle-Ala-Ile-OMe (de Koning, van Rooijen and Visser, 1978) or H.Leu-Ser-Phe(NO$_2$)-Nle-Ala-Leu-OMe (Raymond and Brias, 1979).

The enzymatic hydrolysis of κ-casein continues, although at a reducing rate, down to 0°C (Berridge, 1942); the $Q_{10^\circ C}$ is 1.8-2.0 over the temperature range 0-37°C. The temperature optimum in milk cannot be determined because of overlap with the highly temperature-dependent second stage of coagulation. The K_{cat} for the hydrolysis of κ-casein by chymosin increased up to 40°C but the Km remained essentially constant over the range 25-40°C (Garnier, 1963). The tempeaature optimum for the hydrolysis of small peptides does not appear to have been determined but 30 or 35°C have normally been used with these substrates.

The primary phase of rennet action on κ-casein is stimulated by CaCl$_2$, MgCl$_2$, NaCl, EDTA and orthophosphates (Kato and co-workers, 1970) and inhibited by α_s- and β-caseins (Mikawa, Kato and Yasui, 1973). Heating milk prolongs both the primary and secondary phases of rennet coagulation (Kannan and Jenness, 1961; Morrissey, 1969) and if the treatment is sufficiently severe, prevents coagulation. The inhibition of the primary phase is due to interaction between κ-casein and β-lactoglobulin (cf. Sawyer, 1969; Shalabi and Wheelock, 1977) which possibly causes physical shielding of the susceptible bond of κ-casein by denatured β-lactoglobulin.

The Phe-Met (105-106) bond of κ-casein is several orders of magnitude more sensitive to proteolysis than any other bond in the milk protein system. [It is also very sensitive to cleavage by LiBH$_4$ (Jolles, Alais and Jolles, 1963) and heat (Alais, Kiger and Jolles, 1967)]. The Phe-Met bond in Phe-Met-OMe, Phe-Met-Ala-OMe or Leu-Phe-Met-Ala-OMe is not hydrolysed at an appreciable rate (Hill, 1968) but it is hydrolysed in Ser-Leu-Phe-Met-Ala-OMe although 2-3 orders of magnitude more slowly than in κ-casein. The presence of adjacent serine and histidine residues appears to be essential for the susceptibility of the Phe-Met bond and it is further sensitized by increasing polypeptide chain length (Hill, 1968, 1969; Raymond and co-workers, 1972; Raymond and Brias, 1979). Seryl and histidyl residues are at the active center of many hydrolases and Hill (1968) suggests that in κ-casein these catalytic residues are located on the substrate itself; the coagulant might serve only to induce conformational changes in the substrate necessary for the activation of these residues.

Methionine may be replaced by Nle without significant effect on chymosin activity (Raymond and co-workers, 1972) but photo-oxidation of methionine reduces the sensitivity of the Phe-Met bond. Thomas (1972) suggests that methionine-106, which should normally be buried in a non-polar region, occupies an exposed position in κ-casein rendering it accessible to chemical and physical probes.

The use of synthetic peptides has at least two practical aspects:

1) Definition of chymosin units: the activity of rennets is normally determined and expressed as its milk clotting activity measured as described, for example, by the British Standards Institution (1963). Variability of the substrate makes

inter- and even intra-laboratory comparisons of rennet activities difficult; synthetic peptides provide a much more precisely defined substrate (cf. Salesse and Garnier, 1975).

2) Quantitation of individual enzymes in mixtures of coagulants: most commercial rennets consist of mixtures of proteinases and it is desirable to have methods available to quantify such enzymes. Several methods have been proposed (cf. de Koning, 1979), including the relative milk clotting to proteolytic activities on synthetic peptides e.g. the ratio of clotting to proteolytic activities on H.Leu-Ser-Phe(NO_2)-Nle-Ala-Leu-OMe is $\sim$ 30 times higher for chymosin than for bovine pepsin II (Raymond and co-workers, 1973); on H.Leu-Ser-Phe-Nle-Ala-Ile-OMe the ratio is $\sim$ 10 times higher for chymosin than for bovine pepsin (de Koning and co-workers, 1978); H.Phe-Gly-His-Phe(NO_2)-Phe-Ala-Phe-OMe which is a good substrate for (pig) pepsin but not for chymosin and other milk-clotting enzymes (Voynick and Fruton, 1971), has been proposed as means of quantifying pig pepsins in commercial rennets (Salesse and Garnier, 1975).

Second Stage Rennet Coagulation

The actual coagulation of rennet-altered micelles is a non-enzymatic process. Loss of the hydrophilic C-terminal segment of κ-casein reduces micellar zeta potential (Green and Crutchfield, 1971; Green, 1973; Pearce, 1976) and pre-disposes it to coagulation by Ca^{2+}; interaction of micelles with polyvalent cations also pre-dispose micelles to coagulation (Green and Marshall, 1977). The coagulation process is absolutely dependent on Ca^{2+} and colloidal calcium phosphate (Pyne, 1953; Pyne and McGann, 1962). It does not normally occur < $\sim$ 18°C; its very high temperature-dependence ($Q_{10°C} \sim 16$; Berridge, 1942) permits the primary and secondary phases to be isolated for independent study and also forms the basis of most attempts to render cheese curd production continuous (Berridge, 1976). Pyne (1955) claimed that pH had essentially no effect on the coagulation stage but Kowalchyk and Olson (1977) showed a slight inverse relationship to pH, suggesting that coagulation was more dependent on hydrophobic than on electrostatic interactions. Using immobilized rennets, believed to be free of soluble enzymes, Cheryan and co-workers (1975 a) claimed that pH mainly affected the secondary stage of rennet coagulation with a relatively minor effect on the enzymatic phase; however as will be discussed later, it appears likely that coagulation in these experiments was due to free rather than to immobilized enzyme.

The coagulation of rennet-altered micelles can be conveniently monitored by viscometry, (Scott Blair and Oosthuizen, 1961; Green and co-workers, 1978); light scattering (Payens, 1976, 1978; Dalgleish, 1979); electron microscopy (Green and co-workers, 1978) and oscillatory deformation (Kowalchyk and Olson, 1978) and the thermobelastograph (Tarodo, Alais and Frentz, 1969). Viscosity and turbidity decrease initially but beyond a critical stage both parameters increase rapidly with the onset of aggregation and gelation. The gel assembly process, as observed by electron microscopy, has been described by Green and co-workers (1978). It has long been recognised that coagulation of rennet-altered micelles does not commence until the enzymatic stage is substantially complete (Foltmann, 1959; Castle and Wheelock, 1972), although with increasing temperature and decreasing pH the second stage commences before completion of the enzymatic phase. Coagulation does not commence until the enzymatic stage is 86-88% complete overall (Green and co-workers, 1978; Dalgleish, 1979) and an individual micelle will not participate in aggregation until 97% of its κ-casein has been hydrolyzed (Dalgleish, 1979).

The coagulation of rennet-altered micelles appears to occur by a von Smoluchowski-type mechanism (Payens, Wiersma and Brinkhuis, 1977; Payens, 1977, 1979) and the overall rennet coagulation time of milk is thus determined by 3 factors (Dalgleish, 1979):

1) the rate of κ-casein proteolysis by a Michaelis-Menten mechanism,
2) the probability that an individual micelle has sufficient of its κ-casein removed to allow it to aggregate,
3) the rate of the von Smoluchowski aggeegation.

The overall clotting time, t_c, is equal to the sum of the enzymatic phase and the aggregation phase (Dalgleish, 1980):

$$t_c = t_{prot} + t_{agg}$$

$$\sim \frac{Km}{Vmax} \ln\left(\frac{1}{1-\alpha_c}\right) + \frac{\alpha_c}{Vmax} \cdot S_o + \frac{1}{2 \, k_s C_o}\left(\frac{Mcrit}{M_o} - 1\right)$$

where Km and Vmax are the Michaelis-Menten parameters, α_c is the extent of κ-casein hydrolysis, S_o is the initial concentration of κ-casein, k_s is the rate constant for aggregation, C_o = concentration of aggregating material, Mcrit = weight average M.W. at t_c ($\sim$ 10 micellar units), M_o = weight average M.W. at t = o.

Payens (1979) argues that the observed lag in the coagulation of rennet-altered micelles is not due to an energy barrier of the Deryagin, Landan, Verwey, Oberbeek-type but to a "steric factor" which results in a number of unsuccessful collisions of altered micelles. Payens suggests that rennet-altered micelles can aggregate only at certain surface sites and in this way build up to a gel matrix. The nature of these interaction sites is unknown but some indication may be deduced from the results of rather limited studies of the effects of chemical modification of casein on its rennet coagulability.

It has been traditionally held that gelation of rennet-altered micelles occurs through Ca^{2+} acting to cross-link micelles via organic phosphate groups (McFarlane, 1938). However, casein that had been enzymatically dephosphorylated to the extent of 70% was coagulable by rennet provided the [Ca^{2+}] was increased suggesting that additional factors were involved in coagulation (Hsu and co-workers, 1958).

Photooxidation, in which histidine was the prime target, inhibited rennet coagulation (Zittle, 1965). Hill and Laing (1965) confirmed that oxidation of histidine inhibited both the primary and secondary phases but modification of tryptophan, methionine and probably tyrosine affected neither phase. The secondary phase of rennet coagulation, but not the primary phase, was inhibited by modification of 2-3 lysine residues/molecule of κ-casein with dansyl chloride (Hill and Craker, 1968); modification of α_s- or β-caseins did not affect their stabilization by κ-casein or the coagulation of κ-α_s or κ-β-casein systems. Dansylation increases the net negative charge on proteins making it more difficult for the micelles to interact. Modification of 1.5 arginine residues/mole κ-casein by treatment with glyoxal also prevents aggregation of rennet-altered micelles (Hill, 1970). It was suggested (Hill, 1970) that histidine, lysine and arginine form a positively charged cluster in the κ-casein sequence near the rennet-susceptible bond (a cluster of histidine has been demonstrated, Mercier, Brignon and Ribadeau-Dumas, 1973); in intact κ-casein this cluster is shielded by the macropeptide but is exposed following rennet action, allowing electrostatic interaction with negatively charged regions on neighbouring proteins or micelles leading to aggregation.

Rennet-altered (para) κ-casein aggregates in the absence of Ca^{2+} (Wake, 1959) but self-aggregation is strongly inhibited by low levels of α_{s1}- and β-caseins (Lawrence and Creamer, 1969; Kato and co-workers, 1980) and stimulated by very low concentrations of α_{s1}- and β-caseins (Kato and co-workers, 1980). The significance of these observations to the coagulation process is not clear but it probably has significance in gel formation. Kato and co-workers (1980) also showed that glutamic or aspartic acids promote self-association of para-κ-casein supporting the views of Hill (1970) on the importance of interactions between positive and negative groups in the coagulation of rennet-altered casein micelles.

Action of Rennet in Cheese

Most (∿ 90%) of the rennet added to milk in cheese-making is lost in the whey at draining; the actual amount of rennet, especially chymosin, retained in the curd is strongly influenced by the pH of the curd at draining (Holmes, Duersch and Ernstrom, 1977). The rennet retained by the curd is essential for cheese ripening as it is primarily responsible for the formation of large peptides (O'Keeffe, Fox and Daly, 1976, 1978 ; Visser, 1977 a, b). Proteolysis is essential for the development of proper cheese texture (de Jong, 1976). Small peptides and amino acids contribute to cheese flavour (Mabbitt, 1961) but excessive or incorrect proteolysis leads to off-flavour development, especially bitterness (cf. Lowrie and Lawrence, 1972).

Chymosin, the reference milk coagulant, shows a high degree of specificity for peptide bonds adjacent to, preferably between a pair of, hydrophobic residues. Its specificity on the B-chain of insulin is somewhat narrower than that of pig pepsin (Bang-Jensen, Foltmann and Rombauts, 1964) as is that of bovine pepsin (Pedersen, 1977). In fact all the principal coagulants have similar specificites on insulin (Green, 1977). Chymosin sequentially hydrolyses bonds 189-190, 163-164 and 139-140 of β-casein in solution to yield β-I, β-II and β-III respectively (Creamer, Mills and Richards, 1971; Pelissier, Mercier and Ribadeau-Dumas, 1974; Creamer, 1976; Visser and Slangen, 1977). Additional bonds adjacent to these sites (192-193, 165-166, 167-168) may also be hydrolysed to yield peptides indistinguishable from β-I and β-II by gel electrophoresis. At low pH (2-3), bond 127-128 is also hydrolysed to yield β-IV (Visser and Slangen, 1977; Mulvihill and Fox, 1978).

α_{s1}-Casein has at least 25 sites for chymosin action (Pelissier and co-workers, 1974) but many of these are hydrolyzed only slowly. The most susceptible bond is Phe 23-Phe 24 (Hill, Lahav and Givol, 1974) or Phe 24-Val 25 (Creamer and Richardson, 1974) hydrolysis of which yields α_{s1}-I. The specificity of chymosin is markedly dependent on the experimental conditions (Mulvihill and Fox, 1977, 1979 a, 1980). α_{s1}-I Casein is produced under all conditions investigated but its further breakdown is dependent on the environment, Fig. 2. The segments of α_{s1}-casein represented by some of these peptides are shown in brackets.

Para-κ-casein is hydrolyzed only very slowly by chymosin (Mulvihill and Fox, unpublished) and the susceptible bonds have not been identified. The α_{s2}-caseins are also very resistant to proteolysis (Guiney, 1972) and the hydrolysis products have not been characterized. The hydrolysis of the γ-caseins and proteose-peptone has not been investigated.

Bovine and porcine pepsins are more proteolytic than calf chymosin on isolated bovine β-casein but have similar specificities (Mulvihill and Fox, 1979 b). Both pepsins rapidly hydrolyze α_{s1}-casein to α_{s1}-I which is quite resistant to further proteolysis and the peptides that are produced have different electrophoretic mobilities from the corresponding chymosin peptides (Mulvihill, 1978). Possible effects of pH and NaCl on the specificity of pepsins has not been assessed but the activity of pepsins is affected by NaCl in a manner similar to chymosin.

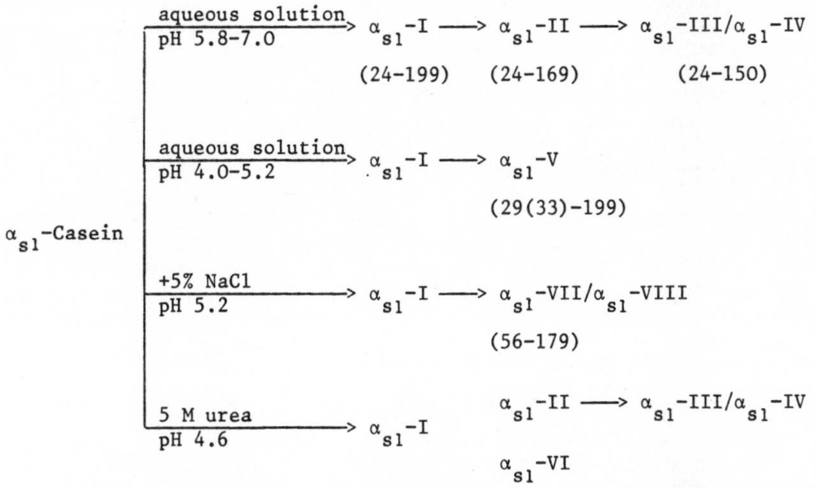

α_{s1}-Casein

$\xrightarrow[\text{pH 5.8-7.0}]{\text{aqueous solution}}$ α_{s1}-I $\longrightarrow$ α_{s1}-II $\longrightarrow$ α_{s1}-III/α_{s1}-IV

(24-199)   (24-169)   (24-150)

$\xrightarrow[\text{pH 4.0-5.2}]{\text{aqueous solution}}$ α_{s1}-I $\longrightarrow$ α_{s1}-V

(29(33)-199)

$\xrightarrow[\text{pH 5.2}]{\text{+5\% NaCl}}$ α_{s1}-I $\longrightarrow$ α_{s1}-VII/α_{s1}-VIII

(56-179)

$\xrightarrow[\text{pH 4.6}]{\text{5 M urea}}$ α_{s1}-I

α_{s1}-II $\longrightarrow$ α_{s1}-III/α_{s1}-IV

α_{s1}-VI

Fig. 2. Proteolysis sequence of α_{s1}-casein by chymosin under various environmental conditions

The proteolytic activity of the microbial rennets is greater than that of chymosin but their specificities on caseins have not been established. However, electrophoretic comparison of the peptides produced from casein by animal and microbial rennets indicates considerable differences in specificity (Edwards and Kosikowski, 1969; Vanderpoorten and Weckx, 1972; El-Shibiny and Abd El-Salam, 1976, 1977).

The susceptibility of caseins to proteolysis by chymosin and pepsins is highly dependent on the degree of aggregation; little proteolysis occurs in intact micelles but disruption of the micelles by removal of colloidal calcium phosphate renders the component caseins susceptible to proteolysis (Fox, 1970; O'Keeffe, Fox and Daly, 1975). The susceptibility of β-casein to proteolysis, relative to that of α_{s1}-casein, is greater at low than at high temperatures (Fox, 1969), probably due to the temperature-dependent aggregation of β-casein (Payens and van Markwijk, 1963; Garnier, 1966). The susceptibility of β-casein to proteolysis by chymosin or pepsins is highly dependent on water activity (A_w) and is inhibited by low (5%) concentrations of NaCl (Fox and Walley, 1971) and high sucrose concentrations (Creamer, 1971) while α_{s1}-casein is optimally hydrolysed at ∿ 5% NaCl.

Chymosin or pepsins do not hydrolyze β-casein in cheese (Ledford, O'Sullivan and Nath, 1966) possibly because of low A_w due in part to its NaCl content (Phelan, Guiney and Fox, 1973). β-Casein is however slowly hydrolyzed by indigenous milk proteinase (Creamer, 1975). α_{s2}-Casein and para-κ-casein are not hydrolyzed in cheese. α_{s1}-Casein in cheese is completely degraded, initially to α_{s1}-I and later to α_s-V and α_{s1}-VII with small amounts of α_{s1}-II (Mulvihill and Fox, 1980). The smaller peptides resulting from the action of the coagulant appear to be degraded further by microbial proteinases and peptidases (O'Keeffe and co-workers, 1976, 1978) but the products have not been characterized.

Rennet Substitutes

As previously indicated, increasing worldwide cheese production with a concomitant reduction of calf slaughterings has led to a shortage of veal rennet and a search for rennet substitutes. Almost all proteinases are capable of coagulating milk but most are unsuitable as rennets because they (1) are too proteolytic or produce the wrong proteolytic pattern thereby causing flavour and textural defects and reduced yields; (2) have pH optima different from that of milk and (3) are too heat stable and survive in processed whey causing problems in foods incorporating whey (Hyslop, Swanson and Lund, 1979; Thunell, Duersch and Ernstrom, 1979; Thunell, Ernstrom and Hartman, 1980). To date only porcine and bovine pepsins and proteinases from *M. pusillus, M. meihei* and *Endothia parasitica* have been used commercially in long-ripened cheeses. Nelson (1975) claims that not more than 15% of all cheese in the U.S. is now made with rennets of bovine origin; rennet-pig pepsin blends were used in 40% of cheese and Mucor-derived enzymes in a further 40%. It is likely that the usage of rennet substitutes, especially those of microbial origin, is considerably less in Europe.

Rennet substitutes have been the subject of several reviews including Nelson (1975), Green (1977) and Martens and Naudts (1978) and as they will be discussed by others at this conference, it is not intended to discuss the matter further. It is worth noting the considerable interest now being generated in the possibility of transferring the genetic information for chymosin biosynthesis from calf cells to microorganisms. If successful, these investigations will have major consequences for both the dairy and enzyme industries. Because the various rennets now available are not equally acceptable to cheese makers or equally priced, several methods have been developed to identify and quantify the type(s) of enzyme (cf. Green, 1977; Martens and Naudts, 1978; de Koning, 1979).

Immobilized Rennets

Not surprisingly there has been considerable recent interest in the development and use of immobilized rennets with one of five objectives in mind: (1) as a technique for the separation of the first and second stages of rennet coagulation to permit their individual investigation; (2) as a technique for identifying the micellar location of κ-casein; (3) as a technique to produce coagulant-free cheese for investigations on the role of starter and non-starter proteinases in cheese ripening; (4) as an enzyme-saving technique in cheese manufacture, and (5) as a component in a system for the continuous production of cheese curd. Reviews on immobilized rennets include Olson and Richardson (1975), Taylor, Richardson and Olson (1976), Olson and Richardson (1979), and Cheryan (1978).

The first attempt to immobilize rennets appears to be that of Green and Crutchfield (1969) who obtained encouraging results with chymosin-agrose but leaching of soluble enzyme from the complex was considerable and the coagulation of milk was attributed entirely to soluble enzyme. This conclusion is supported by Arima and co-workers (1974). Brown and Swaisgood (1975) report the successful immobilization of chymosin on porous glass but supply little information. Jansen and Olson (1999) reported that papain insolubilized by cross-linking with gluteraldehyde could coagulate milk but little information on the milk-clotting properties was supplied. Superpolymerized papain was also used by Ashoor and co-workers (1971) to investigate casein micelle structure.

Pig pepsin immobilized on porous glass was reported by Ferrier and co-workers, (1972) to be capable of catalyzing the enzymic phase of milk coagulation; tests showed that catalysis was not due to soluble enzyme. Milk at pH 5.6 or 5.9 and 15°C was passed through the column and then warmed to 30°C to yield a typical curd. The main problems encountered were plugging of the column due to adsorption of protein and loss of enzymatic activity.

The enzymatic stage of coagulation in a reactor column was further studied by Cheryan and co-workers (1975 a) who ascertained that solubilization of enzyme did not occur although some of the data are somewhat disconcerting in this regard. For example, data are presented showing that renneted (immobilized pepsin) milk mixed with up to 2.5 times its volume of raw milk coagulates slowly which was interpreted as indicating that the κ-casein of only a proportion of the micelles present need by hydrolyzed for gelation to occur. No evidence is presented to show that intact micelles were present in the gel or that enzymic hydrolysis of κ-casein did not occur in the mixture, coagulation of which might have been due to the presence of small amounts of solubilized coagulant. However, a generally similar result was reported by Angelo and Shahani (1979) for rennet-Sepharose-4B.

In contrast to the experience with soluble rennets the release of NPN by immobilized pepsin did not plateau with time (Hicks and co-workers, 1975) and while RCT was inversely related to the level of NPN released by pepsin-glass, coagulation did occur over a wide range of NPN values. Unfortunately the continued release of NPN after the milk had come from the column was not measured. Ultracentrifuged skim milk supernatant (UCSMS) which has been passed through a pepsin-glass column was capable of coagulating untreated skim milk or redispersed casein micelles while simulated milk ultrafiltrate (SMUF) similarly treated was unable to do so. This result was interpreted as supporting the casein micelle model proposed by Parry and Carroll (1969) in which rennet coagulation of milk was considered to be due to para-κ-casein formed from serum κ-casein. The failure of enzyme-treated SMUF appears to preclude solubilized enzyme as responsible but it seems possible that proteins (whey proteins and small micelles) in UCSMS may de-adsorb enzyme from the pepsin-glass column whereas protein-free SMUF is unable to do so. Most of the data reported in the two previous studies were confirmed by Lee, Olson and Richardson (1977).

Porous glass-pepsin performed more satisfactorily in a fluidized bed than in a packed-bed reactor due mainly to freedom from plugging (Cheryan and co-workers, 1975 b). However, loss of enzymic activity persisted; the rapid initial decay appeared to be due to adsorption of κ-casein or macropeptides on the column but the cause of the slower, later decay was not identified. Partial regeneration of the column could be accomplished by treatment with 2 M urea at pH 3.5 but the regenerated column lost activity very quickly on re-use.

The stability of pepsin-glass was improved by using zirconia-coated glass, by precoating glass with protein especially with blood serum albumin and by storing at pH 3 (Cheryan and co-workers, 1976). Limitations of various carrier systems and immobilization techniques were compared by Taylor and co-workers (1977). Best results were obtained by covalent attachment to protein-coated glass or titania or by adsorption to alumina at low pH. When cost and simplicity of preparation are considered, the alumina system was considered best overall. However, rapid loss of activity persisted and while considerable regeneration was possible by acid washing, activity was rapdily lost again on re-use. A generally similar study is reported by Paquot, Thomart and Deroanne (1976).

Chymosin immobilized on porous glass was used by Dalgleish (1979) to study the aggregation of chymosin-treated micelles. Hydrolysis of κ-casein was monitored by measuring the ratio of para-κ-casein: $\alpha_{s1,0}$-casein in electrophoretograms. Hydrolysis was considered to be due mainly to immobilized enzyme although some chymosin appeared to be solubilized. Unlike the results of Hicks and co-workers (1975) hydrolysis of κ-casein had essentially plateaued before micelles were capable of aggregation.

Although α-chymotrypsin, chymosin and Suparen immobilized on glass were capable of hydrolysing whole casein or isolated caseins (although less efficiently than the free enzymes), they were unable to coagulate milk even after prolonged contact time (Beeby, 1979). The data clearly show that sufficient κ-casein is not assessible in intact micelles to immobilized rennets even when long (15-18 nm) spacer were used to attach the enzyme to the support.

Pig pepsin was successfully immobilized on chitosan with acrolein as a linking agent (Hirano and Miura, 1979) but the milk clotting activity was not determined. Immobilization of pepsin in paraffin wax is described by Savangikar and Joshi (1978). The molten wax-enzyme mixture could be coated on various surfaces; the film retained the ability to coagulate milk over extended periods. The technique appears attractive; it is cheap and the film could be formed on, for example, agitator paddles, interior of pipes or storage tanks with which the milk comes in contact.

Adsoprtion of rennets on hydroxylapatite or entrappment in calcium-alginate was investigated by Glogowski and Rand (1978) as a means of increasing the retention of coagulant in cheese curd thereby reducing losses in the whey.

Thus, it is claimed that a number of milk-clotting enzymes have been successfully immobilized by various techniques on various supports. Pig pepsin has been the most ameniable to immobilization and the most widely studied. Porous glass has been the most popular support but better results have been obtained with titania and alumina. All immobilized rennets have had poor stability and although they could be partially regenerated, they rapidly re-loose activity. Although it is claimed that numerous advantages would ensue from the use of an active, stable immobilized coagulant, Taylor and co-workers (1976) concluded that they were not economically viable at that time and there have been no significant developments, technologically or economically, since. Although most of the publications reviewed claim that solubilization of enzyme is insignificant, the coagulation characteristics ascribed to milk treated with immobilized rennets are so different in many respects from those observed with soluble enzymes as to lead this author to have reservations as to the complete absence of soluble enzyme. The data of Beeby (1979) support these reservations.

Even if a stable, economic immobilized enzyme becomes available, many problems remain to be solved:

1) Since enzyme-modified milk will coagulate if the temperature increases > 20°C, there is considerable risk that coagulation will occur in the reactor in the event of a mechanical failure. Should such occur it would be extremely difficult to recover the enzyme.

2) Many cheese factories process up to 10^6 l/day; the amount and cost of immobilized enzyme required would be very considerable.

3) Many factories operate 20 h/day making the maintenance of sanitary conditions in the reactors difficult although the columns could be sanitized by treatment with H_2O_2 solutions.

4) Although it is suggested that the absence of coagulant in the curd would facilitate the more controlled ripening of soft cheeses, it appears to this author that it would be necessary to add proteinase(s) to curd, at least for hard cheeses, since the coagulant retained in the curd in conventional manufacturing methods plays a major role in cheese ripening. This could probably be done by infusion or addition of encapsulated enzymes to cheese milk.

ACCELERATED CHEESE RIPENING

Cheese ripening is a complex process involving proteolysis, glycolysis, lipolysis and a variety of secondary changes catalysed, in the main, by the coagulant, starter bacteria and their enzymes, non-starter bacteria and their enzymes, indigenous milk enzymes, and in some varieties, exogenous lipases. Ripening is a slow process requiring up to 2 years for some varieties; in general the lower the moisture content of the cheese, the longer the ripening time required. In modern dairy technology, most cheeses are matured under controlled conditions of temperature and humidity. Thus cheese ripening is an expensive process because of the large stocks of cheese carried and ripening facilities required. According to Law (1980) it costs £45 per tonne to ripen Cheddar cheese in Britain i.e. 3.6% of the cost of milk and other manufacturing expenses. Furthermore, ripening is to some extent an uncontrolled process and considerable variation is experienced in the quality of the finished product. For these various reasons there is increasing interest in accelerating cheese ripening.

Various approaches, reviewed by Law (1978), might be adopted to accelerate cheese ripening, all of which are very much at the experimental stage:

1) Higher ripening temperatures accelerate both flavour and off-flavour development and are not therefore recommended although it would be the easiest and cheapest approach.

2) Increased microbial population/activity; since the starter plays a key role in ripening, it would appear that ripening could be accelerated by increasing starter numbers; however off-flavour development may also be accelerated. Techniques for increasing starter cell numbers include: addition of modified starters (lysozyme-treated, heat-shocked, aged) or modification of the manufacturing technique (pre-hydrolysis of lactose, higher starter additions with higher manufacturing temperatures).

3) Use of starter mutants selected for fast ripening characteristics.

4) Slurry ripening system (Kristoffersen and co-workers, 1967; Sutherland, 1975).

5) Addition of exogenous enzymes:

    (i)   cheese slurries added to cheese curd,
    (ii) cell-free extracts of starter cultures added to cheese curd,
    (iii) isolated enzymes.

In contrast to the studies of Kristoffersen and co-workers (1967), in which the whole cheese was slurried, Dulley (1976) added cheese slurry to Cheddar curd as a source of microorganisms and enzymes. Slurries were prepared by blending 2 parts of fresh curd with one part of a 5% NaCl brine containing 0.3% potassium sorbate. After ripening for 1 week at $30^{\circ}C$, 600 g slurry was added to 10 Kg Cheddar curd after salting. The curd was then hooped, pressed and ripened as normal. Reductions of up to 6 weeks in ripening time to achieve maturity were regularly achieved. Perhaps surprisingly there was no difference in 12% TCA-soluble N between the experimental cheeses suggesting an absence of proteinases and peptidases from the slurry and also that proteolysis was not a limiting factor in ripening.

Normal flavour and proteolysis fails to develop in chemically acidified, starter-free cheese and the possibility therefore exists of accelerating ripening by the addition of cell-free extracts of starter cultures. Addition of cell free extracts to aseptic cheese curd gave normal proteolysis (Gripon and co-workers, 1977); flavour development was not assessed. Law (1980) estimates that the cells from 200 litres of a constant-pH culture would be required to double the starter population of 1 tonne of cheese; at least an equivalent amount of cell-free extract would be required. Unless there are significant developments in culture production and preparation, high costs would appear to militate against this approach.

The method with the best chance of immediate success appears to be the use of commercially available, cheap enzymes. Chaudhari and Richardson (1971) showed that an extract of lamb gastric tissue which contained gastric lipase and a proteinase (different from chymosin) accelerated lipolysis and proteolysis in a number of cheese varieties and improved flavour. Acceleration of proteolysis in cheese was achieved by addition of *Aspergillus* proteinase to cheese curd (Nakanishi and Itoh, 1973, 1974).

The influence of various combinations of lipases, proteinases and decarboxylases on flavour development in Cheddar cheese and cheese slurries was investigated by Kosikowski and Iwasaki (1975). Low levels of mixed enzymes yielded highly flavoured cheese in 4 weeks at 20°C; higher enzyme levels, especially at higher temperatures, led to flavour and textural defects. The enzymes were added with the salt to reduce enzyme losses. It was concluded that the use of microbial lipases and proteinases held potential as a means of accelerating cheese ripening. The technique was refined by Sood and Kosikowski (1979); as enzyme-treated cheeses stored at 4.5°C showed no pronounced acceleration of ripening, it is suggested that cheese could be ripened at 10°C and then stabilized by cooling to 4.5°C.

Uniform distribution of added enzymes is essential for the success of this approach; a similar problem will have to be solved if immobilized rennets are successful. Uniformity could be ensured by enzyme addition to the cheese milk but most of the enzyme would be lost in the whey and, apart from cost, may cause problems if the whey is used in food products. The salt normally added to Cheddar cheese has been used as a carrier by Nakanishi and Itoh (1973), Kosikowski and Iwasaki (1975) and Sood and Kosikowski (1979). Diffusion of large enzyme molecules in cheese curd is slow and lack of uniformity of ripening may ensue; further, the technique would not be applicable to certain soft varieties where the rennet gel is not drained prior to hooping. Promising techniques, not yet adequately evaluated, include microencapsulation of enzymes in milk fat (Magee and Olson, 1978; Magee, Olson and Linsay, 1979) and high pressure injection (Lee, Olson and Lund, 1978).

PROTEIN HYDROLYZATES

Protein hydrolyzates are well-established products for the utilization of protein-rich wastes and non-conventional proteins e.g. vegetable proteins and course fish. Extensively hydrolyzed proteins have traditional applications in soups, gravies, flavourings and dietetic foods. More recently, limited hydrolysis is being employed as a means of modifying the functional properties of proteins.

Extensive Hydrolyzates

Extensive protein hydrolyzates are generally prepared from soya, gluten, milk proteins, meat scrap and fish proteins by acid hydrolysis. Neutralization of acid hydrolyzates results in a high salt content (30-50%, dry weight basis). For certain applications (soups, gravies, condiments) a high salt content is not objectionable but may not be acceptable for dietetic foods and food supplementation. Further, acid hydrolysis causes total or partial destruction of some amino acids.

Enzymatic hydrolysis is a viable alternative to acid hydrolysis (Adler-Nissen, 1976, 1977). However, proteolysis has the serious disadvantage that combinations of certain proteinases with certain substrates yield bitter peptides. Bitterness is generally considered to be due to high hydrophobicity (Ney, 1971) especially when the C-terminal residue is hydrophobic (Sullivan and Jago, 1972). It is important in the first instance to select the proper enzyme-substrate combination for the enzymatic production of protein hydrolyzates but it is also possible to eliminate bitterness or at least to reduce it to an acceptable level by treatment with activated carbon (Murray and Baker, 1952), carboxypeptidase A (Arai and co-workers, 1970), leucine amino peptidase (Clegg and McMillan, 1974) or ultrafiltration (Roozen and Pilnik, 1973).

Milk proteins are not widely used as substrates for extensive proteolysis because casein is notorious for its propensity to yield very bitter hydrolyzates. Bitterness is common even at the relatively low level of proteolysis that occurs in Cheddar and Gouda cheese if proteinase and peptidase activity is excessive or imbalanced (Lowrie and Lawrence, 1972; Sullivan and co-workers, 1973; Stadhouders, 1978). β-Casein is generally considered to be the principal source of bitter peptides in Cheddar cheese (Fox and Walley, 1971; Sullivan and Jago, 1972) and Clegg, Kim and Manson (1974) identified a peptide containing residues 53-79 of β-casein as principally responsible for the bitterness of papain digests of casein. However, Pelissier and Manchon (1976) report that with most proteinases, α_{s1}-casein is more susceptible to bitterness than β-casein.

A procedure using hog kidney leucine amino peptidase was developed by Clegg and McMillan (1974) to debitter ficin/pepsin or papain hydrolyzates of egg white protein or casein; pronase was partially successful. The process was scaled up by Clegg, Smith and Walker (1974) who showed that the biological value of the hydrolyzate was comparable with that of undigested casein and considered it suitable for patients with digestive disorders.

A product with beef-extract flavour was produced by fermenting reconstituted skim milk powder with Ps. fluorescens (Claydon and co-workers, 1968). The preparation performed well in nutritional trials and it is suggested that it would be suitable for use in soups and bouillon. No attempt was made to isolate the enzyme(s) responsible but an enzyme preparation from this organism may also yield a satisfactory product.

Limited Proteolysis

The principal objective of limited proteolysis is to modify and improve functional properties, e.g. solubility, foamability, emulsifying capacity, viscosity, etc. (cf. Kinsella, 1976). Cheese manufacture is probably the outstanding example of the exploitation of limited proteolysis in food processing but discussion in this section will be restricted to modification of "isolated", functional proteins.

Worldwide utilization of functional proteins is increasing rapidly (Wolnak, 1974).

Soya proteins, caseinates and skim milk powders are the principal functional
proteins at present with considerable interest being exhibited in fish and several
seed proteins. Blood is potentially a very important source of protein with high
biological value and excellent functional properties but it is not widely used for
astetic reasons.

Enzymatic modification of the functional properties has been most widely applied
to fish protein concentrate possibly because of its widespread availability in
large quantities and low susceptibility to bitterness (Hale, 1969; Cheftal and co-
workers, 1971; Spinelli and co-workers, 1972, 1975; Hevia, Whitaker and Olcott,
1976; Yanez and co-workers, 1976; Ballester and co-workers, 1977).

Milk is the source of three types of functional proteins with distinctly different
properties: casein, caseinates and casein-whey protein coprecipitates,
lactalbumin (heat denatured whey proteins) and whey protein concentrates
(undenatured whey proteins prepared by ultrafiltration). Casein, like soya
protein, is insoluble in its isoelectric region, a property which limits its use
in acid foods. A method for the production of acid-soluble casein, free of
bitterness, by controlled proteolysis is described by Haggett (1974). Lactalbumin
is essentially insoluble and has, therefore, limited functional properties.
Partial hydrolysis (8-12% α-amino N) of lactalbumin, particularly with trypsin,
yielded a product which was almost completely soluble at pH 6 and 7 and 65%
soluble at pH 4.5 and which was only slightly bitter (Jost and Monti, 1977; Monti
and Jost, 1978); the product appeared promising as a food ingredient.
Proteolysis of whey protein decreased its emulsifying capacity, increased specific
foam volume by ∿ 25% but greatly decreased the stability of the foam (Kuehler and
Stine, 1974).

Unpublished studies by the author have shown that the viscosity of caseinate
solutions can be markedly reduced by very limited proteolysis which did not impair
its flavour. Reduced viscosity facilitates spray-drying and makes caseinate more
suitable for protein fortification of beverages but reduces its functional value
for other applications e.g. in meat products. The fat emulsification capacity and
whippability of the caseinate were drastically reduced by proteolysis.

Plastein Formation

It appears to have been recognised for ∿ 100 years that proteinases are capable of
synthesizing peptide bonds as well as hydrolysing them; the term "plastein" has
been applied to high molecular weight polypeptides synthesised in such reactions.
The mechanism of plastein formation, which occurs optimally in a 30-40% solution
of small peptides at pH 5-7 (regardless of the optimum for hydrolysis by the
enzyme), is generally considered to be via transpeptidization (Horowitz and
Haurowitz, 1959; v. Hofsten and Lalasidis, 1976), with hydrophobic bonds playing
an important secondary role.

The significance of the plastein reaction to food science was not appreciated
until 1970 when it was exploited by a group at the University of Tokyo to debitter
enzymatic hydrolysates of soya protein. Later work has shown that the plastein
reaction may be exploited to improve the nutritional value of plant proteins by
incorporating lysine or methionine into the plastein, to produce protein for
patients suffering from phenylketonuria, and to facilitate the purification of
novel proteins. Reviews include Fujimaki and co-workers (1971, 1977), Arai,
Yamashita and Fujimaki (1975) and Eriksen and Fagerson (1976).

Most of the work on plastein formation has been on soya proteins with lesser
attention paid to milk proteins, single cell proteins, zein, fish proteins and

ovalbumin. Casein is not a good substrate for plastein formation, it is too
hydrophilic; whey proteins are good (Arai and co-workers, 1975).

As far as can be ascertained, assessments of the functional properties of
plasteins have not been published. Since milk proteins have good biological value
and are free of toxins, off-colours and off-flavours, the only significant
attraction of the plastein reaction to dairy technology would be the improvement
of the already good functional properties of milk proteins. In an unpublished
study in this laboratory it was found that, using the alkaline proteinase,
alcalase, for both hydrolysis and resynthesis, a yield of ∿ 65% plastein could be
obtained from casein but only ∿ 40% yield was obtained from lactalbumin. No
polypeptides large enough to be detectable by gel electrophoresis were detectable
in either plastein. The whippability of the casein plastein was markedly less and
the fat emulsification capacity slightly less than those of sodium caseinate but
its solubility in the isoelectric region was considerably improved. It was
concluded that, considering the processing costs and loss of protein, it is very
unlikely that plastein formation from casein would be economically viable.

The solubility and other functional properties of the lactalbumin plastein were far
superior to those of lactalbumin, which it must be remembered are very poor, but
inferior to other available functional proteins. The plastein reaction thus offers
a means of recovering whey protein via thermal denaturation, centrifugation,
enzymatic hydrolysis and plastein formation. In view of the large volumes of
underutilized whey available, it would appear that pilot plant assessment of the
plastein reaction for the utilization of whey protein appears warrented. However,
there are various methods available for the recovery of undenatured whey protein
(cf. Craig, 1979) and the successful application of the plastein reaction
obviously rests on its economic competiveness visa via these alternatives.

MISCELLANEOUS APPLICATIONS

Treatment of milk with trypsin is reported (Shipe, Lee ans Senyk, 1975) to improve
its stability to oxidative rancidity presumably by modification of the fat globule
membrane and improved copper-binding capacity of trypsin-modified milk proteins.

Concentrated milk sterilized by the UHT process (∿ 150°C x 4-5 s) has superior
colour and flavour to the conventional product (∿ 110°C x 20 min). However, its
low viscosity suggests low solids to the consumer and renders it susceptible to
fat separation and salt sedimentation. Very limited proteolysis by pepsin or
rennet is suggested (Tarassuk & Nury, 1952) as a means of increasing viscosity
while retaining colour and flavour advantages; the loss of heat stability
accompanying proteolysis may be counteracted by using orthophosphates. To the
author's knowledge, the technique has not been used commercially possibly because
of the rigid control necessary to ensure the desired result. Furthermore, UHT-
concentrates have not yet become widespread commercially due to a marked tendency
to age-gelation.

Human milk produces a much weaker clot on treatment with rennet than cows' milk,
due, among other factors, to its lower casein content and different casein profile
(Bezkorovainy, 1977). The firm curd produced from cows' milk renders it
unsuitable or at least undesirable as a food for certain babies and for adults
with peptic disorders. There is considerable interest in modifying or "humanizing"
cows' milk for infant food formulations. Early attempts to produce soft-curd milk
showed that a limited treatment of milk with pancreatin produced a soft-curd milk
without concomitant adverse effects on organoleptic or nutritional qualities
(Conquest, Turner and Reynolds, 1938). Excellent clinical results were obtained
with enzyme-treated milk (cf. Storrs and Hull, 1956). The beneficial effects of

enzyme treatment are not exclusively due to the reduction in curd tension (Storrs and Hull, 1956); about ⅓ of the added enzyme survives the pasteurization treatment to which the product was subjected and was considered to contribute to *in vivo* protein digestion. Human milk contains a much higher level of indigenous proteinase and the pasteurized, enzyme-treated milk had about the same level of proteinase activity as raw human milk.

In modern baby food technology, cows' milk is "humanized" by altering the casein: whey protein ratio by addition of demineralized whey to skim milk; the current usage of enzyme treatment for this type of product is unknown.

REFERENCES

Adler-Nissen, J. (1976). Agric. Food Chem., 24, 1090-1093.
Adler-Nissen, J. (1977). Process Biochem., 12 (6), 18-23.
Alais, C. (1956). Proc. XIV Intern. Dairy Congr., II, 823-839.
Alais, C. (1963). Ann. Biol. Anim. Biochem. Biophys., 3, 391-404.
Alais, C., N. Kiger, and P. Jolles (1967). J. Dairy Sci., 50, 1738-1743.
Angelo, I.A., and K.M. Shahani (1979). J. Dairy Sci., 62, Suppl. 1, 64 (abstr.).
Anifantakis, E., and M.L. Green (1980). J. Dairy Res., 47, 221-230.
Arai, S., M. Noguchi, S. Kurosawa, H. Kato, and M. Fujimaki (1970). J. Food Sci., 35, 392-395.
Arai, S., M. Yamashita, and M. Fujimaki (1975). Cereal Foods World, 20, 107-112.
Arai, S., M. Yamashita, K. Aso, and M. Fujimaki (1975). J. Food Sci., 40, 342-344.
Arima, S., K. Shimazaki, T. Yamazumi, and Y. Kanamaru (1974). Jap. J. Dairy Sci., 23, A83-A87.
Ashoor, S.H., R.A. Sair, N.F. Olson, and T. Richardson (1971). Biochem. Biophys. Acta, 229, 423-430.
Ballester, D., E. Yanez, O. Brunser, P. Stekel, P. Chadud, G. Castano, and F. Monckeberg (1977). J. Fd. Sci., 42, 407-409.
Bang-Jensen, V., B. Foltmann, and W. Rombauts (1964). Comptrend. travs. lab. Carlsburg, 34 (13), 326-245.
Beeby, R. (1979). N.Z.J. Dairy Sci. and Technol., 14, 1-11.
Berridge, N.J. (1942). Nature, 149, 194-195.
Berridge, N.J. (1976). J. Dairy Res., 43, 337-356.
Bezkorovainy, A. (1977). J. Dairy Sci., 60, 1023-1037.
Bingham, E.W. (1975). J. Dairy Sci., 58, 13-18.
British Standards Institution (1963). No. 3624.
Brown, R.J., and H.W. Swaisgood (1975). J. Dairy Sci., 58, 796 (abstr.).
Castle, A.V., and J.V. Wheelock (1972). J. Dairy Res., 39, 15-22.
Chaudhari, R.V., and G.H. Richardson (1971). J. Dairy Sci., 54, 467-471.
Cheftel, C., M. Ahern, D.I.C. Wang, and S.R. Tannenbaum (1971). J. Agric. Food Chem., 19, 155-161.
Cheryan, M. (1978). AIChE Symposium Series, 74 (172), 47-52.
Cheryan, M., P.J. van Wyk, N.F. Olson, and T. Richardoon (1975a). J. Dairy Sci., 58, 477-481.
Cheryan, M., P.J. van Wyk, N.F. Olson, and T. Richardson (1975b). Biotechnol. Bioeng., 17, 585-598.
Cheryan, M., P.J. van Wyk, T. Richardson, and N.F. Olson (1976). Biotechnol. Bioeng., 18, 273-279.
Claydon, T.J., R. Mickelsen, P.J. Pinkston, and N.L. Fish (1968). Food Technol., 22, 215-218.
Clegg, K.M., and A.D. McMillan (1974). J. Food Technol., 9, 21-29.
Clegg, K.M., C.L. Kim, and W. Manson (1974). J. Dairy Res., 41, 283-287.
Clegg, K.M., G. Smith, and A.L. Walker (1974). J. Food Technol., 9, 425-431.

Cogan, T.M. (1977). Ir. J. Fd. Sci. and Technol., 1, 95-105.
Conquest, V., A.W. Turner, and H.J. Reynolds (1938). J. Dairy Sci., 21, 361-367.
Craig, T.N. (1979). J. Dairy Sci., 62, 1695-1702.
Creamer, L.K. (1971). N.Z. J. Dairy Sci. and Technol., 6, 91 (abstr.).
Creamer, L.K. (1975). J. Dairy Sci., 58, 287-292.
Creamer, L.K. (1976). N.Z. J. Dairy Sci. and Technol., 11, 30-39.
Creamer, L.K., and B.C. Richardson (1974). N.Z. J. Dairy Sci. and Technol.
  9, 9-13.
Creamer, L.K., O.E. Mills, and E.L. Richards (1971). J. Dairy Res. 38, 269-280.
Dalgleish, D.G. (1979). J. Dairy Res., 46, 653-661.
Dalgleish, D.G. (1980). J. Dairy Res., 47, 231-235.
Davies, D.T., and C. Holt (1979). Symposium in the Physics and Chemistry of
  Milk Proteins. J. Dairy Res., 46, No. 2 (special issue).
de Jong, L. (1976). Neth. Milk and Dairy J., 30, 242-253.
de Koning, P.J. (1979). Special Address to Annual Session of IDF Commission F;
  Montreux, Switzerland.
de Koning, P.J., R. Jenness, and H.P. Wijnand (1963). Neth. Milk and Dairy J.,
  17, 352-363.
de Koning, P.J., P.J. van Rooijen, and S. Visser (1978). Neth. Milk and Dairy J.,
  32, 232-244.
Delfour, A., J. Jolles, C. Alais, and P. Jolles (1965). Biochem. Biophys. Res.
  Commun., 19, 452-455.
Dulley, J.R. (1976). Aust. J. Dairy Technol., 31, 143-148.
Edwards, J.L., and F.V. Kosikowski (1969). J. Dairy Sci., 52, 1675-1678.
El-Shibiny, S., and M.H. Abd El-Salam (1976). J. Dairy Res., 43, 443-448.
El-Shibiny, S., and M.H. Abd El-Salam (1977). J. Dairy Sci., 60, 1519-1521.
Eriksen, S., and I.S. Fagerson (1976). J. Food Sci., 41, 490-493.
Ernstrom, C.A. (1974). In B.H. Webb, A.H. Johnson and J.A. Alford (Eds.),
  "Milk-clotting enzymes and their action"; "Fundamentsls of Dairy Chemistry",
  Avi Publishing Co. Inc., Westport, Conn. pp. 662-718.
Farrell, H.M. Jr., and M.P. Thompson (1974). In B.H. Webb, A.H. Johnson and
  J.A. Alford (Eds.), "Fundamentals of Dairy Chemistry", Avi Publishing Co. Inc.,
  Westport, Conn. pp. 442-473.
Ferrier, L.K., T. Richardson, N.F. Olson, and C.L. Hicks (1972). J. Dairy Sci.,
  55, 726-734.
Fish, J.C. (1957). Nature, 180, 345
Foltmann, B. (1959). Proc. XV Inter. Dairy Congr., 2, 655-661.
Foltmann, B. (1964). Compt. Rend. trav. lab. Carlsberg, 34, 319-325.
Foltmann, B. (1966). Compt. Rend. trav. lab. Carlsberg, 35, 143-231.
Foltmann, B. (1971). In H.A. McKenzie (Ed.), "Milk Proteins: Chemistry and
  Molecular Biology", Vol. II. Academic Press, New York. pp. 217-254.
Foltmann, B. (1978). Proc. XX Intern. Dairy Congr., Paris; Scientific and
  Technical Session - 33ST.
Foltmann, B., and V.B. Pedersen (1977). In J. Tang (Ed.), "Acid Proteases:
  Structure, function and biology", Plenum Press, New York. pp. 3-22.
Foltmann, B., V.B. Pedersen, D. Kaufmamm, and G. Wybrandt (1979). J. Biol. Chem.,
  254, 8447-8456.
Fox, P.F. (1969). J. Dairy Sci., 52, 1214-1218.
Fox, P.F. (1970). J. Dairy Res., 37, 173-180.
Fox, P.F., and B.F. Walley (1971). J. Dairy Res., 38, 165-170.
Fruton, J.S. (1971). In P.D. Boyer (Ed.), "The Enzymes", Vol. 3. Academic Press,
  New York. pp. 119-164.
Fujimaki, M., S. Arai, and M. Yamashita (1977). In R.E. Feeney and J.R. Whitaker
  (Eds.), "Food Proteins: Improvement through chemical and enzymatic
  modification", Advances in Chemistry Series No. 160, American Chemical Society,
  Washington, D.C.
Fujimaki, M., H. Kato, S. Arai, and M. Yamashita (1971). J. Appl. Bact.
  34, 119-131.

Garnier, J. (1963). Biochim. Biophys. Acta, 66, 366-377.

Garnier, J. (1966). J. Mol. Biol., 19, 586-590.

Garnier, J., G. Mocquot, B. Ribadeau-Dumas, and J.-L. Maubois (1968). Ann. Nutr. L'Alim., 22, B495-B552.

Glogowski, J.A., and A.G. Rand (1978). J. Dairy Sci., 61, Suppl. 1, 105 (abstr.).

Green, M.L. (1973). Neth. Milk and Dairy J., 27, 278-285.

Green, M.L. (1977). J. Dairy Res., 44, 159-188.

Green, M.L., and G. Crutchfield (1969). Biochem. J., 115, 183-189.

Green, M.L., and G. Crutchfield (1971). J. Dairy Res., 38, 151-164.

Green, M.L., and R.J. Marshall (1977). J. Dairy Res., 44, 521-531.

Green, M.L., D.G. Hobbs, S.V. Morant, and V.A. Hill (1978). J. Dairy Res., 45, 413-422.

Gripon, J.-C., M.J. Desmazeaud, D. Le Bars, and J.-L. Bergere (1977). J. Dairy Sci., 60, 1532-1538.

Guiney, J. (1972). M.Sc. Thesis, National University of Ireland.

Haggett, T.O.R. (1974). Proc. IXX Intern. Dairy Congr., 1E, 339 (abstr.).

Hale, M.B. (1969). Food Technol., 23, 107-110.

Harboe, M.K., and B. Foltmann (1975). FEBS Letters, 60, 133-136.

Hevia, P., J.R. Whitaker, and H.S. Olcott (1976). J. Agric. Food Chem., 24, 383-385.

Hicks, C.L., L.K. Ferrier, N.F. Olson, and T. Richardson (1975). J. Dairy Sci., 58, 19-24.

Hill, R.D. (1968). Biochem. Biophys. Res. Commun., 33, 659-663.

Hill, R.D. (1969). J. Dairy Res., 36, 409-415.

Hill, R.D. (1970). J. Dairy Res., 37, 187-192.

Hill, R.D., and B.A. Craker (1968). J. Dairy Res., 35, 13-18.

Hill, R.D., and R.R. Laing (1965). J. Dairy Res., 32, 193-201.

Hill, R.D., D. Givol, and E. Lahav (1974). J. Dairy Res., 41, 147-153.

Hirano, S., and O. Miura (1979). Biotechnol. Bioeng. 21, 711-714.

Hofmann, T. (1974). In J.R. Whitaker (Ed.), "Food Related Enzymes", Advances in Chemistry Series, No. 136. American Chemical Soc., Washington, D.C. pp. 146-185.

v. Hofsten, B., and G. Lalasidis (1976). J. Agric. Food Chem., 24, 460-465.

Holmes, D.G., J.W. Duersch, and C.A. Ernstrom (1977). J. Dairy Sci., 60, 862-869.

Horowitz, J., and F. Haurowitz (1959). Biochim. Biophys. Acta, 33, 231-237.

Hsu, R.Y.H., L. Anderson, R.L. Baldwin, C.A. Ernstrom, and A.M. Swanson (1958). Nature, 182, 798-799.

Hsu, I.-N., L.T.J. Delbaere, M.N.G. James, and T. Hofmann (1977). Nature, 266, 140-145.

Humbert, G., and C. Alais (1979). J. Dairy Res., 46, 559-588.

Humme, H.E. (1972). Neth. Milk and Dairy J., 26, 180-185.

Hyslop, D.B., A.M. Swanson, and D.B. Lund (1979). J. Dairy Sci., 62, 1227-1232.

Jansen, E.F., and A.C. Olson (1969). Arch. Biochem. Biophys., 129, 221-227.

Jolles, J., C. Alais, and P. Jolles (1968). Biochim. Biophys. Acta, 168, 591-593.

Jolles, P., C. Alais, and J. Jolles (1963). Biochim. Biophys. Acta, 69, 511-517.

Jost, R., and J.C. Monti (1977). J. Dairy Sci., 60 1387-1393.

Kannan, A., and R. Jenness (1961). J. Dairy Sci., 44, 808-822.

Kato, I., K. Ando, K. Mikawa, and T. Yasui (1980). J. Dairy Sci., 63, 25-31.

Kato, I., K. Mikawa, Y.M. Kim, and T. Yasui (1970). Mem. Fac. Agric., Hikkaido Univ., 7 (4), 477-481.

Kinsella, J.E. (1976). CRC Crit. Revs. Fd. Sci. and Nutr., 7 (3), 219-280.

Kosikowski, F.V. and T. Iwasaki (1975). J. Dairy Sci., 58, 963-970.

Kowalchyk, A.W., and N.F. Olson (1977). J. Dairy Sci., 60, 1256-1259.

Kowalchyk, A.W., and N.F. Olson (1978). J. Dairy Sci., 61, 1375-1379.

Kristoffersen, T., E.M. Mikolajcik, and I.A. Gould (1967). J. Dairy Sci., 50, 292-297.

Kuehler, C.A., and C.M. Stine (1974). J. Food Sci., 39, 379-381.

Law, B.A. (1978). International Dairy Federation, Annual Bulletin, Doc. 108.
Law, B.A. (1979). J. Dairy Res., 46, 573-588.
Law, B.A. (1980). International Dairy Federation, F-Doc. 79.
Lawrence, R.C., and L.K. Creamer (1969). J. Dairy Res., 36, 11-20.
Ledford, R.A., A.C. O'Sullivan, and K.R. Nath (1966). J. Dairy Sci., 49,
  1098-1101.
Lee, H.J., N.F. Olson, and T. Richardson (1977). J. Dairy Sci., 60, 1683-1688.
Lee, H.J., N.F. Olson, and D.B. Lund (1978). Process Biochem., 13 (12), 14-18.
Lindqvist, B. (1963). Dairy Sci. Abstr., 25, 257-264, 299-308.
Lowrie, R.J., and R.C. Lawrence (1972). N.Z. J. Dairy Sci. and Technol., 7,
  51-53.
Lyster, R.L.J. (1970). J. Dairy Res., 37, 233-243.
Lyster, R.L.J. (1972). J. Dairy Res., 39, 279-318.
Mabbitt, L.A. (1961). J. Dairy Res., 28, 303-318.
Mackinlay, A.G., and R.G. Wake (1965). Biochim. Biophys. Acta, 104, 167-180.
Mackinlay, A.G., and R.G. Wake (1971). In H.A. McKenzie (Ed.), "Milk Proteins:
  Chemistry and Molecular Biology", Vol. II. Academic Press, N.Y. pp. 175-215.
Magee, E.L., and N.F. Olson (1978). J. Dairy Sci., 61, Suppl. 1, 114 (abstr.).
Magee, E.L., N.F. Olson, and R.C. Lindsay (1979). J. Dairy Sci., 62, Suppl. 2,
  65 (abstr.).
Martens, R., and M. Naudts (1978). IDF Annual Bulletin, Doc. 108, 51-63.
McFarlane, A.S. (1938). Nature, 142, 1023-1025.
McKenzie, H.A. (1971). "Milk Proteins: Chemistry and Molecular Biology",
  Vols. I and II. Academic Press, N.Y.
Mercier, J.C., G. Brignon, and B. Ribadeau Dumas (1973). Eur. J. Biochem., 35,
  222-235.
Mikawa, K., I. Kato, and T. Yasui (1973). Mem. Fac. Agric., Hikkaido Univ.,
  9 (1), 110-115.
Monti, J.C., and R. Jost (1978). J. Dairy Sci., 61, 1233-1237.
Morrissey, P.A. (1969). J. Dairy Res., 36, 333-341.
Mulvihill, D.M. (1978). Ph.D. Thesis, National University of Ireland.
Mulvihill, D.M., and P.F. Fox (1977). J. Dairy Res., 44, 533-540.
Mulvihill, D.M., and P.F. Fox (1978). Ir. J. Fd. Sci. and Technol., 2, 135-139.
Mulvihill, D.M., and P.F. Fox (1979a). J. Dairy Res., 46, 641-651.
Mulvihill, D.M., and P.F. Fox (1979b). Milchwissenschaft, 38, 680-683.
Mulvihill, D.M., and P.F. Fox (1980). Ir. J. Fd. Sci. and Technol., 4, 13-23.
Murray, T.K., and B.E. Baker (1952). J. Sci. Food Agric., 3, 470-475.
Nikanishi, T., and M. Itoh (1973). Jap. J. Dairy Sci., 22, A110-A113.
Nakanishi, T., and M. Itoh (1974). Jap. J. Dairy Sci., 23, A121-A127.
Nelson, J.H. (1975). J. Dairy Sci., 58, 1739-1750.
Ney, K.H. (1971). Z. Lebenson u. Forsch., 147, 64-68.
Nitschmann, Hs., and H.U. Bohren (1955). Helv. Chim. Acta, 38, 1953-1963.
Nitschmann, Hs., H. Wissmann, and R. Henzi (1957). Chimia, 11, 76-128.
O'Keeffe, R.B., P.F. Fox, and C. Daly (1975). J. Dairy Res., 42, 111-122.
O'Keeffe, R.B., P.F. Fox, and C. Daly (1976). J. Dairy Res., 43, 97-107.
O'Keeffe, A.M., P.F. Fox, and C. Daly (1978). J. Dairy Res., 45, 465-477.
Olson, N.F., and T. Richardson (1975). J. Dairy Sci., 58, 1117-1122.
Paquot, M., Ph. Thomart, and C. Deroanne (1976). Le Lait, 56, (553/554),
  154-163.
Parry, R.M., and R.J. Carroll (1969). Biochim. Biophys. Acta, 194, 138-150.
Payens, T.A.J. (1976). Neth. Milk and Dairy J., 30, 55-59.
Payens, T.A.J. (1977). Biophysical. Chem., 6, 263-270.
Payens, T.A.J. (1978). Neth. Milk and Dairy J., 32, 170-183.
Payens, T.A.J. (1979). J. Dairy Res., 46, 291-306.
Payens, T.A.J., and B.W. van Markwijk (1963). Biochim. Biophys. Acta, 71, 517-530.
Payens, T.A.J., A.K. Wiersma, and J. Brinkhuis (1977). Biophysical. Chem.,
  6, 253-261.
Pearce, K.N. (1976). J. Dairy Res., 43, 27-36.

Pedersen, U.D. (1977). Acta Chem. Scand., B31, 149-156.
Pelissier, J.P., and P. Manchon (1976). J. Food Sci., 41, 231-233.
Pelissier, J.P., J.-C. Mercier, and B. Ribadeau Dumas (1974). Ann. Biol. Anim. Biochim. Biophys., 14, 343-362.
Phelan, J.A., J. Guiney, and P.F. Fox (1973). J. Dairy Res., 40, 105-112.
Pyne, G.T. (1953). Chem. and Ind., pp. 302-303.
Pyne, G.T. (1955). Dy. Sci. Abstr., 17, 531-533.
Pyne, G.T. and T.C.A. McGann (1962). Proc. XVI Int. Dy. Congr., 13, 611-616.
Raymond, M.N., E. Bricas (1979). J. Dairy Sci., 62, 1719-1725.
Raymond, M.N., J. Garnier, E. Bricas, S. Cilianu, M. Blasnic, A. Chaix, and P. Lefrancier (1972). Biochim., 54, (2), 145-154.
Raymond, M.N., E. Bricas, R. Selesse, J. Garnier, P. Garnot, and P. Ribadeau-Dumas (1973). J. Dairy Sci., 56, 419-422.
Roozen, J.P., and W. Pilnik (1973). Process Biochem., 8 (7), 24-25.
Salesse, R., and J. Garnier (1975). J. Dairy Sci., 59, 1215-1221.
Savangikar, V.A., and R.N. Joshi (1978). J. Food Sci., 43, 1616-1618.
Sawyer, W.H. (1969). J. Dairy Sci., 52, 1347-1355.
Schmidt, D.G., and T.A.J. Payens (1976). In E. Matijevic (Ed.), "Surface and Colloid Science", 9. John Wiley and Sons, N.Y. pp. 165-229.
Scott Blair, G.W., and J.C. Oosthuizen (1961). J. Dairy Res., 28, 165-173.
Sepulveda, P., J. Marciniszyw Jr., D. Liu, and J. Tang (1975). J. Biol. Chem., 250, 5082-5088.
Shalabi, S.I., and J.V. Wheelock (1977). J. Dairy Res., 44, 351-355.
Shipe, W.F., E.C. Lee, and G.F. Senyk (1975). J. Dairy Sci., 58, 1123-1126.
Skinner, K.J. (1975). Chem. and Eng. News, Aug. 22-41.
Slattery, C.W. (1976). J. Dairy Sci., 59, 1547-1556.
Sood, V.K., and F.V. Kosikowski (1979). J. Dairy Sci., 62, 1865-1872.
Spinelli, J., B. Koury, and R. Miller (1972). J. Fd. Sci., 37, 604-608.
Spinelli, J., B. Koury, and R. Miller (1975). Process Biochem., 10 (10), 31-36 (42).
Stadhouders, J. (1978). XXth Intern. Dairy Congr. (Paris). Scientific and Technical Sessions, 39 ST.
Storrs, A.B., and M.E. Hull (1956). J. Dairy Sci., 39, 1097-1103.
Sullivan, J.J., and G.R. Jago (1972). Aust. J. Dairy Technol., 27, 98-104.
Sullivan, J.J., L. Mou, J.I. Rood, and G.R. Jago (1973). Aust. J. Dairy Technol., 28, 20-26.
Sutherland, B.J. (1975). Aust. J. Dairy Technol., 30, 138-142.
Swaisgood, H.E. (1973). CRC Critical Revs. Food Technol., 3, 374-414.
Tang, J. (1977). "Acid Proteinases: Structure, function and biology". Plenum Press, New York.
Tarassuk, N.P., M.S. Nury (1952). J. Dairy Sci., 35, 857-867.
Tarodo, B., C. Alais, and R. Frentz (1969). Le Lait, 49, 400-416.
Taylor, M.J., M. Cheryan, T. Richardson, and N.F. Olson (1977). Biotechnol. and Bioeng., 19, 683-700.
Taylor, M.J., N.F. Olson, and T. Richardson (1979). Process Biochem., 42 (12), 10-16.
Taylor, M.J., T. Richardson, and N.F. Olson (1976). J. Milk Food Technol., 39, 864-871.
Thomas, M.A.W. (1972). Neth. Milk and Dairy J., 27, 273.
Thunell, R.K., J.W. Deursch, and C.A. Ernstrom (1979). J. Dairy Sci., 62, 373-377.
Thunell, R.K., C.A. Ernstrom, and G.H. Hartmann (1980). J. Dairy Sci., 63, 32-36.
Vanderpoorten, R., and M. Weckx (1972). Neth. Milk and Dairy J., 26, 47-59.
Visser, F.M.W. (1977a). Neth. Milk and Dairy J., 31, 188-209.
Visser, F.M.W. (1977b). Neth. Milk and Dairy J., 31, 210-239.
Visser, F.M.W., and A.E.A. de Groot-Mostert (1977). Neth. Milk and Dairy J., 31, 247-264.
Visser, S., and K.J. Slangen (1977). Neth. Milk and Dairy J., 31, 16-30.

Voynik, I.M., and J.S. Fruton (1971). Proc. Natl. Acad. Sci. U.S.A., 68,
    257-259.
Wake, R.G. (1959). Aust. J. Biol. Sci., 12, 479-489.
Waugh, D.F., and P.H. von Hippel (1956). J. Am. Chem. Soc., 78, 4576-4582.
Whitney, R.McL., J.R. Brunner, K.E. Ebner, H.M. Farrell Jr., R.V. Josephson,
    C.V. Morr, and H.E. Swaisgood (1976). J. Dairy Sci., 59, 785-815.
Wolnak, B., and Associates (1974). "Functional Proteins: U.S.A., Europe and
    Japan, 1972-1980". Reports I and II.
Yanez, E., D. Ballister, F. Monckeberg, W. Hiemlich, and M. Rutman (1976).
    J. Fd. Sci., 41, 1289-1292.
Zittle, C.A. (1965). J. Dairy Sci., 48, 1149-1153.

PEPSIN—LIKE PROTEINASES AND THEIR PRECURSORS
RENNET SUBSTITUTES FOR CHEESE MAKING

J. Kay*, M.J. Valler*, H. Keilová** and V. Kostka**

*Department of Biochemistry, University College, Cardiff, Wales, U.K.
**Institute of Organic Chemistry and Biochemistry, Czech. Academy of Sciences,
Prague, Czechoslovakia

ABSTRACT

Rennet substitutes for cheese production fall into two categories - 1) acidified extracts from animal stomachs, e.g. pig and cow pepsins, 2) proteinases from microorganisms.

From a comparison of the biochemical properties of the substitute proteinases with those of calf chymosin (rennin), it seems that even allowing for extensive homologies in amino acid sequence, pig and cow pepsins and the microbial proteinases are more similar to each other than to calf chymosin. None of them approaches having the "specialised" activity of chymosin to clot milk.

By contrast, the pepsin obtained from chicken proventriculae resembles calf chymosin much more closely. Cheese is now being manufactured commercially using chicken pepsin.

KEYWORDS

Aspartate proteinases; precursor activation; inhibitor peptides, enzyme stability; species differences.

Proteolytic enzymes are classified on the basis of their catalytic mechanism into four groups - the serine, cysteine, metallo and aspartate proteinases (Barrett, 1980a). Enzymes belonging to the last category have been identified from three major sources - microbial products, intracellular hydrolases in the lysosomes of many cell types and the pepsin-like proteinases in the stomachs of various species (Barrett, 1980a). Our current knowledge of these different types of aspartate proteinase reflects the ease with which they can be isolated for study. Much is known in detail about the gastric enzymes, e.g. complete amino acid sequences, active site residues and mechanisms, the nature of isoenzymic variations (Kay, 1980a) while somewhat less information is available concerning the microbial proteinases and relatively little is known about the intracellular aspartate proteinases, cathepsins D & E (Barrett, 1980b). The gastric proteinases are synthesised as precursors (mol. wt. approx. 40,000). On secretion into the acid conditions of the lumen of the stomach, these undergo activation to produce the active enzyme. There is no evidence for the existence of zymogens of microbial or lysosomal aspartate proteinases (Kay, 1980a).

The importance of these enzymes in maintaining nitrogen balance in the whole organism cannot be overemphasised (Kay, 1980b). However, this article is concerned with the commercial significance of aspartate proteinases. For many years, an assay has been in use that permits a convenient method of quantitation of these enzymes, i.e. the ability to hydrolyse κ-casein thereby bringing about the clotting of milk (McPhie, 1976). However, long before biochemists needed a convenient measure of activity, mankind had utilised the ability of aspartate proteinases to clot milk for a different purpose; reports on acidified extracts of animals stomachs being used to coagulate milk to produce cheese date back hundreds if not thousands of years.

For the purposes of the present discussion, cheese production can be simplified into a two stage reaction

$$\text{Milk} \xrightarrow[\text{pH 6.3}]{\text{Coagulation}} \text{Curd} \xrightarrow{\text{Maturation}} \text{Milling, Salting, Pressing, Ripening}$$

Proteolysis is required (and to different extents) in both stages.

Traditionally, acidified extracts from the stomachs of milk-fed calves have been used in the coagulation process. The major component of calf rennet is the aspartate proteinase, rennin or chymosin. Prochymosin is the precursor synthesised in the mucosal lining of the stomach of the calf and this converts itself into the more active form of chymosin on acidification (Foltmann, 1966). However, in recent years a world-wide shortage of calf rennet has developed, largely as a result of a decreasing tendency to slaughter calves at a young enough age for the neonatal prochymosin gene to be expressed rather than that for the adult protein, cow pepsinogen. Consequently, a search has ensued to find an adequate substitute and those now being used commercially fall generally into two classes   1) extracts from the stomachs of other animals, e.g. pig and adult cow pepsins and 2) proteinases from microorganisms (Green, 1977).

From a comparison of the biochemical properties of the substitute proteinases with those of chymosin, it might be possible to obtain an indication of which might prove to be the most acceptable replacement enzyme for commercial exploitation. If, in addition, an examination is made of the precursors purified from animal stomachs, extra information might be derived about the ease of preparation and the suitability of the product enzymes generated in acidified extracts.

The best preparations of adult cow pepsinogen and pig pepsinogen are obtained with a final step of purification on the affinity resin, polylysine-Sepharose (Kay, 1972; Harboe, et al. 1974). Even forms of bovine pepsinogen differing only in their serine phosphate content can be resolved on this matrix (Fig. 1); the three peaks of activity are identical in all other respects.

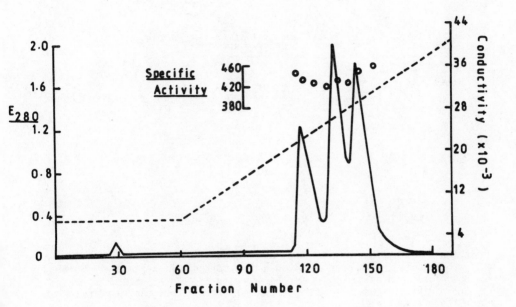

Fig. 1. Chromatography of bovine pepsinogen on polylysine-
Sepharose 4B in 0.05M sodium phosphate buffer,
pH 6.5. A linear NaCl gradient (0-1M) was used
for desorption.

By contrast, calf prochymosin is not retarded on polylysine-Sepharose nor is the
pepsinogen isolated from chicken proventriculae (Dykes, C.W. & Kay, J. - unpublished observations). This might indicate a fundamental difference between the
pairs, adult cow/pig pepsinogens and calf prochymosin/chicken pepsinogen.

The complete amino acid sequences of pig pepsinogen (Stepanov and co-workers, 1973; Moravek and Kostka, 1974) and calf prochymosin (Foltmann and colleagues, 1977) are known and those of cow pepsinogen (Harboe, M. and Foltmann, B. - personal communication) and chicken pepsinogen (Kostka, Keilova and Baudys - see the preceeding chapter - this volume) near completion. While extensive homologies in sequence have been found to exist, nevertheless it is in the functional organisation of the molecules that their importance lies. Comparison of the amino acid compositions of the precursors does not give a particularly significant insight unless attention is focussed on only a few amino acids.

TABLE 1   Amino Acid Composition (residues/mole) of the Precursors

	Pig Pepsinogen	Cow Pepsinogen	Calf Prochymosin	Chicken Pepsinogen
Lys	10	8	15	19
His	3	2	6	8
Arg	4	6	8	7
Asx + Glx	74	72	77	68
Met	4	4	8	9

While the total content of Asn, Asp, Gln and Glu in the four zymogens is similar, calf prochymosin and chicken pepsinogen have markedly higher amounts of the basic residues (and methionine) and would be expected to be less acidic proteins than cow and pig pepsinogens. This might explain the altered behaviour on the basic resin, polylysine-Sepharose.

When the zymogens are exposed to low pH values, they convert themselves into their respective enzymes by undergoing limited proteolysis (Kay, 1980a). Complete resolution of the peptides of the activation segment from the active enzyme can be best achieved by chromatography of the activation mixture on polylysine-Sepharose at pH 3.5 (Fig. 2a).

For cow/pig pepsinogen the activation peptides are not retarded whereas the active enzymes bind and can be eluted by raising the ionic strength (Fig. 2a). By contrast (Fig. 2b) the prochymosin (and chicken pepsinogen) activation mixtures are not fractionated on this matrix and DEAE-cellulose has to be used instead (Kay and Dykes, 1977).

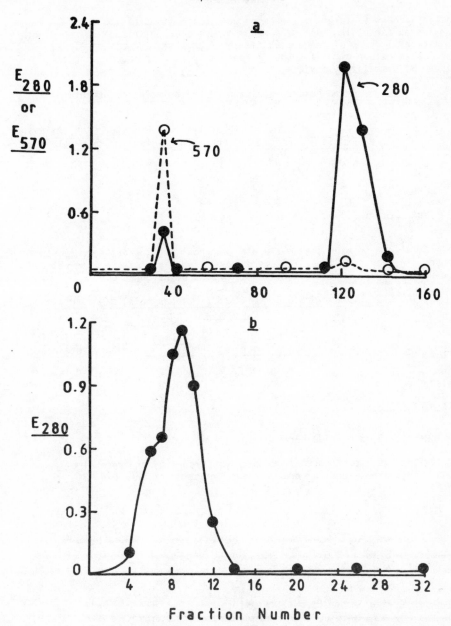

Fig. 2. Chromatography of the activation mixtures of cow
         pepsinogen (a - 360 mg) and calf prochymosin
         (b - 15 mg) on polylysine-Sepharose (2.6 x 43 cm
         and 1 x 15 cm respectively) in 0.05M formate
         buffer, pH 3.5.

If the significant features of the amino acid composition of the active enzymes are considered (including that of a mucor miehei acid proteinase as representative of the microbial group of rennet substitutes (Lagrange, Paquet and Alais, 1980), the more basic nature of calf chymosin, chicken pepsin and mucor miehei proteinase is immediately obvious.

TABLE 2 Amino Acid Composition (residues/mole) of Aspartate Proteinases

	Pig Pepsin	Cow Pepsin	Calf Chymosin	Chicken Pepsin	M. Miehei Proteinase
Lys	1	0	9	9	9
His	1	1	6	4	2
Arg	2	3	4	3	6
Asx + Glx	68	64	69	57	63
Met	4	3	8	8	7

In Table 3, it can be seen that while pig (and cow) pepsins are not glycoproteins, chicken pepsin (Kay and Dykes, 1977) and mucor miehei proteinase (Rickert and McBride-Warren, 1974) contain significant amounts of carbohydrate.

TABLE 3 Carbohydrate Composition (residues/mole) of Aspartate Proteinases

	Pig Pepsin (ogen)	Chicken Pepsin (ogen)	M. Miehei Proteinase[1]
Mannose	0	2.24	7.15
Galactose	0	0.28	0.90
Glucose	0	0.12	1.99
N-Acetyl Glucosamine	0	4.48	1.94

[1]From Rickert, W.S. & McBride-Warren, P.A. (1974).

The active enzymes are produced from their precursors by the removal of about 45 residues from the NH_2-terminal end of the polypeptide chains (Fig. 3).

5 10 15 20 25

PPg: Leu-Val-Lys-Val-Pro-Leu-Val-Arg-Lys-Lys-Ser-Leu-Arg-Gln-Asn-Leu-Ile-Lys-Asp-Gly-Lys-Leu-Lys-Asp-Phe-Leu-Lys-Thr-His

BPg: Ser-Val-Val-Lys-Ile-Pro-Leu-Val-Lys-Lys-Lys-Ser-Leu-Arg-Gln-Asn-Leu-Ile-Glu-Asn-Gly-Lys-Leu-Lys-Glu-Phe-Met-Arg-Thr-His

BpC:Ala-Glu-Ile-Thr-Arg-Ile-Pro-Leu-Tyr-Lys-Gly-Lys-Ser-Leu-Arg-Lys-Ala-Leu-Lys-Glu-His-Gly-Leu-Leu-Glu-Asp-Phe-Leu-Gln-Lys-Gln

30 35 40 45 370

PPg:Lys-His-Asn-Pro-Ala-Ser-Lys-Tyr-Phe-Pro-Glu-Ala-Ala-Ala-Leu | Ile-Gly-Asp ~ PEPSIN ~ Pro-Val-Ala

BPg:Lys-Tyr-Asn-Leu-Gly-Ser-Lys-Tyr-Ile-Arg-Glu-Ala-Ala-Thr-Leu | Val-Ser-Glu ~ PEPSIN ~ Pro-Val-Ala

BpC:Gln-Tyr-Gly-Ile-Ser-Ser-Lys-Tyr-Ser - - Gly-Phe | Gly-Glu-Val-Ala-Ser ~ CHYMOSIN ~ Lys-Ala-Ile

Fig. 3. Sequences of the activation segments of pig (PPg) and cow (BPg) pepsinogens and of calf prochymosin (BpC). The beginning of the pepsin and chymosin sequences are marked by the bars.

However, all of the residues in the activation segments are not released together as one long, intact polypeptide. Instead, it has been shown (Dykes and Kay, 1976; Kay and Dykes, 1976, 1977; Christensen, Pedersen and Foltmann, 1977) that each zymogen converts itself into the active enzyme by a sequential process in which small peptides are released successively from the NH$_2$-terminal end of the polypeptide chain until all of the residues in the activation segment have been removed, i.e.

Precursor → peptide a + intermediate protein A
Intermediate protein A → peptide b + intermediate protein B
Intermediate protein B -----→ peptide n + Enzyme

In the cases of cow/pig pepsinogens, the initial proteolytic split takes place at Leu 16-Ile 17 (see Fig. 3) to release in the first instance a peptide consisting of 16 (pig) or 17 (cow) residues (Dykes and Kay, 1976). For prochymosin and chicken pepsinogen, however, the initial peptide released was much longer and the site of the initial proteolytic attack was between residues Phe 25-Leu 26 (Kay and Dykes, 1977; Keilova, Kostka and Kay, 1977; Pedersen, Christensen and Foltmann, 1979). Nevertheless, the first peptide released in both of these cases was still much shorter than the entire activation segment. The susceptible Leu 16-Ile 17 bond in pig and cow pepsinogens (see Fig. 3) is replaced by a Leu-Lys bond in calf prochymosin (and chicken pepsinogen - Keilova, Kostka and Kay, 1977) that is no longer susceptible to a peptic cleavage. Consequently, the initial proteolytic event has to be deferred to the next susceptible bond which is apparently that between Phe 25-Leu 26.

Thus, it appears that while all of the precursors activate themselves by this sequential removal process, the detailed nature of the activation is slightly different for pig/cow pepsinogens than for calf prochymosin/chicken pepsinogen.

Another intriguing point has emerged from studies of these physiologically important pepsinogen activation systems. It has long been known (Herriott, 1941) that one of the peptides released on activation of pig (cow) pepsinogen(s) binds to pepsin above pH 4.0 to stabilise the enzyme at higher pH values and to inhibit it (usually as measured in the milk-clotting assay at pH 5.3). This "pepsin inhibitor peptide" has now been identified as being derived from the NH$_2$-terminal 16 (17) residues in pig (cow) pepsinogen(s) (Harboe et al. 1974; Kay and Dykes, 1977; Harish-Kumar and Kassell, 1977; Dunn and colleagues, 1978). It inhibits pig (and cow) pepsin(s) very strongly (K_i - 0.25µM, Dunn and colleagues, 1978) but has no inhibitory effect at all on the homologous calf chymosin (Kay and Dykes, 1977) despite the fact that chymosin supposedly operates through a similar (aspartate) active site mechanism to those of pig and adult cow pepsins. It has not been possible to obtain enough of the longer peptide (27 residues released in the first step) from prochymosin activation to permit testing of its inhibitory effects on the proteinases from the various species. However, the very similar 26 residue peptide from the NH$_2$-terminus of chicken pepsinogen was found to act only as a very weak inhibitor of pig pepsin with even less effect on chymosin and chicken pepsin (Keilova, Kostka and Kay, 1977). Thus, it might be postulated that there will be subtle distinctions found between the active sites of pig/cow pepsins and calf chymosin/chicken pepsin when the three-dimensional structures of the proteins are finally determined.

This inhibition effect also introduces possible complications for the commercial usage of unfractionated activation mixtures containing enzyme and peptide. At pH 6.3, chymosin and chicken pepsin are completely unaffected by their own activation peptides. On the other hand, pig/cow pepsin(s) are subject to inhibition by the activation peptide (residues 1-16) in the unfractionated mixture and this effect would be amplified with increasing length of coagulation time. Pig (and cow) pepsins are unstable above pH 6.0 (Fruton, 1976; McPhie, 1975), whereas chymosin is stable at pH 6.5 (Foltmann, 1966) and chicken pepsin retains all of its activity at pH values as high as 8.0 (Bohak, 1969). Thus, the activity of pig/cow pepsins is lost very rapidly as the coagulant is operating in milk at pH 6.3 and as the level of active pepsin declines, then the molar ratio of inhibitor peptide/active pepsin increases and so the effects of the inhibitor peptide will be felt more and more on the diminishing amounts of active pepsin remaining. Consequently, coagulants using pig/cow pepsins are likely to be inconsistent in the time required for the coagulation process and the amount of active enzyme which is maintained into the subsequent (ripening) processes may be variable. A proportion of the coagulant protease is normally incorporated with the curd and survives throughout cheese-making into the ripening period where it is involved in the development of body and flavour (too much proteolysis at this stage can result in the production of bitter-tasting peptides with spoilage of the flavour).

Thus, standardisation of the entire process is most important and from the above considerations, it would seem that this is all too difficult to achieve. Substitutes involving the use of pig/cow pepsins appear to be least similar to calf rennin, despite the apparent homologies in amino acid sequence, since the all-important enzymic and stability properties differ considerably. Chicken pepsin, by contrast, might be better renamed as chicken chymosin since, in its biochemical properties at least, it appears to resemble calf chymosin rather than the animal pepsins.

In a recent comparison of the activities of cow pepsin, calf chymosin and microbial rennets (the enzymes from M. Miehei and M. Pusillus) as general proteinases (measured towards a synthetic peptide at pH 4.7) and as milk coagulants, it was found that while calf chymosin had a ratio of milk clotting/ proteolytic activities of 0.024, the values for the microbial proteinases and cow pepsin were 0.003 and 0.001 respectively (Martin and coworkers, 1980). It would thus appear that the microbial rennets are closer to cow pepsin in their activities and none of the substitutes currently in use can approach the "specialised" activity of calf chymosin in its ability to clot milk. Studies are now in hand to measure the activity of chicken pepsin towards these substrates and to test its suitability for cheese production.

With the enormous market for frozen chickens in modern society, chicken proventri- culae are readily and cheaply available for the large scale production of the chicken enzyme as a potential rennet substitute. Indeed, recent estimates have indicated that 80% of the cheese now being manufactured in Israel has been produced using chicken pepsin as the coagulant (Gutfeld and Rosenfeld, 1975; Gardin and colleagues, 1978).

ACKNOWLEDGEMENTS

The work carried out in the U.K. was supported by grants from the Medical Research Council and the Agricultural Research Council. Our international collaboration was fostered by the Academic Links with Europe Scheme operated by the British Council.

REFERENCES

Barrett, A.J. (1980a). In D.C. Evered and J. Whelan, (eds) Protein Degradation in Health and Disease Vol. 75, CIBA Foundation Symposia. Excerpta Medica pp 1-9.

Barrett, A.J. (1980b). In D.C. Evered and J. Whelan, (eds) Protein Degradation in Health and Disease Vol. 75, CIBA Foundation Symposia. Excerpta Medica pp 37-42.

Bohak, Z. (1969). J. Biol. Chem., 244, 4638-4648.

Christensen, K.A., V.B. Pedersen and B. Foltmann (1977). FEBS Lett., 76, 214-218.

Dunn, B.M., C. Deyrup, W.G. Moesching, W.A. Gilbert, R.J. Nolan and M.L. Trach (1978). J. Biol. Chem., 253, 7269-7275.

Dykes, C.W. and J. Kay (1976). Biochem. J., 153, 141-144.

Foltmann, B. (1966). C.R. Trav. Lab. Carlsberg, 35, 143-231.

Foltmann, B., V.B. Pedersen, H. Jacobsen, D. Kaufmann and G. Wybrandt (1977). Proc. Natl. Acad. Sci. U.S.A., 74, 2321-2324.

Fruton, J.S. (1976). Adv. Enzymol., 44, 1-36.

Gordin, S., I. Rosenthal, S. Bernstein, C. Navrot, H. Balaban and M. Frank (1978). Proc. 20th Int. Dairy Congr. Paris E, pp 441-442.

Green, M.L. (1977). J. Dairy Res., 44, 159-188.

Gutfeld, M. and P.P. Rosenfeld (1975). Dairy Industries, 40, 52-55.

Harboe, M.K., P.M. Andersen, B. Foltmann, J. Kay and B. Kassell (1974). J. Biol. Chem., 249, 4487-4494.

Harish-Kumar, P.M. and B. Kassell (1977). Biochemistry, 16, 3846-3849.

Herriott, R.M. (1941). J. Gen. Physiol., 24, 325-338.

Kay, J. (1972). Proceedings 8th FEBS Meeting, Amsterdam. Abstract No. 458.

Kay, J. and C.W. Dykes (1976). Biochem. J., 157, 499-502.

Kay, J. and C.W. Dykes (1977). In J. Tang (Ed.) Acid Proteases, Structure, Function and Biology Plenum Press N.Y. pp.103-126.

Kay, J. (1980a). In R.B. Freedman and H. Hawkins (Eds.) The Enzymology of Post-Translational Modifications of Proteins Academic Press London pp. 423-456.

Kay, J. (1980b). Biochem. Soc. Trans., 8, 415-417.

Keilova, H., V. Kostka and J. Kay (1977). Biochem. J., 167, 855-858.

Lagrange, A., D. Paquet and C. Alais (1980). Int. J. Biochem., 11, 347-352.

Martin, P., M-N. Raymond, E. Bricas and B. Ribadeau Dumas (1980). Biochim. Biophys. Acta, 612, 410-420.

McPhie, P. (1975). Biochemistry, 14, 5253-5256.

McPhie, P. (1976). Anal. Biochem., 73, 258-261.

Moravek, L. and V. Kostka (1974). FEBS Lett., 43, 207-211.

Pedersen, V.B., K.A. Christensen and B. Foltmann (1979). Eur. J. Biochem., 94, 573-580.

Rickert, W.S. and P.A. McBride-Warren (1974). Biochim. Biophys. Acta, 336, 437-444.

Stepanov, V.M., L.A. Baratova, I.B. Pugacheva, L.P. Belyanova, L.P. Revina and E.A. Timokhina (1973). Biochem. Biophys. Res. Commun., 54, 1159-1164.

ENZYME ACTIONS IN THE CONDITIONING OF MEAT

D.J. Etherington

Muscle Biology Division, Meat Research Institute
Langford, Bristol, BS18 7DY, U.K.

ABSTRACT

The conversion of muscle into meat is an enzymic process beginning with the development of rigor as ATP is depleted. The muscle is subsequently softened by proteolytic disruption of the myofibrils. Cathepsins continue to degrade the muscle proteins during a conditioning period of up to three weeks when full flavour and tenderness are obtained.

KEYWORDS

Muscle; myofibrillar proteins; calcium-activated neutral proteinase; cathepsins.

INTRODUCTION

It is generally reckoned that the most important attribute to the eating quality of meat is its texture. Variations in tenderness depend in part on the quantity and extent of cross-linking of the intramuscular connective tissues which are determined essentially by the age of the animal and have a predictable effect on the texture (Bailey, 1972). In addition there are other factors which have a less certain influence on meat quality. For instance, the pre-slaughter care of the animal and the way the carcass is subsequently handled post-slaughter determine the time of onset of rigor development and whether this is achieved under conditions that will convert the living muscle into tender meat. Finally the progressive improvement in the quality of meat that we can then gain by holding the carcass in storage is dependent on both the temperature and the length of time to retailing (Etherington and Dransfield, 1981). This process of holding meat to enhance its flavour and tenderness is known as conditioning (Lawrie, 1979).

The intramuscular connective tissues may contribute much to the background toughness of meat. In young animals such as calves and lambs the poorly cross-linked collagen is readily converted to soluble gelatine on cooking and the meat should always be tender. However, we have ample evidence to show that if the carcasses are not properly handled then the meat even from young animals can be extremely tough. Here the inducible toughness arises specifically from the muscle fibres, which are found to be strongly contracted. The effect of conditioning on meat quality is described in more detail in a previous publication (Etherington and Dransfield, 1981). In this chapter I shall commence with a brief description of glycolysis and how rigor develops when the post-mortem supply of ATP in the muscle is depleted. This is necessary in order to follow the subsequent enzymic processes that abolish rigor and then contribute to the further development of tenderness in the meat.

PRE-MORTEM FACTORS

Animals that are well-rested before slaughter have a substantial reserve of glycogen in the muscles, whereas animals that are physically exhausted after a long journey or have been badly jostled at the abattoir are quite likely to have depleted these glycogen reserves. Again in animals that are hunted, such as deer, not only is the muscle glycogen depleted at the kill, but the high energy reserves of creative phosphate (CP) may also be greatly reduced (Bendall, 1973). These stressful stimuli promote an increased output of adrenaline, which in turn accelerates the rate of glycolysis. The muscle glycogen level will in consequence progressively fall during the period of stress. Indeed it has been shown that an injection of adrenaline can be used experimentally as a means of depleting the muscle of glycogen (Bendall and Lawrie 1962). Starvation for a short period before slaughter may also cause the muscle glycogen levels to fall and this has been clearly demonstrated in rabbits and pigs, but for cattle the muscle glycogen reserve is normally spared even during prolonged starvation. (Bendall, 1973).

The time of onset of rigor is determined by the levels of glycogen and high energy phosphates present originally in the muscle at the time of death, and the lower these reserves the earlier will be the appearance of rigor. Therefore it can be seen that stress and exercise in lowering the glycogen level in the muscle before slaughter will consequently hasten the development of rigor in the carcass.

POST-MORTEM GLYCOLYSIS

The reader is referred to the review by Bendall (1973) for a detailed description of post-mortem glycolysis. Muscle remains metabolically active for several hours after death of the animal and may be stimulated to contract so long as there are sufficient ATP and CP present to supply the energy. ATP is in equilibrium with CP and in resting muscle ATP is still hydrolysed to ADP by non-specific ATPase activity and by each cell's effort to maintain its internal organisation. ATP is regenerated in the tissue by the breakdown of glycogen with glycolysis being stimulated specifically by an increase in the level of ADP. However, since the oxygen supply to the cells has ceased the pyruvate that is produced from glycolysis cannot be oxidised further. Instead this is rapidly reduced to lactate by NADH, the level of which also rises steadily in post-mortem muscle. Lactate is the normal end-product of anaerobic glycolysis as the equilibrium for pyruvate reduction is to the right with a ΔG value of - 6.0 kcal mole $^{-1}$ (Lehninger, 1975). Each glucose unit of the muscle glycogen allows the resynthesis of 3 ATP molecules from ADP and on balance $2H^+$ ions are generated with the result that the muscle pH is progressively lowered due to an accumulation of free lactic acid in the system. If the muscle glycogen is depleted before slaughter then glycolysis ceases when this has been consumed and in severe depletion there may be only a small fall of pH in the tissues. Where there is adequate glycogen glycolysis will continue until the pH has fallen to a value of about 5.5 and at this point, known as the limit value, no further breakdown of muscle glycogen can occur. The termination of anaerobic glycolysis at this pH value appears to be due to the sudden failure of the enzyme phosphofructokinase (Newbold and Scopes, 1967) and consequently no more ATP can be regenerated even where there is still some residual glycogen in the muscles.

The rate at which glycogen is consumed depends on a number of factors, the principal one probably being the amount of ADP that has accumulated in the muscle. Some years ago Scopes (1974) in this Institute devised a completely in vitro system for the examination of post-mortem glycolysis and concluded that the level of ATPase activity is the rate limiting factor in the early post-mortem regulation of glycogen breakdown. As glycolysis proceeds there is found to be less ATP and ADP available to the system. These nucleotides are progressively lost most probably by way of deaminating side reactions catalysed by the enzyme AMP deaminse, which is very much more active in the acidic muscle. More recently Bendall (1979) has demonstrated mathematical relationships between pH and the levels of lactate, phosphocreatine, ATP and the free level of ADP in the muscle and related these to changes in temperature and work done from contractions.

Post-mortem glycolysis is temperature sensitive and this can be easily demonstrated by measuring the rate of pH change as a function of muscle temperature. Glycolysis is slowed as the temperature is lowered, but below $12^{o}C$ the rate increases markedly. At these lower temperatures there is a cold-induced contraction of the muscle, which consumes more ATP because mechanical work is now done. The detailed mechanism of cold-contraction is not certainly known, but it appears that at low temperature the ATP-driven calcium pump is unable to maintain the normally very low calcium-ion level in the sarcoplasm of the resting muscle. The passive outflow of ions from the sarcoplasmic reticulum then triggers a myofibrillar contraction, which consumes ATP and this in turn stimulates a faster breakdown of glycogen. Raising the temperature can reverse the flow of calcium ions and again relax the muscle until such time as rigor develops, but beyond this point the muscle fibres will become locked in their shortened state.

Post-mortem glycolysis is hastened by stimulating the muscle to contract very shortly after slaughter and electrical stimulation is now becoming a routine procedure in some abattoirs for beef and lamb carcasses. A voltage of up to 1000V is applied ,in regular pulses for 2 minutes, from electrodes attached to the extremities of the carcass, and this causes a generalised contraction of the carcass muscles. Bendall (1980) has shown that a pH fall of 1.0 - 1.5 units is easily achieved during a 2 min stimulation period. The remainder of the muscle glycogen is then rapidly consumed while the muscles are still warm. The rapid depletion of glycogen by stimulation advances the onset of rigor, thus permitting the carcasses to be refrigerated much earlier and without the problem of cold-shortening.

DEVELOPMENT OF RIGOR

Rigor mortis occurs in the carcass when the muscle ATP is nearly depleted and glycolysis is slowing to a halt. When there is insufficient ATP present in the muscle cell, actin and myosin complex together. It is this cross-bridging of the thin and thick filaments in the myofibril that causes the muscle to stiffen and become inextensible. The extent to which these filaments are overlapped in the myofibril determines to a large degree the ultimate quality of the meat. Muscles that are greatly shortened at the time of rigor development therefore contain more attached cross-bridges and this gives an increase in toughness (Marsh and Leet 1966). This inducible toughness due to such fibre shortening is quite different from the previously mentioned connective tissue toughness.

Muscles that contain an adequate reserve of glycogen at the time the animal is slaughtered will not enter rigor until the pH approaches 6.0 and rigor development may not be complete until the muscle is near pH 5.5. However, where the muscle glycogen level is initially low, due perhaps to pre-slaughter stress or exercise, the onset of rigor will occur much earlier and at a correspondingly higher pH. Therefore irrespective of the initial state of the glycogen reserve in the muscle and the temperature changes in the muscle post-mortem, sooner or later the myofibrillar filaments will become permanently cross-bridged and the extent of this cross-bridging will depend on the degree of shortening in each muscle at the time of rigor development. The extent of this shortening can be determined histologically by finding the average sarcomere length in the muscle.

CONDITIONING OF MEAT

The rigidity of the muscle in rigor is not maintained indefinitely. Instead the muscle softens again, a process that was originally described as the "resolution of rigor". We now know that there is no reversal, as such, in the cross-bridging of the thin and thick filaments, but there is a loss of strength in the myofibrils. This time-dependent increase in fragility is of immense importance to the ultimate quality of the meat and the traditional hanging of carcasses was undertaken to gain maximal improvement in the flavour and texture of the meat. The rate of tenderising during this conditioning period is approximately the same for beef, veal and lamb,whereas this is two-fold higher for pork and twenty-fold higher for poultry,which suggests there are marked differences in the levels of muscle enzymes between these species. When beef carcasses are held under refrigeration at 1 - 3°C conditioning may take up to three weeks to complete. As with other enzymic processes the rate of tenderizing, as determined by mechanical sheer measurement, is faster at high temperatures and between 0° and 40° a temperature coefficient (Q_{10}) of 2.4 and activation energy of about 62 kJ/mole have been quoted for beef muscles (Etherington & Dransfield, 1981).

THE ACTION OF MUSCLE PROTEINASES IN CONDITIONING

The softening of the muscles post-rigor is caused by proteolytic disruption of the cross-bridged myofibrillar structures. These effects can be readily observed when post-rigor muscle is examined in the electron microscope. Here the disruption occurs initially at the junction of the I band with the Z disc (Penny, 1980) (Fig 1). If the muscle is stretched then long transverse gaps appear as the thin filaments are detached from their points of anchorage in the Z discs. As conditioning proceeds there is also degradation of the actual Z discs and in fully conditioned meat both the Z-disc and M-line structures are substantially degraded.

When the myofibriller proteins were separated by sodium dodecyl sulphate - polyacrylamide - gel electrophoresis then the most noticeable change that occurs during conditioning was found to be the progressive loss of troponin T (Penny, Voyle & Dransfield, 1974)(Fig 2). Studies in several laboratories have now confirmed that there is also a decrease in the content of desmin, C-protein, M-line protein and tropomyosin. At the same time there was shown to be an accumulation of smaller polypeptides with one particularly strong band of 30,000 daltons,possibly derived from troponin T (Azanza and others, 1979; Penny, 1980). The major component of the Z disc, α-actinin,was not apparently degraded and the disruption of the Z disc may therefore arise from the degradation of a minor protein that is involved in stabilizing the α-actinin lattice in this structure (Azanza and others, 1979; Penny, 1980).

Fig 1.   Electron micrograph of conditioned meat. Bovine
         M. semitendinosus 7 days post-mortem showing
         fractures at the junctions of the I band with the
         Z disc (from Penny, 1980).

As with other tissues, muscle contains both lysosomal cathepsins and neutral non
lysosomal proteinases, but generally the level of proteinase activity is lower
than for other soft tissues. A neutral serine proteinase of muscle that has
previously received much attention appears to reside specifically in the mast
cells and is not now considered to be a muscle cell enzyme (Bird and others, 1980).
The proteinases that may participate in the conditioning of meat, together with
the main pH range in which each is active, are listed in Table 1. This list is
still tentative as not all the enzymes have been certainly identified in muscle.
In addition there is less certainty concerning the actual location of these
enzymes in the muscle cells per se.

TABLE 1   Proteinases of muscle

Proteinase	Main pH range for proteolysis
Non-lysosomal	
Calcium-activated neutral proteinases (CANP)	6.0 - 8.5
Lysosomal	
Cathepsin B	3.0 - 6.0
Cathepsin D	2.5 - 4.5
Cathepsin H	5.0 - 7.0
Cathepsin L	3.0 - 6.5
Cathepsin N	3.0 - 6.0

Storage   Time   (days)

0   1   2   5   7   14

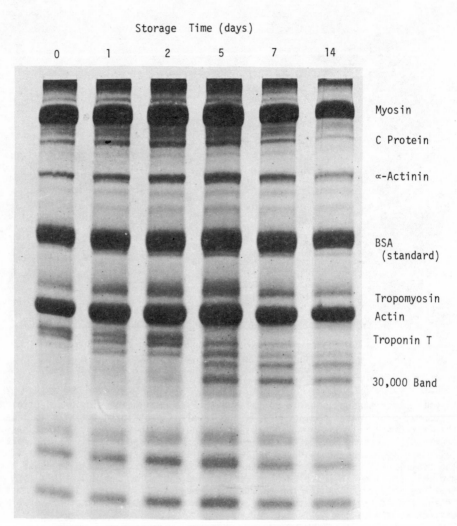

Myosin

C Protein

α-Actinin

BSA
(standard)

Tropomyosin

Actin

Troponin T

30,000 Band

Fig 2.   Electrophoretic analysis of muscle proteins.
         Myofibrils from bovine M. semitendinosus were
         stored at 5°C and the protein components examined
         up to 14 days by sodium dodecyl sulphate-
         polyacrylamide - gel electrophoresis. During
         conditioning there was a loss of troponin T and
         several other proteins were reduced in amount.
         Of the new bands to be formed   the one of
         30,000 daltons was the most noticeable (from
         Penny, 1980).

The initial destruction at the Z disc while the muscle pH is still high is most probably caused by the calcium-activated neutral proteinase (CANP) as this enzyme is active at higher pH values than the lysosomal cathepsins and in muscle that remains at high pH it may be the only effective enzyme. CANP is present in many tissues but appears to be more abundant in muscle . CANP has been purified from several mammalian species and its action on myofibrillar proteins studied in many laboratories. It was reported that the enzyme from pig or beef muscle consists of two dissimilar subunits of MW 80,000 and 30,000 daltons (Dayton and others, 1975). However, rabbit CANP was found to exist as a single polypeptide chain of 73,000 daltons (Azanza and others, 1979). The enzyme requires 1-5 mM Ca^{2+} for full activity and exhibits negligible action at 0.1 mM Ca^{2+} or less. On incubating myofibrils with CANP the Z discs were found to disappear rapidly and free α-actinin was released into the supernatant, (Dayton and others 1975, Azanza and others, 1979, Penny, 1980). CANP also degraded desmin and the loss of this protein from the myofibrils was followed using an immuno-histochemical technique (Lazarides, 1980). Desmin is believed to be attached to the periphery of each myofibrillar Z disc. A possible role for this protein could be to establish the myofibrils in correct register by transversly linking these structures across the muscle fibre at the positions of the Z disc.

CANP appears to be a sarcoplasmic enzyme (Dayton and others, 1975), but from immuno - histochemical evidence it is located close to the Z discs (Ishiura and others, 1980). In the living cell the activity of the enzyme is clearly under very rigid cellular control,which is regulated partly by inhibitors and partly by the level of available calcium ions. The concentration of Ca^{2+}in the sarcoplasm is normally low, of the order of 10^{-8}M, but with a rise to about 10^{-5}M when the muscle is stimulated. An ATP - driven Ca^{2+} pump maintains the low level in relaxed muscle by concentrating the ions in both the mitochondria and in the sarcoplasmic reticulum. With the post-mortem depletion of muscle ATP, the calcium pump fails and the sarcoplasmic level of Ca^{2+} rises, but it remains uncertain whether this may reach a millimolar level as required by CANP in vitro. However recent studies have indicated that in the presence of a physiological concentration of Mg^{2+} (approximately 10mM) there was a diminished requirement for Ca^{2+} and at 0.1 mM 50% of full activity was achieved (Cottin and Ducastaing, personal communication). Unpublished experiments in this laboratory by Penny with washed fibre fragments have demonstrated that autolytic breakdown of the Z disc at pH 7.0 can occur in a medium containing as little as 0.1 mM calcium chloride. Furthermore it was shown that desmin was degraded under these conditions. The reason for the lower requirement of calcium is unclear. One possibility is that the physiologically-effective form of CANP in the muscle cell is much more sensitive to Ca^{2+} ions, whereas the form normally isolated is possibly an autolytic modification of the true enzyme and this is less able to bind calcium. CANP also possesses an essential thiol group that must be in the reduced form for the enzyme to be active. In post-mortem muscle the very low oxygen tension will automatically promote thiol activation and thus the enzyme is unlikely to be blocked at this site in its active centre.

From the many studies that have been made on CANP it is clear that its activity in vitro is consistent with many of the observed changes that take place in muscle when this is converted into meat. However, what is far less certain is the extent to which this enzyme can act on myofibrils if the carcass has been electrically stimulated soon after slaughter. During the 2 mins of stimulation the pH can fall as low as 6.0 and this is also at the lower end of the effective range of action for CANP. Therefore in stimulated carcasses we may anticipate that there is a lesser contribution from this particular enzyme.

There are claims to a measurable improvement in meat quality if carcasses are stimulated. The reasons for this are unclear, but recent data indicate that the lysosomal cathepsins may be participating to a greater degree in myofibrillar degradation (Dutson, Smith and Carpenter, 1980). In contrast to our knowledge of CANP we have less information on the action of the lysosomal enzymes during meat conditioning. However, the sudden fall in pH during electrical stimulation while the carcass is still hot will promote an earlier disruption of the muscle lysosomes. Presumably after their release from the lysosomes the enzymes can diffuse readily into the myofibrillar structures. It has been known for some years that the lysosomes lose their characteristic latency in the post-rigor muscle tissue due to the low pH (Dutson and Lawrie, 1974). Now we have evidence to show that for electrically-stimulated carcasses, more than half of the total measurable lysosomal activity is released within the first hour following stimulation, much sooner than for non-stimulated carcasses. These enzymes therefore are not only released earlier but they will exhibit greater proteolytic activity than in non-stimulated carcasses as the pH is reduced while the temperature of the muscles is still elevated. (Dutson, Smith and Carpenter, 1980).

At the limit pH of 5.5 the muscle enzymes promote a further improvement in the quality of the meat while this is held in refrigerated, but not frozen storage. Since enzyme action is temperature-dependent, proteolytic degradation can be accelerated if the carcasses are held for a while above or near ambient temperatures. Furthermore, there is an early release of enzymes from the lysosomes under these conditions (Moeller and others 1976).

The effect of pH on myofibrillar proteolysis was studied in controlled autolysis of muscle homogenates (Penny and Ferguson-Pryce, 1979). Above pH 6.0 CANP was the most effective enzyme but below pH 6 proteolysis was greatly enhanced in the presence of EDTA and thiol activators, which indicates that the cysteine cathepsins of the muscle lysosome are the more effective enzymes during the main conditioning period. Autolysis is comparatively slow near the final post-mortem pH of 5.5 and in these experimental studies autolytic breakdown was found to be most rapid near pH 4. In the carcass these enzymes act secondarily to the neutral-proteinase activity and the myofibrillar structures are already partially damaged, but in most laboratory studies intact myofibrils or myofibrillar proteins have been employed as test substrates. Therefore it remains uncertain which of the lysosomal proteinases are the more important in the conditioning process when the muscle pH has reached a limit value of about 5.5. The contribution of the major aspartate proteinase, cathepsin D, to conditioning would seem to be small as this enzyme generally requires a pH below 4.5 for activity. However, when myofibrils were incubated with purified cathepsin D at 0.3 or 1.2 units/ml for 24 h at 25⁰ and near the post-mortem limit pH the Z discs were found to be substantially degraded especially at the higher concentration (Robbins and others, 1979). There was also some breakdown of the myosin heavy chains. Rat muscle cathepsin D, MW 41,000 daltons and cathepsin B, MW 27,000 daltons, were purified and their ability to degrade myofibrillar proteins studied (Bird, Schwartz and Spanier, 1977; Schwartz and Bird, 1977). Cathepsin B is an endopeptidase that will hydrolyse various protein substrates in the pH range 5 - 6. With extracted myosin as substrate at pH 5.5 cathepsin B exhibited 60-70% of its full activity. Cathepsin B also degraded actin and for both proteins the degradation products were still somewhat large polypeptide chains. Cathepsin D degraded both actin and myosin, but now smaller peptide fragments were generated although cathepsin D was less active than cathepsin B at the conditions prevailing in post-mortem muscle.

Like cathepsin B, cathepsins H, L and N are all cysteine proteinases, Cathepsin L, MW 22,000 daltons, has been purified from rat liver and other tissues and substantially characterized by Kirschke and co workers (1977a). This enzyme is also an endopeptidase exhibiting maximal activity in the region of pH5.5-6 for most substrates. Experimentally it is readily identified from its ability to degrade azocasein at pH 6.0, in contrast to the other lysosomal cathepsins,which do not cleave the substrate at this pH. Rat cathepsin L was reported to degrade actin in vitro at pH 5.5 - 6.0 (Bird and others, 1980). An enzyme resembling cathepsin L has been purified to near homogeneity from extracts of rabbit muscle (Okitani and others, 1980a). This enzyme caused substantial breakdown of the Z disc α-actinin in isolated myofibrils at pH5.0. The enzyme also cleaved myosin, but only slowly above pH5.0. Proof of identity for cathepsin L is lacking as the enzyme was not tested against azocasein. However, a caseinolytic activity, resembling the present enzyme, was identified in rabbit muscle by the same laboratory some years previously (Arakawa and others, 1967).

Cathepsin H, MW 26,000 daltons, is both an endopeptidase and an exopeptidase. The enzyme has also been purified from rat liver by Kirschke and co workers (1977b), but it remains uncertain if the enzyme exists in muscles. It was reported that cathepsin H can degrade myosin in vitro at pH 5.5 - 6.0 (Bird and others, 1980). An enzyme similar to cathepsin H, but with a much larger MW, 340,000 daltons was purified from rabbit muscles (Okitani and others, 1980b). The activity of the enzyme against myofibrillar proteins was not described.

As stated previously intramuscular connective tissue contributes to the background toughness of meat and this toughness increases with the age of the animal. How much hydrolysis of the connective tissues does occur during conditioning is difficult to assess and at the present time is still not easily demonstrated. The collagen is not solubilized, but there is evidence that the connective tissues may be weakened presumably by a limited number of proteolytic 'clips' in the structure. This can be seen from an increase in the amount of soluble gelatin that is subsequently released during cooking (Herring, Cassens and Briskey, 1967). These breakages could be caused by cathepsins and certainly a crude enzyme preparation did produce histological changes to the endomysium when treated muscle fibres were examined in the electron microscopes (Robbins and Cohen, 1976). The strength of the connective tissues can now be estimated independently of the myofibrils. If a suitable tissue containing collagen fibres is heated under isometric conditions along the fibre axis then a tension is generated as the collagen component denatures. The size of this tension is determined both by the quantity of the collagen in the sample and the extent to which this collagen is cross-linked. In preliminary experiments it has been demonstrated that if intramuscular collagen is first incubated with a lysosomal lysate preparation at pH 5.5 then the isometric tension value is reduced (Kopp and Valin, 1980). Thus these enzymes have caused a sufficient number of breakage points in the connective tissue such that the structure now fails more readily during denaturation.

Three lysosomal enzymes are known to possess collagenolytic activity; these are cathepsins B, L and N (Etherington, 1980; Kirschke, 1980). This enzymic property has been investigated mainly from in vitro testing of the enzymes against defined collagenous substrates. With insoluble fibrous collagen as substrate there was negligible solubilization above pH 4.5. When, however, monomeric collagen in solution was examined under non-denaturing conditions the non-helical telopeptides, on which the cross-links originate, were readily cleaved from the main non-helical part of the molecule at pH values up to 6.5 (Etherington 1980; Etherington and Evans 1977). It is not yet known how far these individual cathepsins can weaken the connective tissues at pH 5.5, but here we may also anticipate that the enzymic 'clips' occur in these telopeptide regions. Now that we have the technique for measuring isometric tension during thermal denaturation, it should be possible to examine this aspect of conditioning in some detail.

As already discussed, cathepsin B and L may be involved in the breakdown of myofibrillar structure, but cathepsin N appears to have a very limited action on proteins other than collagen. The enzyme has been purified from both bovine and human tissues with molecular-weight values of 20,000 and 35,000 daltons respectively. In its action against collagen cathepsin N was almost indistinguishable from cathepsin B (Etherington 1980). For these enzymes to have any action on muscle connective tissue post-mortem they must be released first into the extra-cellular space and the time course for this may be significantly longer than the time required initially to exit from the lysosomes. Furthermore these enzymes need not necessarily come from the muscle cells per se. There are other sources of lysosomal enzymes, such as fibroblasts, mast cells and leucocytes, that could release these proteinases into the connective tissues. At present we have no way of knowing how soon post-slaughter these enzymes permeate into the connective tissues.

CONCLUDING REMARKS

From the many studies that have been made of the conversion of muscle into meat we can say that most of our knowledge of the conditioning process indicates that a substantial part of the improvement is due to the breakdown of the myofibrillar structures, especially at the junctions of the Z disc with the I band. It is technically much more difficult to define the extent of post-mortem weakening of the connective tissues and it remains uncertain how far conditioning does indeed affect the extracellular matrix. Modern meat handling techniques are attempting to reduce the time between slaughter and the retail sale of the meat as wholesale storage is expensive. The use of electrical stimulation to reduce rapidly the reserves of muscle glycogen is permitting an earlier action of the muscle enzymes in the conditioning process. However, although the overall affect appears to be similar, the relative contribution of the lysosomal enzymes to this process may well be enhanced following the rapid depletion of glycogen and early acidification of the muscles.

REFERENCES

Arakawa, N., S. Fujiki, C. Inagaki, and M. Fujimaki (1967). A catheptic
protease active in ultimate pH of muscle. Agr. Biol. Chem., 40, 1265-1267.
Azanza, J-L., J. Raymond, J-M. Robin, P. Cottin, and A. Ducastaing (1979).
Purification and some physico-chemical and enzymic properties of a calcium
ion-activated neutral proteinase from rabbit skeletal muscle. Biochem. J.,
183, 339-347.
Bailey, A.J. (1972). The basis of meat texture. J. Sci. Food Agric., 23,
995-1007.
Bendall, J.R. (1973). Post-mortem changes in muscle. In E.J. Briskey,
R.G. Cassens and B.B. Marsh (Eds.), The Structure and Function of Muscle.
Vol. 2, 2nd ed. Academic Press, New York, pp. 243-309.
Bendall, J.R. (1979). Relations between muscle pH and important biochemical
parameters during the post-mortem changes in mammalian muscles. Meat Sci.,
3, 143-147.
Bendall, J.R. (1980). The electrical stimulation of carcasses of meat animals.
In R. Lawrie (Ed.), Developments in Meat Science, Vol. 1. Applied Science
Publishers, Barking. pp. 37-59.
Bendall, J.R., and R.A. Lawrie (1962). The effect of pre-treatment with various
drugs on post-mortem glycolysis and the onset of rigor mortis in rabbit
skeletal muscle. J. Comp. Pathol., 72, 118-130.
Bird, J.W.C., J.H. Carter, R.E. Triemer, R.M. Brookes, and A.M. Spanier (1980).
Proteinases in cardiac and skeletal muscle. Fed. Proc., 39, 20-25.
Bird, J.W.C., W.N. Schwarz, and A.M. Spanier (1977). Degradation of myofibrillar
proteins by cathepsins B and D. Acta Biol. Med. Germ., 36, 1587-1604.
Dayton, W.R., D.E. Goll, M.H. Stromer, W.J. Reville, M.G. Zeece, and R.M. Robson
(1975). Some properties of a Ca^{++}- activated protease that may be involved
in myofibrillar protein turnover. In E. Reich, D.B. Rifkin and E. Shaw (Eds),
Proteases and Biological Control. Cold Spring Harbor Laboratory. pp. 551-577.
Dutson, T.R., and R.A. Lawrie (1974). Release of lysosomal enzymes during
post-mortem conditioning and their relationship to tenderness. J. Food Technol.
9, 43-50.
Dutson, T.R., G.C. Smith, and Z.L. Carpenter (1980). Lysosomal enzyme
distribution in electrically stimulated ovine muscle. J. Food Sci., 45,
1097-1098.
Etherington, D.J. (1980). Proteinases in connective tissue breakdown. In
D. Evered and J. Whelan (Eds.), Ciba Foundation Symposium No 75: Protein
Degradation in Health and Disease. Excerpta Medica, Amsterdam.pp. 87-100.
Etherington, D.J., and E. Dransfield (1981). Enzymes in the tenderisation of
meat. In G.G. Birch, K.J. Parker, and N. Blakebrough (Eds.), Enzymes and
Food Processing, Applied Science Publishers, Barking. (In press).
Etherington, D.J., and P.J. Evans (1977). The action of cathepsin B and
collagenolytic cathepsin in the degradation of collagen. Acta Biol. Med.
Germ., 36, 1555-1563.
Herring, H.K., R.G. Cassens, and E.J. Briskey (1967). Factors affecting collagen
solubility in bovine muscles. J. Food Sci., 32, 534-538.
Ishiura, S., H. Sugita, I. Nonaka, and K. Imahori (1980). Calcium-activated
neutral proteinase: its localization in the myofibril, especially at the
Z-band. J. Biochem., 87, 343-346.
Kirschke, H., J. Langner, S. Siemann, B. Wiederanders, S. Ansorge, and P. Bohley
(1980). Lysosomal cysteine proteinases. In D. Evered and J. Whelan (Eds.),
Ciba Foundation Symposium No 75: Protein Degradation in Health and Disease.
Excerpta Medica, Amsterdam. pp. 15-31.
Kirschke, H.,J. Langner, B. Wiederanders, S. Ansorge, and P. Bohley (1977a).
Cathepsin L. Eur. J. Biochem., 74, 293-301.

Kirschke, H., J. Langner, B. Wiederanders, S. Ansorge, and P. Bohley (1977b). Cathepsin H: an endoaminopeptidase from rat liver lysosomes. Acta Biol. Med. Germ., 36, 185-199.

Kopp, J., and C. Valin (1981). Can muscle lysosomal enzymes affect muscle collagen post-mortem? Meat Sci., 5, (in press).

Lawrie, R.A. (1979). Meat Science, 3rd ed. Pergamon Press, Oxford. pp.132-162.

Lazarides, E. (1980). Intermediate filaments as mechanical integrators of cellular space. Nature, 283, 249-256.

Lehninger, A.L. (1975). Biochemistry. Worth Publishers, New York.pp.417-439

Marsh, B.B., and N.G. Leet (1966). Studies in meat tenderness: III the effects of cold shortening on tenderness. J. Food Sci., 31, 450-459.

Moeller, P.W., P.A. Fields, T.R. Dutson, W.A. Landmann, and Z.L. Carpenter (1976). Effect of high temperature conditioning on sub-cellular distribution and levels of lysosomal enzymes. J. Food Sci. 41, 216-217.

Newbold, R.P., and R.K. Scopes (1967). Post-mortem glycolysis in ox skeletal muscle. Biochem. J., 105, 127-136.

Okitani, A., U. Matsukura, H. Kato, and M. Jujimaki (1980a). Purification and some properties of a myofibrillar protein-degrading protease, cathepsin L, from rabbit skeletal muscle. J. Biochem., 87, 1133-1143.

Okitani, A., T. Nishimura, Y. Otsuka, U. Matsukura, and H. Kato (1980b). Purification and properties of BANA hydrolase H of rabbit skeletal muscle, a new enzyme hydrolysing α N-benzoyl-arginine-β-naphthylamide. Agric. Biol. Chem. 44, 1705-1708.

Penny, I.F. (1980). The enzymology of conditioning. In R. Lawrie (Ed), Developments in Meat Science, Vol. 1. Applied Science Publishers, Barking. pp. 115 - 143.

Penny, I.F., and R. Ferguson-Pryce (1979). Measurement of autolysis in beef muscle homogenates. Meat Sci., 3, 121-134.

Penny, I.F., C.A. Voyle, and E. Dransfield (1974). Tenderizing effect of a muscle proteinase on beef. J. Sci. Food Agric., 25, 703-708.

Robbins, F.M., and S.H. Cohen (1976). Effects of catheptic enzymes from spleen on the microstructure of bovine semimembranosus muscle. J. Text. Stud., 7, 137-142.

Robbins, F.M., J.E. Walker, S.H. Cohen, and S. Chatterjee (1979). Action of proteolytic enzymes on bovine myofibrils. J. Food Sci., 44, 1672-1677.

Scopes, R.K. (1974). Studies with a reconstituted muscle glycolytic system: the rate and extent of glycolysis in simulated post-mortem conditions. Biochem. J., 142, 79-86.

Schwartz, W.N., and J.W.C. Bird (1977). Degradation of myofibrillar proteins by cathepsins B and D. Biochem. J., 167, 811-820.

PREPARATION AND APPLICATION OF IMMOBILIZED PROTEINASES

J. Turková

Institute of Organic Chemistry and Biochemistry, Czechoslovak Academy of Science, 166 10 Prague 6, Czechoslovakia

ABSTRACT

Chymotrypsin, trypsin, papain, pepsin and acid proteinase of A. oryzae were immobilized on different supports developed in Czechoslovakia (hydroxyalkyl methacrylate gels - Spheron or Separon, glycidyl methacrylate copolymer, periodate-oxidized glucosehydroxyalkyl methacrylate gel and cellulose in bead form, polyamides - APA, poly(ethylene terephthalate) - Sorsilen and aluminium oxide activated with titanic chloride) using different coupling methods. Immobilized proteinases were studied from the theoretical point of view (the dependence of the activity on the surface area and porosity of support and on the amount of coupled enzyme, the influence of the nature of supports and coupled proteins on the coupling procedure, the stability in dependence on the nature of supports and the number of the bounds between the enzyme and matrix) and in connection with their use in organic synthesis (esterification of acetyl-L-tryptophan by ethanol in an immiscible organic solvent, chloroform or toluene) and from technological aspects (chill-proofing of beer or wine). From the comparison of different derivatives. it is concluded that neither a universal support nor a generally most convenient method of coupling of proteinases to a solid support exist. When a certain proteinase is immobilized an individual method of coupling must be selected according to its future use.

KEYWORDS

Immobilized proteinases; advantages of their use; preparations; influence of nature of supports; esterification of amino acids; chill-proofing of beer or wine; criteria for selection of support and coupling procedure.

INTRODUCTION

The amount of the enzyme bound to solid support and the activity of immobilized biocatalyst, but also its stability, temperature or pH optimum of catalysis and other factors can be affected by the

choice of the support and the method of coupling the enzyme. Therefore it is very important to study these relationships.

In our experiments - both from the practical and the theoretical
viewpoint - we used exclusively supports developed in Czechoslovakia. It is a great advantage that in Prague many polymer chemists
are interested in the development of solid supports. A collaboration with them gives us a possibility to influence some properties
of the developed supports. In Table 1 a survey of supports is presented which have been developed in Czechoslovakia and are discussed in this article.

HYDROXYALKYL METHACRYLATE GELS

The copolymerization of hydroxyalkyl methacrylate with alkylene
dimethacrylates gives rise to heavily cross-linked microparticles
of a xerogel which subsequently aggregate and yield macroporous
structures of spheroids (Turková and Seifertová, 1978). Because
of this structure the gels share many properties in common with
the most commonly used support, agarose. Thus, e.g., the hydroxyl
groups of the gel can be activated with cyanogen bromide (Turková
and co-workers, 1973) in analogy to the hydroxyl groups of agarose.
Amino acids, peptides, and proteins can be bound to the activated
gels through their amino groups. At the same time, however, the
gels resemble-because of their macroreticular structure - inorga-
nic supports. They do not practically change their volume with
changes in pH or after the addition of organic solvents.

The Dependence of the Amount of Coupled Chymotrypsin on the Sur-
face Area and Porosity of Hydroxyalkyl Methacrylate Gels

Table 2 shows by way of example the quantities of chymotrypsin, a
high molecular weight product, and the quantities of glycin, a low
molecular weight compound, attached to seven types of Spheron
after their BrCN activation (Turková and co-workers, 1973). The
individual types are listed in order of decreasing exclusion molecular weights. As can be seen in the Table 2, the number of the
Spheron type times 10^3 gives the molecular weight of the product
emerging in the hold-up volume. The individual gels considerably
differ in their specific surface. The quantity of attached chymotrypsin is directly proportional to this surface, whereas the
quantity of attached glycine indicates only small differences in
the number of reactive hydroxyl groups. Table 2 shows also the
proteolytic and esterolytic activities of the individual enzymatic
preparations. These values were obtained by testing with denatured
hemoglobin and acetyl-L-tyrosin ethyl ester as substrates.

The Dependence of the Activity of Immobilized Pepsin on the Amount
of Coupled Enzyme

The coupling of proteins to cyanogen bromide activated supports
occurs in alkaline or at least in neutral media. It cannot be
therefore used for binding of carboxylic proteinases most of which
are inactivated at pH 5 or higher. We used therefore supports with

TABLE 1. Survey of Supports Developed in Czechoslovakia and Discussed in this Paper

Support	Developed by:	at:	Commercial name	Method of Coupling	Price
Hydroxyalkyl methacrylate gels	Čoupek, Křiváková and Pokorný (1973)	Inst.Macromol. Chem., Czech. Acad.Sci.	Spheron Separon	BRCN activation Carbodiimide activation	High
Glycidyl methacrylate copolymer	Švec and co-workers (1975)	"	—	Suspension	High
Cellulose in bead form	Peška, Štamberg and Hradil (1976)	"	—	Periodate - oxidation	Low
Glucosehydroxyalkyl methacrylate gel	Filka, Kocourek and Coupek (1978)	Laboratory Instruments Works	Separon H 1000-glc	"	High
Hydroxyalkyl methacrylate gel modified by epichlorohydrin	Gulaya and co-workers (1979)	"	Separon 300 E	Suspension	High
Polyamides	Kubánek and co-workers (1978)	Inst.Chem.Technology, Dept. Polymer.Chem.	APA	Partial hydrolysis and activation with glutaraldehyde	Low
Poly(ethylene terephthalate)	Budín and Kubánek (1980)	"	Sorsilen	Adsorption	Very low

TABLE 2. Dependence of the Amount of Chymotrypsin Bound to Hydroxyalkyl Methacrylate Gels (Spheron) on their Surface Areas

Proteolytic and esterolytic activities are indicated.

Gel	Mol.wt. exclusion limit	Specific surface (m2/ml)	Amount of bound glycine (mg/ml)	Amount of bound chymotrypsin (mg/ml)	Proteolytic activity A_{280} nm/min. ml	Relative proteolytic activity (%)	pH optimum of esterase activity	Esterase activity (moles/min·ml)	Relative esterase activity (%)
Spheron-10^5	10^8	0.96	0.5	0.73	–	–	–	–	–
Spheron-10^3	10^6	5.9	3.1	7.8	1.23	44	9.4	305	29
Spheron-700	700 000	3.6	2.8	6.7	1.17	48	9.1	392	43
Spheron-500	500 000	23.0	2.6	17.1	2.28	37	9.2	810	35
Spheron-300	300 000	19.5	3.15	17.7	2.8	44	9.1	1320	55
Spheron-200	200 000	0.6	2.3	6.9	1.33	53	9.0	626	67
Spheron-100	100 000	0.2	2.6	4.3	0.58	38	9.1	354	61

spacers containing amino or carboxyl groups for the attachment of
pepsin (Valentová and co-workers, 1975). Amino-Spheron was prepared by coupling 1,6-hexamethylenediamine to Spheron, carboxy-
Spheron by coupling ε-aminocaproic acid. Using soluble N-ethyl-N´-
(3-dimethylaminopropyl)carbodiimide methoiodide, we were able to
attach pepsin to these supports through its carboxyl and amino
groups respectively.

TABLE 3. Properties of Pepsin Immobilized on Hydroxyalkyl Methacrylate Gels

Support	Quantity of Pepsin Bound (mg/g)	Bed Volume (ml/g)	Proteol.Acti-vity of Bound Pepsin (A_{280}/ /min.mg bound enzyme)	Relative Pro-teol.Activity of Bound Pep-sin (%)
NH_2-Spheron P-300	13	3.3	4.0	92.8
NH_2-Spheron 300	46.8	3.3	2.8	65.7
COOH-Spheron 300	50.8	3.3	2.0	45.3
NH_2-Spheron 1000 – 100% BTD	65	4.0	1.6	37.8

As shown in Table 3, in both cases active preparations of immobilized pepsin were obtained. It is also evident, that the relative
proteolytic activity of immobilized pepsin preparations decreases
with the increasing quantity of enzyme bound.

EPOXIDE-CONTAINING SUPPORTS

The use of supports bearing epoxide groups offers the main advantage in the simple coupling reaction, which involves only the
treatment of the dry gel with a solution of the protein that should
be coupled. The epoxide group of the carrier can react with a number of groups occurring in proteins as amino, thiol, hydroxyl,
carboxyl, phenol, imidazole or indole groups.

Under analogous conditions chymotrypsin, papain and the polyvalent
natural trypsin inhibitor antilysine were coupled to glycidyl methacrylate gel in dependence on pH. When the pH increased up to
pH 6 the amount of the chymotrypsin attached decreased gradually,
while above pH 7 it slightly increased. The dependence of the
amount of attached antilysine on pH showed just the opposite course
with a maximum of binding at pH 10. The maximum of papain attachment to a glycidyl methacrylate gel was achieved at pH 7. From
what has been said it follows that the optimum pH for the binding
of individual proteins differs considerably.

In order to determine to what extent the nature of the solid matrix is involved, serumalbumin was coupled onto (2,3-epoxypropoxy)-
propyl glass, two types of (2,3-epoxypropoxy)-propyl-silica gel,
Epoxy-activated Sepharose 6B and oxirane-acrylic beads (Röhm
Pharma) in dependence on pH. The amount of the attached protein
as a function of pH of the reaction mixture is considerably affected by the nature of the solid matrix. While the largest amount
of serumalbumin was bound on the derivatives of glass and silica
gel in the acid region (optimum pH 4-5), in the case of Sepharose
the maximum was at alkaline pH (pH 9). The amount of the protein
coupled to oxirane acrylic gel practically did not change with pH.
In order to explain these differences in greater detail the study
of the coupling with model peptides has been studied (Turková,
1980; Turková and co-workers, 1978).

IMMOBILIZATION VIA REDUCTIVE ALKYLATION

The advantages of the immobilization of enzymes via reductive alky-
lation on solid carriers containing diols have been demonstrated
by Royer and co-workers (1975). The use of their method for the
coupling of trypsin onto periodate oxidized glucose-hydroxyalkyl
methacrylate gel and cellulose beads permits a comparison of the
methacrylate matrix with a natural polysaccharide support. The
first step of the reaction sequence for the coupling of protein
to polysaccharides is the oxidation of the saccharide molecules
with $NaIO_4$. Aldehyde groups are formed which bind the amino groups
of the proteins as Schiff s bases. The bond is further stabilized
by addition of $NaBH_4$ which also eliminates the unreacted aldehyde
groups.

Optimum conditions for the coupling of trypsin onto hydroxyalkyl
methacrylate gel containing glucose on its surface (Separon H1000-
glc) after the oxidation of the gel with periodate were developed
by Vančurová and co-workers (1979). The amount of trypsin coupled
on periodate-oxidized Separon H1000-glc in dependence on pH increases first moderately with the increasing pH, but above pH 8 the
increase is steep up to pH 10. The amount of the immobilized trypsin and its activity determined by means of benzoyl-L-arginine
p-nitroanilide as substrate increases with increasing concentration of free trypsin in the reaction mixture during the coupling.
For hydroxyalkyl methacrylate gel the initial high yield of the
bound protein determined from mass balance is characteristic; it
decreases steeply with the increasing concentration of the enzyme
in the reaction mixture, from 90 to 20%.

Optimum conditions for the coupling of trypsin onto cellulose in
bead form after its oxidation with periodate have been developed
by Turková and co-workers (1979). Equally as in the case of the
preceding gel the rate of binding of trypsin increases with the
increasing pH. However, the solubilization of cellulose derivatives also increases with pH, due to alkaline degradation yielding
soluble immobilized trypsin fractions and soluble oxidized oligo-
saccharides. Equally as with glucose-hydroxyalkyl methacrylate gel
the amount of bound trypsin and also its activity increase with
increasing concentration of free trypsin in the reaction mixture
during the coupling. The yield of the bound protein, determined

from mass balance, does not exceed 40% even at very low trypsin
concentration in the reaction solution during the coupling.

The Thermal Stability of Immobilized Chymotrypsins in Dependence on the Nature of Supports

In order to check the effect of the nature of the support (cellulose, Separon-glc and Separon E) on the thermal stability of immobilized chymotrypsins, we followed the decrease in activity after
10 min incubation at pH 7.3 and temperatures in the 45-90°C interval. The experiment was carried out in collaboration with the co-
workers of the Bioorganic Institute of the USSR Academy of Sciences in Moscow (Turková and co-workers, 1980). From Fig.1 it is
evident that the dependence of the decrease of activity on temperature differs in individual preparations. The course of the

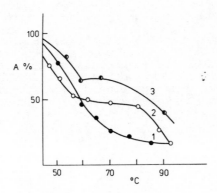

Fig. 1. Activity of chymotrypsin immobilized on
        Separon-glc (1), cellulose (2), and Se-
        paron-E (3) in dependence on temperature
        after 10 min incubation at pH 7.3

curves indicates that the preparations are not homogeneous with
respect to their thermal denaturation. The differences observed
may be explained by the differences in the number of bonds between
the protein surface and the solid support. Therefore, evidently,
chymotrypsin bound to hydroxyalkyl methacrylate gel modified with
epichlorohydrin (Separon E) displays the highest thermal stability,
due to the high number of groups in proteins which can react with
the epoxide groups of the support. The number of bonds between
the protease and the solid support should not be excessively large,
however, in order to prevent the deformation of the native structure of the enzyme.

APPLICATION OF IMMOBILIZED PROTEINASES

As regards the application of immobilized proteases, two main lines
are followed at our Institute. Primarily we take advantage of the
fact that we are an Institute of organic chemistry and biochemistry

where not only proteases have been investigated for a number of
years, but also the synthesis of peptides. As an example of the
use of proteases in organic synthesis let me mention the esterification of N-acetyl-L-tryptophan with ethanol, catalyzed with immobilized chymotrypsin. The second line is classical proteolysis. In
collaboration with co-workers from applied research we are studying the use of immobilized proteases for the cleavage of proteins
in beer and in wine, in order to increase their stability.

Synthesis of Ac-L-TrpOEt from Ethanol and Ac-L-Trp in Biphasic Aqueous-chloroform Medium Catalyzed by Chymotrypsin Adsorbed on Sorsilen

We have already learned an important rule from our co-workers in
the field of applied research: for each application always use
the cheapest possible support and the cheapest possible method of
immobilization. Since we carry out the esterification of N-acetyl-
-L-tryptophan according to Klibanov and co-workers (1977), i.e.
in an organic solvent immiscible with water (chloroform), simple
sorption to a very cheap poly(ethylene terephthalate) - Sorsilen -
suffices for the immobilization of chymotrypsin. From the determination of the dependence of the amount of the adsorbed chymotrypsin, papain, pepsin and serumalbumin on pH it is again evident
that the individual proteins are bonded in completely different
ways. The amount of the chymotrypsin adsorbed decreases with increasing pH.

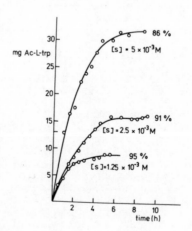

Fig. 2. Kinetics of chymotrypsin-catalyzed synthesis of N-acetyl-L-tryptophan ethyl ester.
% of esterification for substrate solutions
of different concentration are indicated.

The esterification of Ac-L-Trp was carried out in the following
manner: Chymotrypsin adsorbed on Sorsilen was placed in 30 ml of
chloroform containing 5×10^{-3}M (or 2.5×10^{-3}M or 1.25×10^{-3}M)
N-acetyl-L-tryptophan and 1M ethanol. The system was continuously
stirred at room temperature. Aliquots taken in 30 min intervals
were analyzed by HPLC.

The kinetics of the esterification of Ac-L-Trp, catalyzed by immobilized chymotrypsin, is shown in Fig. 2. From this it is evident
that the % of esterification increase with decreasing concentration of the substrate.

The kinetics of esterification was studied with a preparation of
chymotrypsin adsorbed onto Sorsilen at pH 8, possessing an activity of 9.6 units A_{280}/min,g of wet preparation. When a preparation
showing the activity 11 units/min,g (chymotrypsin adsorbed at pH 7)
was used under analogous conditions the yield of esterification
increased from 86% to 95%. If a two-phase system was used, i.e.
water-immiscible organic solvent, the enzyme remained in the
aqueous phase on the surface of the solid support, where it is protected against denaturation by the organic solvent. Evidently the
catalysis takes place in the interpspace.

Derivatives of Immobilized Proteinases for Chill-Proofing of

Beer or Wine

We prepared the first samples of immobilized papain by coupling
of papain onto the hydroxyalkyl methacrylate gel, activated with
cyanogen bromide (Turková and co-workers, 1973). The immobilization process was much facilitated by the use of p-nitrophenol ester
derivatives of hydroxyalkyl methacrylate gel (NPAC) (Turková, 1976;
Čoupek and co-workers, 1977). Both these preparation were proofed
good not only in laboratory tests for the chill-proofing of beer
(Basařová and Turková, 1977)but also in pilot plant experiments.
In connection with the development of other supports, papain was
also coupled onto glycidyl methacrylate gels (Turková and co-workers, 1978), hydroxyalkyl methacrylate gel treated with epichlor-
hydrin, hydroxyalkyl methacrylate gel containing glucose and oxidized with periodate (Vančurová and co-workers, 1979) and onto
cellulose in bead form, also after periodate oxidation (Turková
and co-workers, 1979). Since the economic viewpoint is an important factor in industrial applications, a new preparation of papain coupled on the cheap polyamide support has been developed.
This preparation was obtained by binding papain onto a polyamide
support after its partial hydrolysis and activation with glutar-
aldehyde. All these preparations were tested and evaluated at the
Beer-Brewery Institute both from the point of view of their activity, stability, behaviour during the catalysis operated on a continuous basis and also from the economic viewpoint which cannot
be neglected in industrial applications

However, none of the methods used in the binding of papain was
found suitable for the binding of carboxyl proteinase from A.ory-
zae for the chill-proofing of wine. Here the binding of enzyme
to alumina activated with titanium tetrachloride was found best.
After annealing the chloride is converted to titanium dioxide
which forms stable complexes with proteins.

CONCLUDING REMARKS

Enzymes as biokatalysts have an exceedingly high catalytic activity in mild conditions and a unique substrate specificity. These
properties should have provided them a wide application in technological catalytic processes. The immobilization of enzymes facilitates their application. Immobilized proteinases may be easily
separated from soluble substrates and products and reused. Their
stability increases, because first of all their autolysis is further impossible. When the catalytic process is carried out in a
water-immiscible organic solvent, the proteinases localized in the
aqueous phase on the solid support, are prevented against the inactivation by a nonaqueous solvent.

From the comparison of different derivatives of immobilized proteinases we may conclude that neither a universal support nor a
generally most convenient method of coupling of proteinases on
a solid support exist. When a certain proteinase is immobilized
an individual method of coupling must be selected, with respect
to its future use.

REFERENCES

Basařová, G. and Turková, J (1977). Eigenschaften, Bindungsweise
    und Applikation von gebundenem Papain an Hydroxyalkyl-metha-
    crylategele. Brauwissenschaft, 30, 204-209.
Budín, J. and V.Kubánek (1980). Powdered poly(ethylene terephtha-
    late). Angew. Makromol.Chem. 84, 37-49.
Čoupek, J.M., M.Křiváková and S. Pokorný (1973). New hydrophilic
    materials for chromatography: glycol methacrylates. J.Polymer.
    Sci.Symp., 42, 185-190.
Čoupek, J., J. Labský, J.Kálal, J.Turková and O.Valentová (1977).
    Reactive carriers of immobilized compounds. Biochim.Biophys.
    Acta, 481, 289-296.
Filka, K., J.Kocourek and J.Čoupek (1978) Studies on lectins XL.
    O-Glycosyl derivatives of Spheron in affinity chromatography
    of lectins. Biochim.Biophys.Acta, 539, 518-528.
Gulaya, V.E., J.Turková, V.Jirků, A.Frydrychová, J.Čoupek and
    S.N.Ananchenko. Immobilization of yeast cells onto hydroxyal-
    kyl methacrylate gels modified by spacers of different lengths.
    (1979). Eur.J. Appl.Microbiol.Biotechnol. 8, 43-47.
Klibanov, A.M., G.P. Samokhin, K.Martínek and I.V.Berezin (1977).
    A new approach to preparative enzymatic synthesis. Biotechnol.
    Bioeng., 19, 1351-1361.
Kubánek, V., B.Veruovič, J.Králíček and Z. Cimburek (1978). Pro-
    duction of microporous polymers of high active surface. Czech.
    Patent Appl. PV-1648-78.
Peška, J., J.Stamberg and J.Hradil (1976). Chemical transformation
    of polymers XIX. Ion exchange derivatives of bead cellulose.
    Angew. Makromol.Chem., 53, 73-80.
Royer, G.P., F.A. Liberatore and G.M. Green (1975). Immobilization
    of enzymes on aldehydic matrices by reductive alkylation.
    Biochem.Biophys.Res.Commun., 64, 478-484.

Švec, F., J.Hradil, J.Čoupek and J.Kálal (1975). Reactive poly-
    mers I. Macroporous methacrylate copolymers containing epoxy
    groups. Angew. Makromol. Chem. 48, 135-143.
Turková, J. (1976). Immobilization of enzymes on hydroxyalkyl
    methacrylate gels. Methods Enzymol., 44, 66-83.
Turková, J. (1980). Enzymes covalently bound to solid supports.
    In Lj. Vitale and V. Simeon (Eds.) Industrial and Clinical
    Enzymology, FEBS Vol. 61, Pergamon Press, Oxford. pp. 65-76.
Turková, J., O.Hubálková, M. Křiváková and J.Čoupek (1973).
    Affinity chromatography on hydroxyalkyl methacrylate gels I.
    Preparation of immobilized chymotrypsin and its use in the
    isolation of proteolytic inhibitors. Biochim. Biophys.Acta,
    322, 1-9.
Turková, J., K.Bláha, M.Malaníková, D.Vančurová, F.Švec and
    J.Kálal (1978). Methacrylate gels with epoxide groups as
    supports for immobilization of enzymes in pH range 3-12.
    Biochim.Biophys.Acta 524, 162-169.
Turková, J. and A.Seifertová (1978) Affinity chromatography of
    proteases on hydroxyalkyl methacrylate gels with covalently
    attached inhibitors. J.Chromatogr. 148, 293-297.
Turková, J., J.Vajčner, D.Vančurová and J.Stamberg (1979). Immo-
    bilization on cellulose in bead form after periodate oxidation
    and reductive alkylation. Collect.Czech.Chem.Commun. 44,
    3411-3417.
Turková, J., L.V.Kozlov, L.Ya. Bessmertnaya, L.V. Kudryavtseva,
    V.M. Krasilnikova and V.K. Antonov (1980). Study of neutral
    matrices effects on stability of immobilized trypsin and
    chymotrypsin. Bioorg. Khim., 6, 108-115.
Valentová, O., J.Turková, R.Lapka, J.Zima and J.Čoupek (1975).
    Pepsin immobilized by covalent fixation to hydroxyalkyl metha-
    crylate gels: preparation and characterization. Biochim.
    Biophys. Acta, 403, 192-196.
Vančurová, D., J.Turková, J.Čoupek and A.Frydrychová (1979).
    Immobilization of trypsin on periodate-oxidized glucosehydro-
    xyalkyl methacrylate gel Separon-H 1000-glc. Collect. Czech.
    Chem.Commun. 44, 3405-3410.

Proteinase Inhibitors

THIOL INHIBITORS OF PROTEASES — THE REVERSIBLE DISULPHIDE EXCHANGE IN THE REGULATION OF TRYPSIN—LIKE ENZYMES

F.S. Steven

Dept. Medical Biochemistry, Stopford Building, Univ. of Manchester, Manchester M13 9PT, England

ABSTRACT

Cytoplasm from human polymorphonuclear leucocytes and mouse Ehrlich ascites tumour cells has been shown to contain a protein inhibitor of trypsin and trypsin-like enzymes. This inhibitor possesses a reactive thiol group which forms an intermolecular disulphide bond with a significant disulphide bond responsible for maintaining the conformation of the active site of trypsin. The kinetics of this disulphide exchange reaction have been studied in model systems in which trypsin was first inhibited with a thiol and the resultant latent form of trypsin re-activated by the subsequent addition of disulphides, oxidising agents, organo-mercurials and chemically inactivated trypsin or chymotrypsin.

This control of enzyme activity by disulphide exchange has been shown to apply to chymotrypsin, elastase, plasmin and collagenase. Disulphide exchange could have biological significance in that it might offer temporal control of enzymic activity without requiring any net change in thiol or disulphide content but rather a temporary relocation of disulphide bonds with consequent modification of enzyme activity.

KEYWORDS

Disulphide exchange; enzyme activity; regulation; organomercurials; latency; re-activation; reversible; control.

INTRODUCTION

The concept of disulphide exchange is simple. The scheme below typifies a disulphide exchange in which an intermolecular disulphide is formed between two distinct protein molecules (P_1 and P_2).

$$P_1 \begin{matrix} S \\ | \\ S \end{matrix} \quad + \quad P_2 - SH \rightleftharpoons P_1 - S - S \quad P_2 \\ \qquad\qquad\qquad\qquad\qquad\qquad\qquad SH$$

Intermolecular disulphide exchange is common practice in the formation of permanant waves in the hairdressing industry. Given the correct conditions disulphide exchange is easy to demonstrate in the laboratory if this change can be linked to a corresponding change in a physical property or the biological activity of the system under study. It has taken me a long time to get this concept accepted in the literature of enzyme control mechanisms, much frustration has been experienced in getting journal editors round to the idea that disulphide exchange plays any role in the control of enzymes. I am grateful to have this opportunity to develop the concept which has been developed from experimental results obtained over the last few years in Manchester.

As far as I am concerned my interest in this subject began as the result of some observations on a time-dependent re-activation of a latent form of trypsin (Steven, Milsom and Hunter, 1976). This data was presented at an enzymology meeting and the audience clearly disbelieved the data and indicated that I was talking rubbish. Having spent twenty years working on the organisation of monomers within connective tissue polymers rather than study enzymology, perhaps I should have taken the audience's advice and left quietly. I persisted in this empirical research even though I had no idea where it would lead. It was much later that thiols, disulphides and an exchange mechanism were organised into this simple concept of enzyme control by disulphide exchange. Recently other enzyme systems have been shown to be activated by intermolecular disulphide exchange (Macartney and Tschesche. 1980) or inhibited (Aull and Daron, 1980). Pedersen and Jacobsen (1980) report an intermolecular disulphide exchange involving serum albumin and there must be many more such exchanges. Enzymes may be involved in catalysing disulphide-disulphide exchanges (Rafter and Harmison, 1979), and the effect of disulphides on the health of cattle and man is now under study (Smith, 1978, 1980).

The evidence for the data presented in this review is largely kinetic data which has already been published in the form of graphs. In order to save space these graphs will not be reproduced here but will be referred to in the literature at the appropriate place in this text.

Substrates

One of the substrates which we have used extensively has been fluorescein-labelled polymeric collagen fibrils (Steven and Lowther, 1975; Steven, Torre-Blanco and Hunter, 1975). Polymeric collagen fibrils (Steven, 1967) consist of millions of intermolecularly cross-linked monomeric collagen molecules (tropocollagen molecules). The individual tropocollagen molecules consist of a rigid triple helix with short N- and C- terminal non-helical peptides attached to the ends of the helical region. These non-helical peptide regions are cleaved by trypsin and chymotrypsin without solubilising the helical regions, the latter remains covalently attached to one another in the insoluble partially digested collagen fibrils. Polymeric collagen fibrils have been uniformly labelled with fluorescent dyes, eg fluorescein, rhodamine and dansyl chloride, such that four fluorescent labels are located within the helical region and three in the non-helical peptide regions of each tropocollagen molecule within the polymeric collagen fibrils. Attack by trypsin and trypsin-like enzymes results in the solubilisation of a maximum of two fluoresein-labelled peptides per tropocollagen molecule. On the other hand, attack by chymotrypsin leads to the solubilisation of peptides from the non-helical regions which lack fluorescein-label; thus chymotrypsin appears to have no detectable proteolytic action on fluorescein-labelled polymeric collagen when assayed for the release of fluorescein-labelled peptides. Fluorescein-labelled polymeric collagen fibrils provide a useful substrate for trypsin and trypsin-inhibitor assays in which fluorescent peptides may be assayed throughout the course of a 3h digestion period simply by removing 100 μl aliquots from the supernatant fluid at suitable time intervals prior to

fluorimetry.   This substrate has been particularly useful for the study of crude
enzyme preparations containing latent enzyme, zymogen and inhibitors.   (Steven,
Podrazky and Foster, 1978).   The advantages of this substrate are that no
precipitation step is needed prior to estimating the quantity of solubilised
peptides, the high sensitivity of fluorescent measurements and the fact that
small aliquots of the digestion products can be sampled at suitable times without
disturbing the equilibrium of the reaction.   These advantages enable the
researcher to obtain a great deal of reliable kinetic data in a relatively short
time.

We have also employed casein as a substrate, for example, when chmotrypsin is
under study (Steven and Al-Habib, 1979).   The solubilised peptides were
estimated by fluorimetry after coupling with fluram (Weigele and co-workers,
1972).   It is sometimes convenient to employ very short incubation periods for
the determination of initial rates of enzyme reactions, for this purpose we
employed carbobenzoxy-L-tyrosine-nitro phenyl ester (Martin, Golubow and Axelrod,
1959) as an esterase substrate for proteolytic enzymes.   For similar reasons we
also employed the β-naphthylamidase substrate, α-N-benzoyl-DL-arginine-β-
naphthylamide and assayed the product by fluorimetry (MacDonald, Ellis and
Reilley 1966).

Evidence for Trypsin Inhibitor in Polymorphonuclear Leucocytes.

Human polymorphonuclear leucocytes were shown to possess a granule enzyme similar
to trypsin and a cytoplasmic inhibitor of both trypsin and their granule enzyme.
(Steven, Torre-Blanc and Hunter, 1975).   The inhibitor could be readily
quantitated against an internal standard of added trypsin, there being a linear
relationship between the quantity of cytoplasm added and the degree of trypsin
inhibition observed.   It was noticed however that if the totally inhibited
trypsin was allowed to stand at 37° for a few hours in the presence of the
substrate, much of the original trypsin activity was regained.   This chance
observation lead to an investigation of the time-dependent re-activation of this
trypsin-inhibitor complex and also a study in which the inhibitor was preincubated at 37° prior to mixing with the trypsin.   In both these studies it was
found that the inhibitor lost activity on standing whether it was originally
combined with typsin as a trypsin-inhibitor complex or whether the inhibitor was
incubated prior to adding trypsin.   This re-activation of trypsin and
simultaneous destruction of inhibitor was later shown to be due to oxidation of a
reactive thiol on the inhibitor protein molecule derived from the cytoplasm
(Steven, Podrazky and Itzhaki, 1978).

Evidence for a Trypsin Inhibitor in Mouse Ehrlich Ascites Tumour Cells.

Ehrlich ascites tumour cells may be grown in large numbers within the
intraperitoneal cavity of mice (Itzhaki, 1972).   These cells were washed in
isotonic saline and sonicated prior to centrifugation according to the method
described by Kopitar and Lebez (1975) to produce a granule fraction and a post-
granule supernatant fraction.   The granule fraction contained the zymogen of a
trypsin-like neutral protease which was activated by trypsin.   The post-granule
supernatant fraction contained a latent form of a trypsin-like enzyme and an
excess of a trypsin inhibitor similar to the one found in human polymorphonuclear
leucocyte cytoplasm.   The quantitation of all the components in this system was
achieved by the technique which we refer to as incremental analysis (Steven,
Podrazky and Foster, 1978) and we feel strongly that this was only possible
because we used this technique rather than standard assays of enzymic activity
or of inhibitory power.   The complex biphasic kinetics of this type of plot

should make the above points self evident to a first year student of enzymology.

Re-activation of Latent Collagenase by Organomercurials.

Whilst we were attempting to explain the time dependent re-activation of our latent trypsin systems, Sellers and co-workers (1977) published their observations that latent collagenase could be activated by organomercurials such as mersalyl.   These agents are known to bind thiol compounds although Sellers and co-workers did not implicate a reactive thiol as the functional group of their inhibitor nor a disulphide exchange as the mechanism of collagenase-inhibitor complex formation.   Since tumour cells also possessed latent collagenase (Steven and Itzhaki, 1977) as well as a latent trypsin-like enzyme (Steven, Podrazky and Itzhaki, 1978) we examined the effect of incremental additions of mersalyl on our preformed trypsin-inhibitor complex obtained from tumour cells.   We observed a biphasic re-activation of trypsin-like activity followed by a subsequent inhibition when the concentration of mersalyl increased above a critical level.   This evidence provided the vital clue that a thiol protein was involved in the inhibition of trypsin and that trypsin-thiol complexes could be used as a simple model for the mersalyl reactivation of latent enzymes (Sellers and co-workers, 1977).   Studies on the thiol inhibition of trypsin and similar enzymes and their re-activation with disulphides were designed to elucidate a general mechanism for the control of enzyme action by disulphide exchange between the enzyme and its potential inhibitor.   These studies will be described in the following sections.   It is appropriate to end this section on collagenase re-activation with the heartening publication of Macartney and Tschesche (1980) in which they show latent collagenase to be formed by disulphide exchange between a thiol bearing inhibitor obtained from leucocytes and a significant disulphide bond in collagenase.   The mechanism of disulphide exchange elucidated by these authors corresponded exactly to the mechanism defined for trypsin-like enzymes (Steven and Pdrazky, 1978, 1979) and later shown to apply to mersalyl re-activation of latent trypsin (Steven and Griffin, 1980(a).

Re-activation of Latent Enzymes by Disulphides.

The idea that the cytoplasmic inhibitor of trypsin and trypsin-like neutral proteases might possess a reactive thiol group which was necessary for the interaction was the result of the observed re-activation of latent trypsin by incremental additions of mersalyl.   It was first necessary to add an appropriate amount of crude inhibitor to a fixed quantity of trypsin to achieve partial inhibition (eg 70%) of this enzyme in a series of tubes.   To each tube was then added an incremental addition of cystine and the regain in trypsin activity followed by fluorimetric analysis.   (Steven and Podrazky, 1978).   It was further demonstrated, by similar kinetic analysis, the cystine could be replaced as re-activating agent by TPCK-chymotrypsin, ie chymotrypsin lacking enzymic activity but retaining its significant disulphide group for exchange with trypsin in the trypsin-inhibitor complex.   TLCK-trypsin also re-activated the latent trypsin (see Fig. 1) by disulphide exchange.

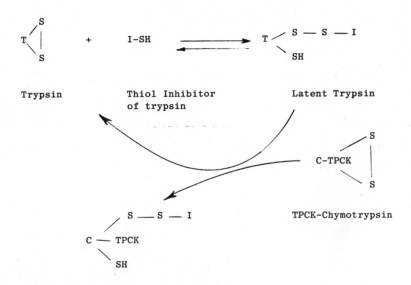

Fig. 1.   Re-activation of latent trypsin by enzymically
          inactive chymotrypsin through intermolecular
          disulphide exchange.

We were able to follow this exchange reaction by gel electrophoresis in which the
inhibitor was observed as a high molecular weight band which failed to enter the
gel when complexed with trypsin but which could be shown to reappear when cystine
was added to the enzyme-inhibitor complex.

Since TLCK-trypsin and TPCK-chymotrypsin were shown to complex with the inhibitor
by the gel electrophoretic analysis and also by kinetic studies described above
in the re-activation of latent trypsin it was concluded that the active centres
of these enzymes were not directly involved in the interaction with the thiol
inhibitor.   It followed that the interaction between the thiol of the inhibitor
molecule and a disulphide bond in trypsin resulted in a conformational change in
the active centre of trypsin with a consequent loss of enzymic activity.   The
studies of Knights and Light (1976) would strongly suggest that this significant
disulphide bond is the one which they found to be most reactive and which spans
part of the active centre of trypsin (cystine 191-220).   This bond would be most
capable of stabilising the active centre in the biologically functional form.
The evidence of Macartney and Tschesche (1980) confirmed this suggestion by
demonstrating a single disulphide bond in trypsin is responsible and that
selective reduction followed by substitution of one or both of the resultant
thiol groups (formed by selective reduction) lead to complete failure of this
modified trypsin to re-activate latent collangease.   On the other hand
TLCK-trypsin, cystine and oxidised glutathione were all capable of this
re-activation.

The realisation that enzymic activity was not necessary for an exchange reaction
to take place (Steven and Podrazky, 1978) provided an opportunity to make well
known biological molecules behave in an unconventional manner in order to
illustrate the disulphide exchange mechanism.   We all take for granted the   ·
reaction between trypsin and chymotrypsinogen in the production of the enzyme
chymotrypsin.   If we consider that the structure of the zymogen is similar to
chymotrypsin with respect to its disulphide bond organisation, then it should be
possible to employ the significant disulphide bond of chymotrypsinogen to
re-activate trypsin from its latent form by disulphide exchange.   It is necessary
to choose a substrate which does not produce products in the assay as a direct
result of chymotrypsin enzymic activity and fluorescein-labelled polymeric
collagen fibrils meet these requirements admirably.   With this system, we had all
the reagents necessary to make the well known conventional activation of
chymotrypsinogen work backwards in the sense that we re-activated trypsin
(Steven and Podrzky, 1979) from the preformed trypsin-inhibitor complex.

Inhibition of Pancreatic Proteases by Thiols.

Simple thiols such as cysteine, reduced glutathione thioglycolate and
dithiothreitol readily inactivate trypsin (Steven and Podrazky, 1978) and their
ability to inhibit this enzyme was shown to vary greatly according to the
structure of the thiols used.   The data suggested that inhibition by disulphide
exchange was an equilibrium reaction which was strongly concentration dependent.
This is not a stoichiometric reaction and is unlike an active site titration of
trypsin or the inhibition of trypsin by ovamucoid.   The molar concentration of
thiol was greatly in excess of the molar concentration of trypsin.   Chymotrypsin
was also shown to be inhibited by thiols (Steven and Al-Habib, 1979).   Elastase
and plasmin behaved similarly (Steven and Griffin, 1980(b)), thus the class of
pancreatic proteases have this property of regulation of enzymic activity by
disulphide exchange as a common feature.

Re-activation of Thiol-Trypsin Complexes.

We had established that thiol compounds inhibited trypsin-like enzymes by means of
a disulphide exchange.   Mild oxidation should be capable of re-activation of the
preformed latent enzyme, since the oxidation of excess thiol should drive this
equilibrium reaction towards regeneration of the active enzyme (Fig. 2).

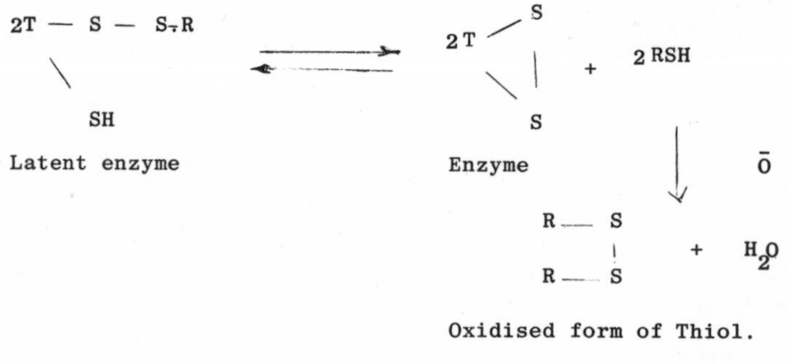

Fig. 2. Re-activation by oxidation.

Trypsin and chymotrypsin were first complexed with dithiothreitol with
consequent loss in enzymic activity.   These latent enzymes were then re-activated
by incremental addition of sodium periodate (Steven and Al-Habib, 1979).   It was
observed that as the concentration of sodium periodate increased a biphasic
kinetic plot resulted, due to the direct action of the periodate on the activity
of the newly re-activated trypsin or chymotrypsin.   Periodate oxidation of
trypsin resulted in inhibition of enzymic activity with simultaneous oxidation of
disulphide groups which appeared on hydrolysis as cysteic acid.

This biphasic plot was only obtained when the initial inhibition was produced by
a thiol and the re-activation was by oxidation, the latter stages of disulphide
oxidation were irreversible (Fig. 3).

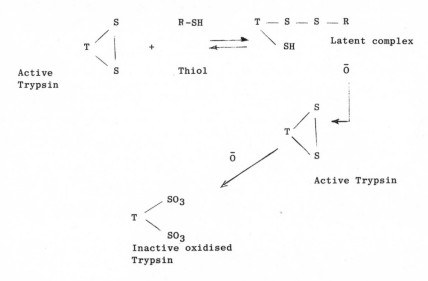

Fig. 3.   Steps in Oxidation of Trypsin-Thiol-Complex
           Leading to Biphasic Kinetics.

The evidence obtained from the oxidation of the model system helped to explain
the time-dependent re-activation of enzymic activity initially observed with the
polymorphonuclear leucocyte inhibitor system.   The validity of this explanation
was confirmed by experiments inolving chemical reduction followed by atmospheric
oxidation of this test system (Steven, Podrazky and Itzhaki, 1978).   At present
it is not known to what extent the partial pressure of oxygen in living tissues
influences the activities of enzymes which may be regulated by disulphide
exchange mechanisms.   Organomercurials re-activate the trypsin-thiol complex as
would be expected from the use of these compounds in re-activating latent
collagenase (Sellers and co-workers, 1977, 1978).   The mechanism we proposed
(Steven and Griffin, 1980 (a)) is outlined in Fig. 4 in which the organomercurial
is represented by R-Hg-OH.

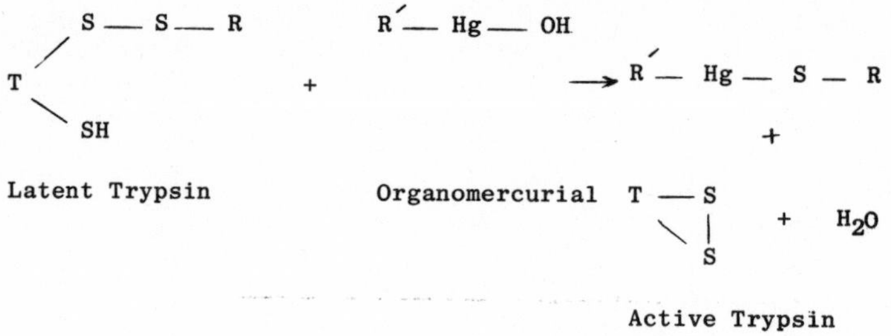

Fig. 4. Organomercurial Re-activation of Latent Trypsin

Location of significant Disulphide close to Enzyme Active Centre

We were able to combine studies on the dithiothreitol inhibition of trypsin with active site titration (Chase and Shaw, 1967) of the remaining active sites on the trypsin (Steven and Griffin, 1980(a)). This showed a similar fall in the number of available active sites and the inhibition of β-naphthylamidase activity with incremental additions of dithiothreitol to a fixed quantity of trypsin. Secondly a series of tubes was set up in which a fixed quantity of trypsin-thiol complex was prepared with 100 µM thiol addition and incremental amounts of mersalyl added to each tube. It was observed that the regain in β-naphthylamidase activity corresponded to the regain of active sites available for titration. Thus, disulphide exchange, promoted by a thiol and resulting in trypsin inhibition, was directly linked by experimentation to a site which controlled the conformation of the enzymes active centre of trypsin and by inference the other pancreatic proteases.

Re-activation of Trypsin-Thiol Complexes by Metal Ions.

Martineck and co-workers (1971) reported that silver ions inhibited trypsin by combining with a histidine residue within the active site of this enzyme. Metal ions also bind thiol groups and it was therefore of interest to observe the competitive events which took place when incremental additions of Ag^+, Cu^{++} and Hg^{++} were added to the trypsin-dithiothreitol complex (Steven, Podrazky, Al-Habib and Griffin, 1979). At low concentrations of metal ions typsin re-activation took place. As the concentration of metal ion increased, a sharp fall in enzymic activity was observed as would be predicted from the results obtained by Martineck and co-workers (1971). So it can be concluded that although Ag^+ and other metal ions inhibit free trypsin, these same ions at low concentration re-activate the latent form of trypsin by displacing the inhibitory thiol group from the intermolecular disulphide bond as shown in Fig. 5.

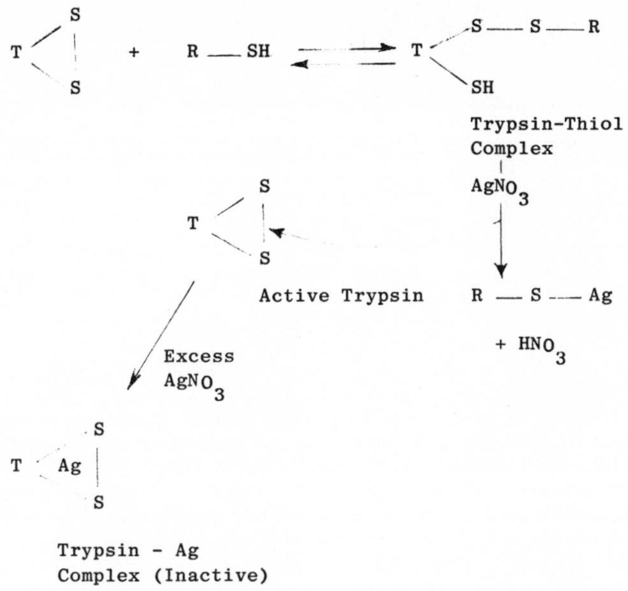

$$T \overset{S}{\underset{S}{<}} \quad + \quad R — SH \quad \underset{\longleftarrow}{\overset{\longrightarrow}{\quad}} \quad T \overset{S — S — R}{\underset{SH}{<}}$$

Trypsin-Thiol
Complex

AgNO$_3$

$$T \overset{S}{\underset{S}{<}} \quad \quad \text{Active Trypsin} \quad R — S — Ag$$

+ HNO$_3$

Excess
AgNO$_3$

$$T < Ag \overset{.S}{\underset{.S}{|}}$$

Trypsin - Ag
Complex (Inactive)

Fig. 5.   Re-activation of Trypsin-Thiol Complex by Silver
Ions.

General Comments

This review presents the evidence for a reversible disulphide exchange mechanism
involved in the biological control of extracellular proteases.   In such a thiol-
disulphide intermolecular reaction there is no change in the total number of thiol
groups or disulphide groups, only a re-location of these groups.   This type of
reaction has two important properties as a consequence of this exchange,
(i) the reactions cannot be measured unless a change in physical or biological
properties accompanies this disulphide exchange, and (ii) temporal control of
enzymic reactions may be possible by means of disulphide exchange at times at
which it is advantageous to the cell to possess a certain enzyme activity which
may remain latent at other times.

This review has considered only protease inhibition with reference to enzymes
acting extracellularly.   Many intra-cellular enzymes (eg. lysosomal enzymes) are
known to contain a reactive thiol in their active centres.   It may well be that
thiol enzymes can be inhibited by disulphides and re-activated by thiols, through
disulphide exchange reactions;   for example Aull and Daron (1980) have reported
the inhibition of thymidylate synthase by disulphides.

When considering the location of an enzyme and its potential inhibitor it is worth
drawing attention to a trypsin-like enzyme located on the surface of tumour cells
(and also fibroblasts) which is probably derived from the granule zymogen found in
these cells.   This cell surface enzyme behaves as a membrane bound trypsin being
inhibited by low molecular weight inhibitors of trypsin but is only very weakly
inhibited by high molecular weight protein inhibitors of trypsin in free solution
(Steven, Griffin and Itzhaki. 1980).   This unusual behaviour becomes significant

when it is known that these cells export a zymogen of collagenase which is activated by the cell surface neutral protease and also that the cell surface enzyme possesses plasminogen activation activity. Both these activities can take place under experimental conditions in whcih trypsin in free solution would be inhibited and both these activities are associated with tumour invasive properties.

Finally, let us consider the whole animal rather than isolated cells in vitro. We could consider what is likely to happen if the balance of circulating thiols or disulphides is distrubed. In the first case, it is easy to increase the level of thiol by injection, dithiothreitol injected into mice is lethal (Al-Habib, 1980). In the second case there is the physiological example of cattle fed on cabbages or kale which may lead to kale poisoning. Thiols in the diet were shown to be converted to dimethyldisulphide by the bacterial flora. The dimethyl disulphide passed into the blood and caused two very noticeable effects (i) anaemia and (ii) Heinz body formation due to oxidation of the haemoglobin; cattle having a very low level of erythrocyte glutathione reductase activity exhibit these symptoms more readily than other animals fed dimethyl disulphide (Smith, 1978, 1980). It may be significant that the disulphide could also interact with membrane thiols thought to be essential for the flexibility of the red cell membrane (Smith, 1980). It was also observed that circulating dimethyl disulphide caused a marked reduction in circulating cholesterol and Augusti and Mathew (1974) postulated that the disulphide might bind to the thiol of coenzyme A or possibly to a thiol enzyme directly involved in the biosynthesis of cholesterol. The involvement of a disulphide in a disease of cattle suggests that disulphide exchange and the control of chemical reactions may be much more widespread than those in which this process may be easily demonstrated in vitro, such as the pancreatic proteases. Once one has observed these reactions at the laboratory bench and elucidated a simple concept of enzyme control by disulphide exchange it is easier to tackle the harder problems such as the control of enzymes on the surface of cells and the apparently disasterous consequences of cattle grazing extensively on kale.

ACKNOWLEDGEMENTS

I wish to express my thanks to my colleagues for their help over the years and most importantly for their continued faith in the rationale behind these experiments even when we appeared to make no progress. I also wish to thank the Wellcome Research Foundation for enabling me to participate in this conference.

REFERENCES

Al-Habib, A. (1980). M.Sc Thesis, University of Manchester.
Augusti, K.T. and Mathew, P.T. (1974). Experimentia 30, 468-470.
Aull, J.L. and Daron, H.H. (1980). Biochim. Biophys. Acta 614, 31-39.
Chase, T. and Shaw, E. (1967). Biochem. Biophys. Res. Comm. 29, 508-514.
Itzhaki, S. (1972). Life Sci. 11 part II, 649-655.
Knights, R.J. and Light, A. (1976). J. Biol. Chem. 251, 222-235.
Kopitar, M. and Lebez, D. (1975) Eur. J. Biochem. 56, 571-581.
MacDonald, J.K., Ellis, S. and Reilley, T.J. (1966). J. Biol. Chem., 241 1494-1501.
Macartney, H.W. and Tschesche, H. (1980). Febs Letters, (in press).
Martin, C.J., Golubow, J. and Axelrod, A.E. (1959). J. Biol. Chem. 234, 1718-1725.
Pedersen, A.S. and Jacobsen, J. (1980). Eur. J. Biochem. 106, 291-295.
Rafter, G.W. and Harmison, G.G. (1979). Biochim. Biophys. Acta 567, 18-23.

Sellers, A. Cartwright, E., Murphy, G. and Reynolds, J.J. (1977).   Biochem. J.
    163, 303-307.
Sellers, A. and Reynolds, J.J. (1977).   Biochem. J. 167, 353-360.
Smith, R.H. (1978).   Veterinary Science Comm. 2, 47-61.
Smith, R.H. (1980).   Veterinary Record 107, 12-15.
Steven, F.S. (1967).   Biochim. Biophys. Acta 140, 522-528.
Steven, F.S. and Al-Habib, A. (1979).   Biochim. Biophys. Acta 568, 408-415.
Steven, F.S. and Griffin, M.M. (1980(a)).   Eur. J. Biochem. (in press).
Steven, F.S. and Griffin M.M. (1980(b)).   Biochem. Soc. Trans. 8, 80-81.
Steven, F.S., Griffin, M.M. and Itzhaki, S. (1980).   Brit. J. Cancer (in press).
Steven, F.S. and Itzhaki, S. (1977).   Biochim. Biophys. Acta 496, 241-246.
Steven, F.S. and Lowther, D.A. (1976).   Conn. Tiss. Res. 4, 7-10.
Steven, F.S., Milsom, D.W. and Hunter, J.A.A. (1976).   Eur. J. Biochem. 67,
    165-169.
Steven, F.S., and Podrazky, V. (1978).   Eur. J. Biochem. 83, 153-161.
Steven, F.S., and Podrazky, V. (1979).   Biochim. Biophys. Acta 568, 49-58.
Steven, F.S., Podrazky, V., Al-Habib, A. and Griffin, M.M. (1979).   Biochim.
    Biophys. Acta 571, 369-373.
Steven, F.S., Podrazky, V. and Foster, R.W. (1978).   Anal Biochem. 90, 183-191.
Steven, F.S., Podrazky, V. and Itzhaki, S. (1978).   Biochim. Biophys. Acta 524,
    170-182.
Steven, F.S., Torre-Blanco, A. and Hunter, J.A.A. (1975).   Biochim. Biophys. Acta
    405, 188-200.
Weigele, M., De Bernardo, S.L., Tengi, J.P. and Leimgruber, A.A. (1972)
    J. Am. Chem. Soc. 94, 5927-5928.

MULTIPLE FORMS OF RABBIT ALPHA–1–ANTITRYPSIN AND DIFFERENTIAL INACTIVATION OF SOME SERINE PROTEINASES

A. Koj, E. Regoeczi* and A. Dubin

Institute of Molecular Biology, Jagiellonian University, 31-001 Krakow, Poland
*Department of Pathology, McMaster University, Hamilton, Ontario, Canada

ABSTRACT

Rabbit plasma α-1-antitrypsin /α-1-AT/ exhibits heterogeneity both in respect of charge and of mol.wt. By employing preparative poly-acrylamide gel electrophoresis /PAGE/ two principal forms, F and S, were separated and exposed to bovine trypsin and chymotrypsin, equine leucocyte proteinases, porcine pancreatic elastase and rabbit plasmin. Kinetic measurements or analyses by crossed immuno-electrophoresis and SDS-PAGE showed that the F form reacted more rapidly with all tested enzymes. However, this was not related to molar inhibitory capacities, i.e. moles of enzyme inhibited by one mole of α-1-AT in a complete reaction. The inhibitory capaci-ties ranged from 1 for trypsin to 0.26 for leucocyte proteinases and showed profound differences between the two antitrypsins. The highest F:S inhibitory ratio was recorded with chymotrypsin /1.88/ and the lowest with elastase /0.69/. The differences result pro-bably from a dual nature of the reaction between the inhibitor and a proteinase, i.e. either complex formation or inactivation of α-1-AT without enzyme inhibition.

KEYWORDS

Antitrypsin variants, inactivation of serine proteinases, kinetics of inhibition, inhibitory capacities

INTRODUCTION

Alpha-1-antitrypsin is a polyvalent proteinase inhibitor of mammalian blood plasma exhibiting a marked polymorphism during electrophoresis. Physicochemical properties responsible for this polymorphism, and biological significance of α-1-AT variants, have not been fully explained. In rabbit blood α-1-AT occurs in two principal forms, designated according to their electrophoretic mobility as F /fast/ and S /slow/. They are immunologically iden-tical, show a rather similar amino acid and carbohydrate compo-sition /Koj and others, 1978/ but appear to be separately synthe-sized in the liver and independently metabolized in the circulation /Regoeczi, Koj and Lam, 1981/. Here we present further evidence on

molecular heterogeneity and differential reactivity of the two
forms of rabbit α-1-antitrypsin with various proteinases.

MATERIALS AND METHODS

Rabbit α-1-antitrypsin was routinely isolated from citrated plasma
by a procedure described previously /Koj and others, 1978/. In two
experiments α-1-AT was obtained by the thiol-disulphide interchange
method of Laurell on a Sepharose-cysteine column following the
procedure described elsewhere /Regoeczi, Koj and Lam, 1981/. The
two forms of α-1-AT were separated by preparative polyacrylamide
gel electrophoresis /Koj and others, 1978/ except that each gel
was pre-electrophoresed for 60 min at 200 V.

Crystalline bovine trypsin and α-chymotrypsin were from Koch-Light,
Colnbrook, England, porcine pancreatic elastase III from Sigma
Chemical Co., St.Louis, Mo.,U.S.A., while horse leucocyte proteina-
ses 1 and 2A were isolated by a published method /Dubin, Koj and
Chudzik, 1976/ and rabbit plasmin was prepared as described by
Hatton and Regoeczi /1977/.

Absorbance of protein solution was measured at 280 nm and protein
concentrations were calculated using the following A /1%-1cm/
values: α-1-antitrypsin, 5.1; chymotrypsin, 20.8;, leucocyte proteinases, 7.0; pancreatic elastase, 20.0; plasmin, 13.9 and trypsin,
15.4. The amount of active trypsin in the commercial preparation
was determined by titration with p-nitrophenyl guanidinobenzoate
/Chase and Shaw, 1970/ and of active chymotrypsin by titration
with p-nitrophenyl acetate /Kézdy and Kaiser, 1970/ and appropriate
corrections were made when calculating inhibitory activity of
α-1-antitrypsin.

To obtain immobilized enzymes approx. 20 mg of trypsin or chymotrypsin were coupled to 2 g of CNBr-activated Sepharose 4B /Phar-
macia Fine Chemicals, Uppsala, Sweden/ according to the instruction
provided by the manufacturer. From estimations of esterolytic activity of trypsin or chymotrypsin in the obtained preparation it was
concluded that 1 ml of the reconstituted gel contained approx. 1.5
mg of the active enzyme.

The activities of trypsin and plasmin were determined with N-α-
benzoyl-L-arginine ethyl ester /Bz-Arg-OEt, Koch-Light, Colnbrook,
England/ and chymotrypsin with N-benzoyl-L-tyrosine ethyl ester
/Bz-Tyr-OEt, Sigma Chemical Co.,St. Louis, Mo.,U.S.A./ using the
procedure of Schwert and Takenaka /1955/. Pancreatic elastase and
leucocyte proteinases were assayed with N-benzyloxy-carbonyl-L-
alanine-4-nitrophenyl ester /Z-Ala-ONp, Fluka A.G., Buchs, Switzer-
land/ as described by Janoff /1969/. To measure inhibition of
proteinases by α-1-AT, 0.2-0.5 nmoles of an enzyme were incubated
with 0.1-0.8 nmoles of the inhibitor in 0.4 ml of 0.05 M Tris-HCl
pH 8.0 at 25°C. Correspondingly, molar concentrations of the reactants were in the range of 0.5-1.2 x 10^{-6}M /enzymes/ and 0.2-2.0 x
10^{-6}M /inhibitor/. After 10 min incubation 2.6 ml of the appropriate
synthetic substrate were added and the remaining esterolytic activity was determined. Titration of a constant amount of an enzyme
with increasing amounts of α-1-antitrypsin enabled to establish
conditions required for complete inhibition of the proteinase. For

kinetic studies the enzyme and inhibitor were incubated with the excess of proteinase at three different pH values /8.0, 7.0 and 6.0/ in 0.05 M Tris-acetate for periods varying from 15 s to 1 h. Inactivation rates were calculated according to second-order kinetics /Downing, Bloom and Mann, 1978/ assuming bi-molecular reaction between enzyme and inhibitor.

The products of interaction of the inhibitor with proteinases were analyzed by polyacrylamide gel electrophoresis in Tris-glycine buffer pH 8.3 /Gordon, 1969/, by polyacrylamide gel electrophoresis with 0.1% sodium dodecyl sulphate /SDS/ at pH 7.0 /Weber, Pringle and Osborn, 1972/ or at pH 9.5 /Laemmli, 1970/, and also by crossed immunoelectrophoresis in polyacrylamide gel - agarose /Koj and Dubin, 1978/. The monospecific guinea pig antiserum to rabbit α-1-antitrypsin and polyvalent goat antiserum to rabbit plasma proteins were a gift from Dr. J.Gauldie /McMaster University/.

RESULTS

Molecular heterogeneity of rabbit α-1-antitrypsin

The inhibitor isolated by either of the two methods migrated on PAGE as two distinct protein bands /Fig.1/.

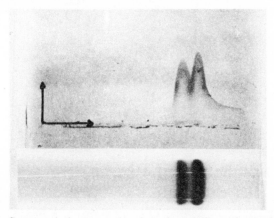

Fig.1. Crossed immunoelectrophoresis of purified unfractionated rabbit α-1-AT. Migration in the first direction /in 7.5% polyacrylamide gel/ was from left to right /gel load 3 μg protein/; migration in the second direction /in 1% agarose with antiserum to rabbit α-1-AT/ was from bottom to top. The real position of the sliced gel cylinder is visible under the double peak of immunoprecipitate. The gel cylinder shown below was run separately with the same preparation of α-1-AT /load 8 μg protein/ and was stained with Coomassie brilliant blue R-250.

Under optimal loading conditions each band showed further heterogeneity in agreement with the results of isoelectric focusing in polyacrylamide gel /Koj and others, 1978/ when a multiple-band

pattern was observed in the region of pH 4.4 - 4.9. However, all
these components must share antigenic determinants since two characteristic peaks were seen on a plate after crossed immunoelectrophoresis with monospecific antiserum to rabbit α-1-AT /Fig.1/. With
the use of polyvalent antiserum to rabbit serum it was found that
the obtained preparation of α-1-AT did not contain detectable
amounts of other plasma proteins. Still, when subjected to electrophoresis in polyacrylamide gel in the presence of SDS the F component migrated always as 2 closely spaced bands of variable relative intensity corresponding in mol.wt. to 56 000 and 60 000, while
the S component showed only one protein band of approx. 56 000
daltons /Fig.2/. Hence rabbit α-1-antitrypsin exhibits a rather
high degree of molecular heterogeneity in comparison with other
examined mammalian species.

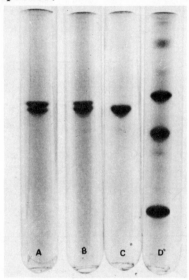

Fig.2. Electrophoresis in 7.5% polyacrylamide gel
with 0.1% SDS in phosphate buffer pH 7.0. A -
α-1-AT F /6 µg/; B - mixture of α-1-AT F and S
/3 µg each/; C - α-1-AT S /6 µg/; D - mixture of
bovine serum albumin, ovalbumin and cytochrome C
/3 µg each/.

Kinetics of interaction with proteinases

At micromolar concentrations of enzymes and the inhibitor and at pH
8.0 inactivation of trypsin is very fast while reaction with plasmin
proceeds slowly, especially in case of the S form /Fig.3/. Plasmin
inactivation is not accelerated in the presence of heparin which
contrasts sharply with another plasmin inhibitor, antithrombin III.

The reaction with trypsin could be slightly retarded at pH 7 or 6
and then differences between the two forms of antitrypsin were
observed, especially during the first two minutes of incubation.
Again, as in case of plasmin, the S form reacted more slowly than
F at identical protein concentration. From the early section of
the inactivation curves the second-order rate constants could be

calculated for plasmin at pH 8 and for trypsin at pH 6 and 7
/Table 1/. The reactions with chymotrypsin, pancreatic elastase
and leucocyte proteinases were too fast to enable direct comparison.

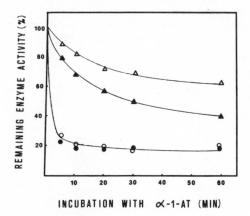

Fig.3. Residual esterolytic activity /Bz-Arg-OEt/
of rabbit plasmin /triangles/ and bovine trypsin
/circles/ after incubation at pH 8.0 with rabbit
α-1-AT F /full symbols/ or S /open symbols/. The
molar ratio of the proteinase to the inhibitor was
approximately 5:4.

TABLE 1   Second-Order Reaction Constants for the

Inactivation of Rabbit Plasmin and Bovine Trypsin

with Rabbit α-1-Antitrypsins F and S

Enzyme	pH	K M^{-1} sec^{-1}		K_F/K_S
		F form	S form	
Plasmin	8.0	0.99×10^3	0.50×10^3	2.0
Trypsin	7.0	0.96×10^5	0.23×10^5	4.2
Trypsin	6.0	0.74×10^5	0.12×10^5	6.1

Enzyme and inhibitor were incubated for variable periods of time
as indicated in Fig.1 before the estimation of residual esterolytic /Bz-Arg-OEt/ activity. Enzyme-to-inhibitor ratios were approx.
5:4. The rate constants for trypsin were calculated from the early
sections of the inactivation curves.

However, differential reactivity of the two α-1-antitrypsins could
also be concluded from the appearance in crossed immunoelectrophoresis of unfractionated α-1-AT, purified or contained in plasma,
after a brief exposure to sub-saturating quantities of a proteinase. As shown for trypsin and chymotrypsin in Fig.4 the F form was

depleted to a larger extent under these conditions than the S form.
Similar results were obtained also with elastase, leucocyte proteinases and plasmin.

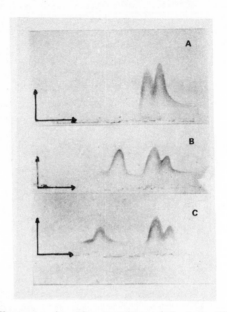

Fig.4.   Changes in the pattern of α-1-AT examined
by crossed immunoelectrophoresis /polyacrylamide
gel - agarose + antibodies/ after 10 min incubation of purified unfractionated rabbit α-1-AT /A/
with chymotrypsin /B/ and trypsin /C/. The molar
ratio of the proteinase to the inhibitor was
approximately 1:2. Gel   load - approx. 3 μg α-1-AT.

When purified F and S α-1-AT were exposed to proteinases and the
products analyzed by SDS-PAGE a more complex picture was obtained
due to the fact that the F form includes two sub-components /cf.
Fig.2/. However, both trypsin and chymotrypsin were found to react
preferentially with the slower protein band, characteristic for the
F α-1-antitrypsin /Fig.5/. The presence of 2-mercaptoethanol in the
samples subjected to SDS-PAGE had very little effect on the protein
pattern except bands corresponding to complexes of the inhibitor
with chymotrypsin.

Inhibitory capacities of rabbit α-1-antitrypsin

Kinetic observations in the previous section have established
that the interaction of all enzymes tested, with the notable exception of plasmin, was practically complete in less than one min at
pH 8 by either form of rabbit α-1-AT. Hence by a gradual increase
of the inhibitor -to-enzyme ratio it should be possible to calculate
the stoichiometry of inactivation reaction. Such typical titration
curves of trypsin and chymotrypsin are shown in Fig.6. The results
indicate that the extrapolated equivalence points correspond to
approximately 1:1 molar ratio of enzyme:inhibitor for both forms

of α-1-antitrypsin and trypsin, as well as for α-1-AT F and chymotrypsin. On the other hand, in the conditions used in the present
experiments over 1.8 moles of α-1-AT S is required to inactivate
1 mole of chymotrypsin. Similar experiments were also carried out
with porcine pancreatic elastase and equine leucocyte proteinases,
and in all cases more than 1 nmole of α-1-AT /58 µg/ was required
to inactivate 1 nmole of the proteinase /calculated from the protein content of the preparation and mol.wt. of the enzyme/.

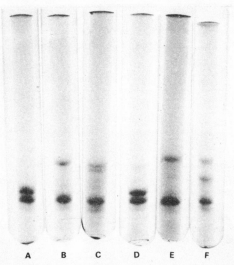

Fig. 5.   Electrophoresis in 7.5% polyacrylamide
gel with 0.1% SDS in Tris-HCl buffer pH 9.5. All
gels contained 5.2 µg of α-1-AT F but in gels B
and E the inhibitor was first incubated for 10
min with trypsin /approximate molar ratio 2:1/
and in gels C and F with chymotrypsin /approxima-
te molar ratio 3:1/. Gels D, E and F - samples
were reduced with 2-mercaptoethanol before loading.

Apart from titration curves stoichiometry of interaction of α-1-AT
with proteinases could be calculated directly from individual measurements of the remaining esterolytic activity at a single inhi-
bitor:enzyme ratio using the early, linear section of the inhibition curve. Such measurements enabled to determine µg of enzymes
inhibited by 1 mg of α-1-AT F or S. This value is referred to as
inhibitory capacity and, after its conversion to moles of enzyme
neutralized by 1 mole of α-1-AT, as molar inhibitory capacity.
These calculations revealed marked differences between the two
forms of rabbit α-1-antitrypsin with regard to their inhibitory
capacities towards various proteinases /Table 2/. Thus, while the
difference in molar inhibitory capacities was negligible in case
of trypsin, the S form possessed a comparatively lower capacity
for chymotrypsin as well as for the two leucocyte proteinases,
and the F form for pancreatic elastase.

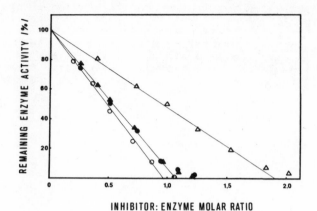

Fig.6. Titration curves of trypsin /circles/ and chymotrypsin /triangles/ by α-1-AT F /closed symbols/ or S /open symbols/. Trypsin /0.51 nmoles/ or chymotrypsin /0.34 nmoles/ were incubated in 0.05 M Tris-HCl pH 8 at 25°C for 10 min with variable amounts of α-1-AT /0.1-0.7 nmoles/ and the remaining esterolytic activity was determined with Bz-Arg-OEt /trypsin/ or Bz-Tyr-OEt /chymotrypsin/.

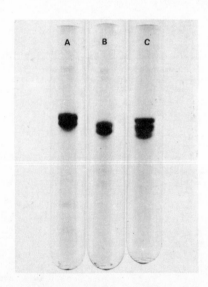

Fig.7. Electrophoresis in 7.5% polyacrylamide gel with 0.1% SDS in phosphate buffer pH 7.0. All gels contained approx. 8 μg protein. A - α-1-AT F; B - the effluent from a Sepharose-trypsin column; C - - mixture of /A/ and /B/.

TABLE 2 Comparison of the Inhibitory Capacities of Rabbit α-1-Antitrypsins F and S for Five Serine Proteinases

Proteinase		α-1-Antitrypsin inhibitory capacity			
Name and source	Assumed mol.wt.	Form	µg/mg	moles/mole	F/S ratio
Trypsin, bovine	23 900	F	391	0.95	0.94
		S	416	1.01	
Chymotrypsin, bovine	22 000	F	365	0.96	1.88
		S	194	0.51	
Pancreatic elastase type III, porcine	22 000	F	175	0.46	0.69
		S	257	0.68	
Leucocyte proteinase type 1, equine	38 000	F	270	0.41	1.57
		S	173	0.26	
Leucocyte proteinase type 2A, equine	24 500	F	240	0.57	1.21
		S	200	0.47	

Incubations were in 0.05 M Tris-HCl, pH 8.0 at 25°C for 10 min. Concentration of the reactants were in the range of 0.2-2.0 x 10^{-6}M with approximate molar ratios of enzyme-to-inhibitor ranging from 2:1 to 5:4. Inhibitory capacity, defined in the text, was determined from the remaining enzymic activity at a given inhibitor concentration. In the column, µg/mg, inhibitory capacities are shown as µg of enzyme inhibited by 1 mg of inhibitor, and the values obtained after molar conversion are listed under mole/mole /assuming mol.wt. of α-1-AT as 58 000/. Values are means from nine experiments with trypsin and chymotrypsin using five batches of α-1-AT, and from 3 - 4 determinations with the other enzymes using one batch of the inhibitor.

Complexing of one enzyme molecule with one molecule of inhibitor
should yield a value of 1.0 for molar inhibitory capacity; yet most
of the corresponding values in Table 2 were considerably lower than
that. One possible explanation is that **portions of the α-1-AT are**
being inactivated by some proteinases during incubation without
inhibition of the enzyme. This supposition is indirectly supported
by the experiments with immobilized enzymes and SDS-PAGE analysis
of the products of interaction of proteinases and α-1-AT.

The columns filled with Sepharose-trypsin or Sepharose-chymotrypsin
were loaded with purified active α-1-AT F or S. The effluents from
both columns were completely devoid of both antitrypsic and anti-
chymotryptic activity but on crossed immunoelectrophoresis showed
significant amounts of the antigen. When the effluent was analyzed
by SDS-PAGE two bands were detected, corresponding to mol.wt. of
approx. 52 000 and 56 000 /Fig.7/. Such a new fast band of the mobility corresponding to mol.wt. of 52 000 was also observed after
exposure of α-1-AT F or S to chymotrypsin in solution /cf. Fig.5/.
Notwithstanding the likely limitations of immobilized enzymes in
the studies of the mechanism of inhibitory reactions, these observations suggest that the examined proteinases can, indeed, render
α-1-AT inactive independently of the formation of stable enzyme-
-inhibitor complexes.

DISCUSSION

The data presented above indicate that the two rabbit α-1-antitryp-
sins are functionally distinguishable on at least two accounts,
namely, kinetically and in respect of inhibitory capacities. As to
the former, the F form emerges as being more reactive than the S
form from comparison with the enzymes tested. Preferential depletion
of the F form in crossed immunoelectrophoresis after a brief exposure of plasma or unfractionated α-1-AT with subsaturating quantities of a proteinase/Fig.4/, is the prime evidence for this view,
supported by the differences in inactivation rate constants when-
ever calculations of this kind were technically feasible.

The different inhibitory capacities of the two forms of α-1-AT for
various proteinases was a surprising finding. Artefacts, such as
the presence of inactive α-1-antitrypsin, or contamination by another inhibitor have been carefully considered and they may be dis-
counted on the following grounds. There is neither immunochemical
nor physicochemical evidence for the presence of any other proteinase inhibitor, such as antichymotrypsin or antithrombin III, in
rabbit α-1-AT prepared by two independent methods. Moreover, molar
inhibitory capacities for trypsin were close to unity with several
preparations of α-1-AT F and S. This could not have been the case
had our preparations contained significant proportions of inactive
or denatured inhibitor. On the other hand, preparations of plasmin,
pancreatic elastase or leucocyte proteinases could contain some
proportion of inactive enzymes, since the active site titration was
performed only with trypsin and chymotrypsin. However, since the F
and S forms were always examined simultaneously, inactive enzymes
could not have affected the ratio of molar inhibitory capacities.

For the reasons listed above a more likely explanation is that the
reaction of α-1-AT with proteinases encompasses two different mecha-

nisms, i.e. either complex formation or inactivation of the inhibitor /Scheme 1/. This view is supported by recent studies from several laboratories showing that the interaction of human α-1-AT with
porcine elastase is of dual nature: some inhibitor molecules form
a stable complex with the enzyme whereas others are immediately
inactivated by limited proteolysis /Baumstark, Lee and Luby, 1977;
James and Cohen, 1978; Satoh and others, 1979; Baumstark and others,
1980/. As shown by Laskowski and co-workers /Satoh and others, 1979/
the relative rates of reactions k_1 and k_2 /Scheme 1/ are influenced
by various factors such as temperature and ionic strength. Our
observations suggest that dual reactions of the porcine elastase -
human α-1-AT type can also occur between rabbit α-1-AT and a number
of serine proteinases. Furthermore, it seems that the relative proportion of the reaction products /enzyme-inhibitor complex or inactivated inhibitor/ varies considerably from enzyme to enzyme and is
different for two forms of α-1-AT.

Scheme 1. Postulated general reaction mechanism

between some serine proteinases and plasma α-1-

antitrypsin or antithrombin III

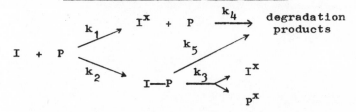

I - inhibitor, P - proteinase, x denotes modified
form of the inhibitor or proteinase, k - reaction
rate constant. For human α-1-AT and elastase k_1/k_2
ratio appears to be enhanced by low temperature
and low salt concentration /Satoh and others,1979/
while for antithrombin III and thrombin or factor
Xa k_3 is probably increased by Tris pH 9 /Jesty,
1979/.

Inactive, proteolytically modified inhibitor may appear also as the
result of the dissociation of a complex that had already been formed
/k_3, Scheme 1/. This is in the case with antithrombin III - plasmin
complex, although electrophoretic evidence of dissociation only
manifests after several hours of incubation /Lam, Regoeczi and
Hatton, 1979/. Nucleophilic ions /Tris/ at high pH expedite the
dissociation of the complexes of antithrombin III and thrombin or
Factor Xa /Jesty,1979/. Although in our experiments complexes of
α-1-AT with trypsin or chymotrypsin appear to be stable during
electrophoresis in polyacrylamide gel with SDS in the absence of
reducing agent we do not know yet whether dissociation was a contributing factor in the obtained results and, if.so, to what extent.
The additional fast component appearing in the sample of inhibitor
exposed to immobilized trypsin, or to chymotrypsin in solution
/Figs. 5 and 7/ may correspond to the modified inhibitor /reaction k_1/
or dissociation product /reaction k_3/. The pattern observed on SDS-
PAGE is further complicated by degradation products that may be
produced at the excess of a proteinase /k_4 and k_5, Scheme 1/.

The structural features that are responsible for the different reactivities of the two forms of rabbit α-1-antitrypsin, are at present
unknown. Both proteins occur in various strains of rabbits and are
synthesized independently in the liver /Regoeczi, Koj and Lam, 1981/.
Hence it appears that the rabbit has evolved two closely related
proteinase inhibitors. They share antigenic determinants but exhibit
different reactivities towards several proteinases, probably due to
small differences in amino acid composition /Koj and others, 1978/.
Comparison of the amino acid sequences of the F and S forms is
clearly warranted. Future work will also have to decide whether the
complex α-1-antitrypsin system seen in the rabbit, is unique to this
species. The multiple molecular forms of human α-1-antitrypsin type
MM lack the capability of differentially reacting with bovine trypsin, chymotrypsin or equine leucocyte proteinase 1 as can be concluded from observations with isoelectric focusing in agarose gels
/W. Pajdak and A. Koj, unpublished work/.

ACKNOWLEDGEMENTS

These studies were supported by the Polish Academy of Sciences
/grant No 09.7/ and the Medical Research Council of Canada /grant
No MT-4074/. Technical assistance of Miss A. Kurdowska is gratefully acknowledged.

REFERENCES

Baumstark, J.S., C.T. Lee, and R.J. Luby /1977/. Biochim.Biophys.
    Acta,482, 400-411.
Baumstark, J.S., D.R. Babin, C.T. Lee, and R.C. Allen /1980/.
    Biochim.Biophys.Acta,628, 293-302.
Chase, T.Jr., and E.S.Shaw /1970/. Methods Enzymol.,19, 20-27.
Downing, M.R., J.W. Bloom and K.G. Mann /1978/. Biochemistry, 17,
    2649-2653.
Dubin, A, A.Koj and J. Chudzik /1976/. Biochem.J.,153, 389-396.
Hatton, M.W.C., and E. Regoeczi /1977/. Thromb.Res.,10, 645-660.
James, H.J., and A.B. Cohen /1978/. J.Clin.Invest.,62, 1344-1353.
Janoff, A. /1969/. Biochem.J.,114, 157-159.
Jesty, J. /1979/. J.Biol.Chem.,254, 1044-1049.
Kézdy, F.J. and Kaiser,E.T. /1970/ Methods Enzymol.,19, 3-20.
Koj, A. and A. Dubin /1978/. Br.J.Exp.Path.,59, 504-513.
Koj A., M.W.C. Hatton, K.-L.Wong and E. Regoeczi /1978/. Biochem.J.
    169, 589-596.
Laemmli, U.K. /1970/. Nature, 227, 680-685.
Lam, L.S.L., E. Regoeczi and M.W.C. Hatton /1979/. Br,J.Exp.Path.,
    60, 151-160.
Regoeczi, E., A.Koj, and L.S.L. Lam /1981/. Biochem. J. - in press
Satoh, S., T. Kurecki, L.F. Kress, and M. Laskowski, Sr. /1979/.
    Biochem.Biophys.Res.Comm.,86, 130-137.
Schwert, G.E., and Y. Takenaka /1955/. Biochim.Biophys.Acta, 16,
    570-575.
Weber, K., J.R. Pringle and M. Osborn /1972/. Methods Enzymol.,26,
    3-27.

INTRACELLULAR NEUTRAL PROTEINASES AND INHIBITORS

M. Kopitar, M. Drobnič–Košorok, J. Babnik, J. Brzin, F. Steven*
and V. Turk

Department of Biochemistry, J. Stefan Institute, 61000 Ljubljana, Yugoslavia
*Dept. Medical Biochemistry, University of Manchester, Manchester M13 9PT,
England

ABSTRACT

The at present known leucocyte and spleen endogenous inhibitors of proteinases are described. These proteins are located in nuclei and cytosol and they differ in molecular weight, isoelectric point, specificity against the tested proteinases and immunologically.

Inhibition effects of inhibitors are reported: 1) on isolated proteinases, 2) on in vitro colony forming ability of cells, 3) on in vivo tumour growth and metastases formation.

Some of these inhibitors could be degraded by cathepsin D, by hydrolysis of the inhibitor molecule.

KEYWORDS

Leucocyte; spleen; proteinase inhibitors; intracellular distribution; inhibition studies; inactivation – degradation of inhibitors.

INTRODUCTION

Proteinases, as well as their protein inhibitors, are present in multiple forms in numerous tissues of animals and plants, and in microorganisms (Barrett, 1977; Laskowski, 1980). Recent investigations in the field of tissue proteinases have shown that their action is extracellulary(Fritz and co-workers, 1974; Collen and co-workers, 1979) and intracellulary(Kopitar and co-workers, 1980; Lenny and co-workers,1979), regulated by inhibitors. Extracellular inhibitors have been much more studied than intracellular inhibitors.

Of great physiological significance are the inhibitors which are present in the same cells as the enzymes they inhibit. Our interest in this concept originated with investigations of the proteolytic system in various mammalian tissues which have a great preponderance of leucocyte cells. Besides different types of proteinases, leucocyte cells are known to contain their endogenous inhibitors. So far it was found that leucocytes possess inhibitory activity against elastases, chymotrypsin-like neutral pro-

329

teinases, thiol proteinases, plasminogen activator and plasmin, and many groups of
investigators have studied the isolation (Kopitar and co-workers, 1980, 1975; Dubin,
1977), molecular characterization (Kopitar and co-workers, 1978; 1981) and mode
of inhibition (Steven and co-workers, 1976) of these proteins. Until recently, it was
generally accepted that they are located in cytoplasm, but in one of our last studies
we reported on their location and isolation from leucocyte nuclei (Kopitar, 1981).

In the course of our continuing studies, three types of specific inhibitors of proteinases were isolated from leucocytes. These proteins differ significantly in electrophoretic mobility, molecular weight, isoelectric point, immunologically and in their
inhibition ability against tested enzymes. The purified inhibitors are stable in acid
as well as in neutral pH. Two of them can be inactivated - degraded by the action of
cathepsin D.

The present review will emphasize studies of our laboratory and of our co-workers
which have delineated the biochemical and biological properties of endogenous proteinase inhibitors.

Subcellular Distribution of Leucocyte Intracellular Inhibitors

Leucocyte cells were isolated from peripheral pig blood, by the Dextran procedure,
as described previously (Kopitar and co-workers, 1975). According to the evidence
of many studies, leucocyte inhibitors could be isolated only from the soluble phase
of disrupted leucocytes, freed of nuclei and granules, by differential centrifugation,
named the post granular supernatant (PGS) (Kopitar and co-workers, 1975; Dubin,
1977). But so far no work has been published dealing with the extraction of these inhibitors from nuclei.

During the last five years many studies appeared dealing with the extractability of
proteinases and their endogenous inhibitors from tissue homogenates. It was found
that their extractability could be regulated by the ionic strength and pH of salt solutions (Lenny and co-workers, 1979; Fraki and co-workers, 1973; Seppa and Jarvinen,
1978). Acid and thiol proteinases as well as inhibitors are extracted at low ionic
strength, whereas the solubility and activity of neutral proteinases are dependent on
a high concentration of salt. On the basis of the different solubility of proteinases at
low ionic strength, dialysis (against water) caused the precipitation of neutral proteinases, while acid proteinases and inhibitors remain soluble (Lenny and co-workers, 1979 ; Seppa and Jarvinen, 1978). When the extraction is directed towards the
isolation of inhibitors, another important finding has to be considered, namely, the
inactivation of these inhibitors at acid pH, by acid proteinases (Kopitar and co-workers, 1980). In the preparation of inhibitors from nuclei all these important points
were taken into consideration.

A two-step extraction of the inhibitors from the nuclear pellet was performed. The
first was made with 0.9% NaCl of pH 7.0 and the extract obtained showed no inhibitory activity against thiol or neutral proteinases. This extract was centrifuged and
the sediment was used for the second extraction, whereas the inactive supernatant
was further treated to activate it. This supernatant generated inhibitory activity
after dialysis against distilled water, pH 6.0. Dialysis caused precipitation of some
proteins, whereas the inhibitors remained in solution. The precipitate formed during
dialysis was removed by centrifugation and the supernatant was tested against thiol
and neutral proteinases, it showed high inhibitory activity. From these results it can

be concluded that generation of nuclear inhibitory activity under the chosen conditions is a time dependent process, presumably based on dissociation of the enzyme inhibitor complexes and on precipitation of proteinases mainly inactivated under the given conditions.

The second extraction with 0.029 M NaCl, pH 7.0, followed by freezing 4 times in liquid air, released proteins that showed high inhibitory activity against the tested proteinases.

Purification, Biochemical and Biological Characteristics of Intracellular Inhibitors

Inhibitors could be isolated from the soluble cellular fractions which were obtained from the cell homogenates (0.34 M sucrose) by differential centrifugation and from nuclear extracts. In the case of isolation from tissues still another procedure has been developed, especially in the case of inhibitors of thiol proteinases. This isolation procedure is based on the high thermal stability of this type of inhibitors.

Subsequent purification was achieved by ion exchange chromatography, gel chromatography on Sephadex G-100, G-150, G-50 and by affinity chromatography on papain Sepharose.

TABLE 1 Properties of Leucocyte and Spleen Inhibitors.

Inhibitor	$M.W. \times 10^3$ approx.	pI	Spleen	Leucocyte	
				N	PGS
$I-1_A$	12 – 15	4.6–4.8	+	+	+
$I-1_B$	11 – 13	5.8–7.0	+	+	–
I-2	40	5.7–5.8	+	–	+
I-3	68	4.4–4.5	–	+	+
I-4	8	7.0	+		

N – nuclei; + present in; – absent in.

From leucocytes three types of proteinase inhibitors have been isolated (Kopitar and co-workers, 1980, 1981), and from spleen two types identical to those from leucocytes (Brzin and co-workers, 1978), as well as a low molecular weight inhibitor of plasmin (Brzin and co-workers, 1981).

Proteinase inhibitor I-1 could be isolated either from leucocytes or from spleen. This type of inhibitor is present in many multiple molecular forms in various human and rat organs and also in non-mammalian animals (Lenny and co-workers, 1979; Kopitar and co-workers, 1978; Jarvinen, 1978). Inhibitors ($I-1_A$) isolated from cytosol or from spleen have isoelectric points around pH 4.6 – 4.8, whereas inhibitors from nuclei include this type as well as the inhibitors ($I-1_B$) with isoelectric points near pH 7.0. A similar inhibitor was isolated from muscles by Bird (1977). Inhibitors I-1 inhibit thiol proteinases: cathepsin B, H and papain and chymotrypsin- like neutral proteinases.

Proteinase inhibitor I-2 could be isolated only from the soluble cellular fractions of spleen and leucocytes (Kopitar and co-workers, 1975; Dubin, 1977; Brzin and co-

workers, 1978). This inhibitor is a specific inhibitor for leucocyte and spleen elastase, and also inhibits proteinases of the chymotrypsin type. Fig. 1 shows the inhibition
of elastase by increasing concentrations of I-2 in the case of fluorescein labeled polymeric collagen as substrate.

TABLE 2 Leucocyte and Spleen Inhibitors and Related Proteinases

Inhibitor	Inhibition of						
	Elast.	CLNP	Papain	Cath.B	Cath.H	UK	Plasmin
I-1$_A$	-	++	++	+	+	-	-
I-1$_B$	-	+	++	++	++	-	-
I-2	++	+	-	-	-	-	-
I-3	-	-	-	-	-	++	-
I-4	-	-	-	-	-	-	++

CLNP - chymotrypsin-like neutral proteinases;
UK - urokinase; ++ strong , + weaker , - no inhibition.

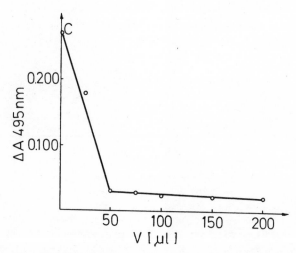

Fig. 1. Inhibition of spleen elastase by increasing concentrations
of leucocyte inhibitor I-2.

Proteinase inhibitor I-3 could be isolated from PGS or nuclei of leucocyte cells. This
inhibitor is a specific inhibitor of the plasminogen activator urokinase (Kopitar,1981).
Inhibitors with similar characteristics have been already isolated from aorta (Okamura and co-workers, 1979) and from plasma (Collen and co-workers, 1979). Fig.2
shows the inhibition of urokinase by increasing quantities of I-3.

Proteinase inhibitor I-4 was isolated from spleen and its location in the cell is presently unknown. This inhibitor is a specific inhibitor of plasmin and also inhibits
trypsin (Brzin and co-workers, 1981). As is evident from Tables 1 and 2, inhibitors
differ in isoelectric point, molecular weight and their specificity against tested enzymes. Another difference is that of their thermal stability. Only inhibitors I-1 are

thermostable; all other types are not. Another important characteristic of these inhibitors was also determined, namely their pH stability. It was found that highly purified inhibitors are quite stable in buffer solutions from pH 3 - 5; tle loss of inhibitory activity at acid pH of PGS, is connected with a specific interaction with acid
proteinases. The mechanism is clarified in a subsequent section of this paper. These
inhibitors belong to the type of so called fast reacting inhibitors of proteinases. That
is, they reach their full inhibitory ability toward the tested proteinases in a very
short preincubation time (1 min), and the complex formed is stable over the tested
time intervals of 1 - 240 min (Kopitar and co-workers, 1978, 1981).

From immunological studies we determined that these inhibitors belong to specific
types. This has been shown already by immunoelectrophoresis (Kopitar and co-
workers, 1980), and in the present paper by double immunodiffusion. Fig. 3 shows
the cross reactivity obtained from samples of I-1$_A$ and I-3. This result confirms
that the inhibitors are unrelated proteins.

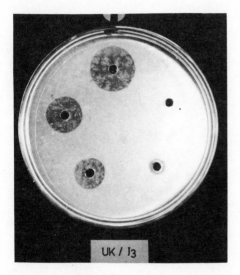
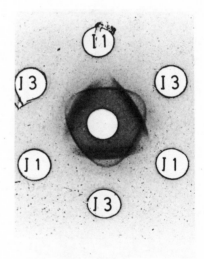

Fig. 2.   Fig. 3.

Fig. 2. Inhibition of urokinase by I-3. Samples: UK, I-3 and
mixtures of UK and I-3 in volume ratios: 1:1, 1:0.5, 1:0.25.

Fig. 3. Immunodiffusion plate in 3% agar, buffer pH 7.1. In
outher well I-1$_A$ and I-3. In center well antibody of I-1$_A$, I-3
and I-2.

The regressive effect of leucocyte as well as of spleen intracellular inhibitors on tumour growth and metastases formation has already been reported (Giraldi and co-
workers, 1977, 1980). It was found that I-1$_A$ and I-2 from both sources do significantly reduce metastases formation in the lungs of mice bearing Lewis lung carcinoma.

The in vitro inhibitory effect of leucocyte inhibitor I-1$_A$ and spleen inhibitor I-4 was first tested on the colony forming ability of V-79-379 A cells (Kopitar and co-workers, 1981B). It was found that leucocyte I-1$_A$ displayed a stronger cytocydal effect than spleen I-4.

Inactivation-Degradation Studies of Intracellular Inhibitors by Cathepsin D

In our first study (Kopitar and co-workers, 1975) of leucocyte intracellular inhibitors we found that the inhibition ability of leucocyte cytoplasm was highly diminished at acid pH, from pH 3 - 5. In further investigations (Kopitar and co-workers,1980, 1980B), the mechanism of the acid inactivation of cytoplasm inhibitor activity was clarified. Highly purified, isolated inhibitors are quite stable over the pH range from pH 3 - 8, but two of them, I-2 and I-3, could be inactivated when treated with cathepsin D at acid pH. We also observed that cathepsin D does not inactivate I-1. In one of our last study we determined that cathepsin D inactivates these inhibitors by hydrolysis of the inhibitor molecule (Drobnič-Košorok and co-workers, 1981). The inactivation as well as the degradation of the inhibitor depends on the concentration of cathepsin D and the pH of the medium. The products resulting from the interaction of inhibitor with cathepsin D were examined by polyacrylamide gel electrophoresis, as shown in Fig. 4. The incubation mixture was made up in a 2:1 M ratio of I-3 and cathepsin D. The reaction was blocked at the time intervals indicated with pepstatin. Examination of the incubation mixtures by polyacrylamide gel electrophoresis indicated the degradation of the inhibitor molecule into several fragments.

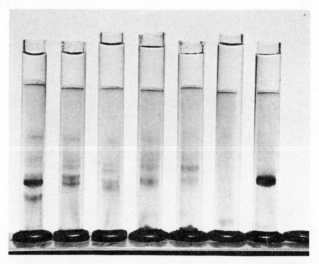

Fig. 4. Polyacrylamide gel electrophoresis of incubation mixtures of I-3 : cathepsin D (pH 4.0, 37 ^{o}C) after different time of incubation. The direction of migration was dawnwards the anode. Gel 1 (1 hr), gel 2 (2 hrs), gel 3 (4 hrs), gel 4 (6 hrs), gel 5 (12 hrs), gel 6 (24 hrs), gel 7 (control of I-3.)

Inactivation of the inhibitor can also be studied immunologically. From Fig. 5 it can be seen that the inhibitor was degraded during incubation with cathepsin D. Degradation was apparent within the first hour of incubation and continued through 4 hrs, while after 6 hrs incubation the mixture failed to form precipitin lines.

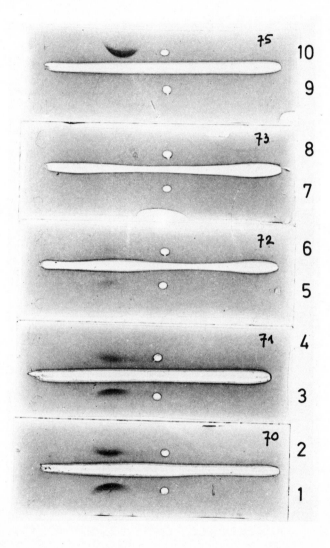

Fig. 5. Immunoelectrophoretic results of degradation of I-3 by cathepsin D demonstrating the degradation of inhibitor. Trough (T) contain rabbit anti(pig leucocyte I-3)serum. Well 1 (5 min), well 2 (15 min), well 3 (30 min), well 4 (1 hr), well 5 (2 hrs), well 6 (4 hrs), well 7 (6 hrs), well 8 (12 hrs), well 9 (cathepsin D), well 10 (I-3).

CONCLUSIONS

From the results presented it is evident that leucocytes and spleen contain different types of protein inhibitors of proteinases. They are located in cytoplasm and in nuclei. In our laboratory we have succeeded so far in isolating 4 different types of such inhibitors: I-1 (leucocyte, spleen), I-2 (leucocyte, spleen), I-3 (leucocyte), I-4 (spleen). These proteins differ in molecular weight, isoelectric point, inhibition ability against tested proteinases and interactions (inactivation) with cathepsin D.

In Laskowski's opinion (1980), the physiological function of proteinase inhibitors in the broad sense is clear – elimination of unwanted proteolysis. However, detailed functions are not clear. What the regulatory role of cellular proteinase inhibitors is, which are located in cytoplasm and nuclei is presently unknown. It is only known that their activity could be regulated by cathepsin D. Another important property of intracellular inhibitors is the ability to inhibit the growth of metastases formation (in vivo), as well as the growth of cells in colonies (in vitro).

ACKNOWLEDGEMENT

The excellent technical assistance of Mrs. M. Božič and Mr. K. Lindič is greatfully acknowledged. Supported by the research grant from the Research Council of Slovenia and in part by the NSF.

REFERENCES

Barrett, A.J. ed.,(1977). Proteinases in mammalian cells and tissues. North Holland Publishing Co., Amsterdam.

Brzin, J., M. Kopitar and V. Turk (1978). Isolation and characterization of inhibitors of neutral proteinases from spleen. Acta Biol.Med.German. 36, 1883–1886.

Brzin, J., M. Kopitar and V. Turk (1981). Isolation and characterization of an low molecular weight plasmin inhibitor from spleen. Submitted for publication.

Collen, D., B. Wiman and M. Verstraete, eds.(1979). The physiological inhibitors of blood coagulation and fibrinolysis. Elsevier North Holland, Amsterdam.

Dubin, A. (1977). Polyvalent proteinase inhibitor from horse blood leucocyte cytosol. Europ.J.Biochem. 73, 429–435.

Drobnič-Košorok, M., M. Kopitar, J. Babnik and V. Turk (1981). Inactivation studies of leucocyte inhibitor of urokinase by cathepsin D. Mol.Cell.Biochem., in press.

Fritz, H., H. Tschesche, L.J. Greene and E. Truscheit, eds. (1974). Bayer Symp. V, Proteinase inhibitors . Springer Verlag, Berlin.

Fraki, J., O. Ruuskanen, V.K. Hopsu-Havu and K. Kouvalainen (1973). Thymus proteases: extraction, distribution, comparison to lymph node proteases and species variation. Hoppe Seyler's Z.Biol.Chem. 354, 933–943.

Giraldi, T., M. Kopitar and G. Sava (1977). Antimetastatic effects of a leucocyte intracellular inhibitor of neutral proteases. Cancer Res. 37, 3834–3837.

Giraldi, T., G. Sava, M. Kopitar, J. Brzin and V. Turk (1980). Neutral proteinase inhibitors and antimetastatic effects in mice. Europ.J.Cancer, 16, 449–454.

Jarvinen, M. (1978). Purification and some characteristics of the human epidermal SH-protease inhibitor. J.Invest.Dermatol. 72, 114–118.

Kopitar, M., and D. Lebez (1975). Intracellular distribution of neutral proteinases and inhibitors in pig leucocytes. Europ.J.Biochem., 56, 571-581.

Kopitar, M., J. Brzin, T. Zvonar, P.Ločnikar, I. Kregar, and V. Turk (1978). Inhibition studies of an intracellular inhibitor on thiol proteinases. FEBS Letters, 91, 355-359.

Kopitar, M., J. Brzin, J. Babnik, V. Turk, and A. Suhar (1980). Intracellular neutral proteinases and their inhibitors. In P. Mildner and B. Ries (eds.), Proc. of the FEBS Special Meeting on Enzymes, Enzyme Regulation and Mechanism of Action, Vol. 60, Pergamon Press, Rotterdam, pp.363.

Kopitar, M., T. Giraldi, P.Ločnikar, and V. Turk (1980) Biochemical and biological properties of leucocyte intracellular inhibitors of proteinases. In M.C.Escobar and H. Friedman (eds.), Macrophages and Lymphocytes, Part A, Plenum Publishing Co., London, pp. 75.

Kopitar, M. (1981). Isolation and characterization of urokinase inhibitors isolated from pig leucocytes, Haemostasis, in press.

Kopitar, M., M. Drobnič-Košorok, J. Babnik, J. Brzin, V. Turk, M. Korbelik, U. Batista, S. Svetina, J. Škrk, T. Giraldi, and G. Sava (1981). Biochemical and biological characteristics of leucocyte proteinase inhibitors. In F. Rossi and P. Patriarca (eds.). Movement, metabolism, microbicidal and tumouricidal mechanisms of phagocytes, Plenum Publishing Co., London, in press.

Laskowski, M.Jr., and K. Ikunoshin (1980). Protein inhibitors of proteinases. Ann. Rev.Biochem., 49, 593-626.

Lenny, J.F., J.R. Tolan, W.J. Sugai, and A.G. Lee (1979). Thermostable endogenous inhibitors of cathepsin B and H. Europ.J.Biochem.,101, 151-161.

Okamura, T., S. Nanno, and K. Tanaka (1979). Inhibitors of fibrinolysis in human arterial wall. Abstr.of 7th International Congress on Thrombosis and Haemostasis. Schattauer Verlag, Stuttgart, 220.

Schwartz, W.N., and J.W. Bird (1977). Degradation of myofibrillar proteins by cathepsin B and D. Biochem.J., 167, 811-820.

Seppa, E.J., and M. Jarvinen (1978). Rat skin main neutral protease: purification and properties. J.Invest.Dermatol., 70, 84-89.

Steven, F.S., P.W. Milsom, and J.A.D. Hunter (1976). Human polymorphonuclear leucocyte neutral protease and its inhibitor. Europ.J.Biochem., 67, 165-169.

INHIBITION OF TRYPSIN AND CHYMOTRYPSIN BY INHIBITORS FROM <u>VIPERA AMMODYTES</u> VENOM

A. Ritonja, V. Turk and F. Gubenšek

*Department of Biochemistry, J. Stefan Institute, E. Kardelj University,
61000 Ljubljana, Yugoslavia*

ABSTRACT

Dissociation constants of the three inhibitors of serine proteinases isolated from the venom of <u>Vipera ammodytes</u> were determined for trypsin and chymotrypsin. The amino acid compositions of inhibitors are presented. It was found that all three inhibitors are single headed and that they are resistant toward proteolysis by trypsin and chymotrypsin.

KEYWORDS

Serine proteinase inhibitors; trypsin, chymotrypsin inhibition; dissociation constant; snake venom.

Takahashi and co-workers (1974) reported that the crude venom of the European viper <u>Vipera ammodytes</u> inhibits a number of serine proteinases. Since we have been working with this venom for many years (Ritonja and co-workers, in preparation) we had only to look where in our fractionation pattern such an inhibitory activity could be detected. Two different inhibitors were found in the fractions eluted from CM – cellulose at pH 8.2. The first (T.i.) inhibits trypsin substantially better than chymotrypsin, the second (Ch.i.) does the opposite. In the course of further purification on Sephadex G-100 and following rechromatography on CM-cellulose, T.i. was separated into two different proteins: T.i.I and T.i.II.

All three inhibitors are proteins with monomer molecular weight of 7,000. They are present as dimers in solutions of moderate ionic strength at neutral pH. The amino acid composition of the three inhibitors is shown in Table 1. Both T.i. forms are relatively similar. The most striking difference between the two is the absence of methionine in T.i.I. Chymotrypsin inhibitor has quite a different composition. All three, however, contain 6 Cys, presumably bound by 3 disulphide bridges. Immunologically T.i.I and T.i.II are identical, but they both differ from Ch.i. The isoelectric points of all three inhibitors are near pH 10.

Laskowski and Kato (1980) classified numerous known protein proteinase inhibitors into several groups according to their properties. Our T.i.I and T.i.II seem to belong to the Kunitz BPTI family of inhibitors, along with inhibitors from the venoms of Vipera russelli, Haemachatus haemachatus and Naja nivea. Ch.i., however, may not fit into the same group of inhibitors. Only the determination of primary structures will permit us to clarify the problem of their exact classification.

TABLE 1 Amino acid composition of inhibitors

	T.i.I	T.i.II	Ch.i.
Asp	9	8	7
Thr	2	2	2
Ser	2	2	3
Glu	3	3	2
Pro	5	6	5
Gly	5	5	4
Ala	6	6	5
Cys	6	6	6
Val	0	0	1
Met	0	1	1
Ile	2	2	2
Leu	1	1	2
Tyr	4	4	5
Phe	4	4	4
His	2	2	1
Lys	6	5	5
Trp	1	1	1
Arg	3	4	6
	61	62	62
M.W.	6869	7010	7192

Protein proteinase inhibitors have extremely low dissociation constants, usually as low as 10^{-10} M. The dissociation constants of our inhibitors are also of the same order of magnitude. Already a simple titration experiment has shown that all three inhibitors inhibit their corresponding enzymes in a molar ratio close to 1:1. In order to obtain more information on the kinetic properties of these inhibitors, we determined the following features of the interaction between a particular inhibitor and the corresponding enzyme: the resistance of the inhibitor to enzymatic degradation; the time needed to complete the formation of the complex in the absence and in the presence of the substrate; displacement of the inhibitor from the EI complex with substrate and displacement of the substrate from the ES complex with the inhibitor; and finally the dissociation constants.

All measurements were made potentiometrically using TAME as a substrate for trypsin and ATEE as a substrate for alpha-chymotrypsin. Prolonged exposure of inhibitors to trypsin or chymotrypsin showed that all the inhibitors are completely resistant toward proteolysis by corresponding enzymes. We established that the time needed to reach the steady state reaction velocity was less than 2 min in the absence of the substrate for all three inhibitors at the pH optimum of the corresponding enzyme. The bottom curve of Fig. 1 shows inhibition of trypsin by T.i.I (or T.i.II) in the case when substrate was added after 2 min of preincubation of the enzyme with the inhibitor. The bottom curve did not band upwards within the first 10 min even when the substrate concentration was doubled. This is the result of the combination of two effects: of a slow rate of dissociation and of a low dissociation constant. A similar effect was also observed by Green (1953) in the inhibition of trypsin by BPTI. Essentially the same behaviour was observed with T.i.II.

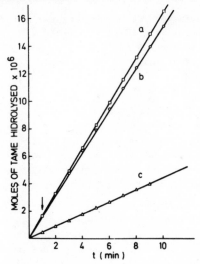

Fig. 1. Inhibition of trypsin by T.i.I (or T.i.II). a. no inhibitor, b. inhibitor was added (where indicated by arrow) to the mixture of E and S, c. substrate was added after 2 min to the mixture of E and I.

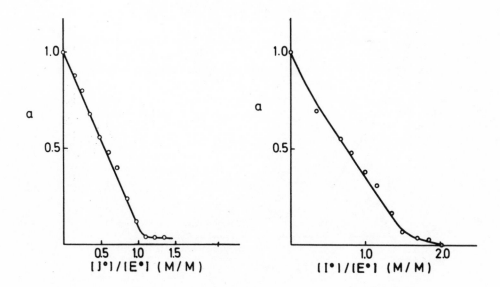

Fig. 2. Inhibition of trypsin by T.i.I (left) and T.i.II (right).

The time course of the reaction after addition of T.i.I (or T.i.II) to the mixture of trypsin and substrate is shown by curve b in Fig. 1. We can see that even after 10 min only a very small amount of enzyme was inhibited, which means that displacement of substrate with T.i.I (or T.i.II) is slow and thus the inhibitor has a low affinity for trypsin.

Chymotrypsin inhibitor behaves entirely differently in inhibiting alpha-chymotrypsin. In this case the substrate could be displaced by the inhibitor within 2 min after its addition, to the mixture of enzyme and substrate, and the reaction velocity becomes the same as in the case when substrate was added to the reaction mixture last. Apparently Ch.i. has a high affinity for alpha-chymotrypsin.

The possibility that the inhibitors have different sites for trypsin or chymotrypsin was ruled out. T.i.I (or T.i.II) already bound to trypsin was without effect on chymotrypsin. Similarly, Ch.i. bound to alpha-chymotrypsin could not inhibit trypsin. For the determination of dissociation constants, the methods of Bieth (1974) and of Greene and Work (1953) were used. As shown in Fig. 2 the binding of T.i.I and T.i.II to trypsin is stoichiometric and thus the method of Greene and Work applies. The dissociation constant at pH 7.8 was essentially the same for both inhibitors of the order of 10^{-10} M.

Binding of Ch.i. to alpha-chymotrypsin is also rather strong (Fig. 3). In this case both methods, the stoichiometric and the non-stoichiometric, could be used since this is a limiting case. The dissociation constants obtained differ, however, by one order of magnitude. A value of 4×10^{-9} M was obtained by Greene's method and 6×10^{-8} M by Bieth's.

We determined also what we call "cross constants", that is dissociation constant of the complex between alpha-chymotrypsin and T.i.I or T.i.II and the constant of the complex of trypsin and Ch.i. In these cases the binding was non-stoichiometric and

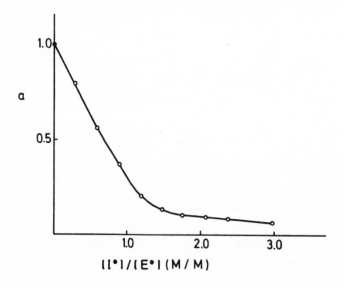

Fig. 3. Inhibition of alpha-chymotrypsin by Ch.i.

only the method of Bieth was appropriate. All the results are summarized in Table 2. Similar results have been already obtained with other known inhibitors. For example, BPTI, which has an inhibition constant for trypsin of 10^{-10} M or even less (recently a value as low as 10^{-14} M has been reported (Vincent, Lazdunski, 1972)) inhibits alpha-chymotrypsin with a K_i of only 10^{-8} M, which is a few orders of magnitude below the K_i for trypsin.

TABLE 2 Dissociation constants of enzyme-inhibitor complexes

Inhibitor	Trypsin	alpha-chymotrypsin
T.i.I	3.4×10^{-10} M	2.7×10^{-7} M
T.i.II	5.6×10^{-10} M	3.0×10^{-7} M
Ch.i.	1.1×10^{-6} M	4.3×10^{-9} M

Besides different dissociation constants, differences in the velocities of binding are also notable. T.i.I and T.i.II bind in the presence of substrate, when the enzymes are already in the ES complex, slowly to trypsin, but quickly to chymotrypsin. Ch.i. binds quickly to both enzymes. The explanation of these differences is at present not possible. Several mechanisms could be responsible for such behaviour, such as local charge effect, structural changes, differences in the formation of intermediates and probably others. Additional investigations will be needed to explain the different behaviour of these otherwise rather related inhibitors.

ACKNOWLEDGEMENT

This work was supported by grant no. 106/3372-80 from the Research Community of Slovenia.

REFERENCES

Bieth, J. (1974). Some kinetic consequences of the tight binding of protein proteinase inhibitors to proteolytic enzymes and their application to the determination of dissociation constants. In H. Fritz, H. Tschesche, L.J. Greene and E. Truscheit (Eds.), Proteinase inhibitors , Springer Verlag, Berlin. pp 463-469.
Green, M. (1953). Competition among trypsin inhibitors. J.Biol.Chem.,205,535-551.
Green, N.M. and E. Work (1953). Pancreatic trypsin inhibitor. Biochem.J.54, 347.
Laskowski, Jr., M., and I. Kato (1980). Protein inhibitors of proteinases. Ann.Rev. Biochem., 49, 593-626.
Ritonja, A., V. Turk, and F. Gubenšek (in preparation). Serine proteinase inhibitors from Vipera ammodytes venom.
Takahashi, H., S. Iwanaga, and T. Suzuki (1974). Distribution of proteinase inhibitors in snake venoms. Toxicon, 12, 193-197.
Vincent, J.P., and M. Lazdunski (1972). Trypsin-pancreatic trypsin inhibitor association. Dynamics of the interaction and role of disulfide bridges. Biochemistry, 11, 2967-2977.

PROTEOLYTIC AND ANTI-PROTEOLYTIC ACTIVITIES
IN CHONDROSARCOMA

H. Levy, I. Shaked and G. Feinstein

*Dept. of Biochemistry, the George S. Wise Faculty of Life Sciences,
Tel Aviv University, Ramat Aviv, Israel*

ABSTRACT

Highly purified proteoglycan monomer (PGM) preparation from Swarm rat chondrosarcoma exhibit self-degradation as well as caselinolytic activity. The proteolytic activity associated with PGM showed two pH optima (7.0 and 8.5). It was inhibited by several protein inhibitors of serine proteases (e.g. LBTI) as well as by low molecular weight microbial inhibitors (e.g. leupeptin). The proteolytic activity was effected by cations and organic solvents.
Five electrophoretic bands also isolated from chondrosarcoma exhibit protease inhibiting activity. These inhibitors, localized in the chondrosarcoma matrix, were shown to be acidic proteins which shared common antigenic determinants (isoinhibitors). The pooled inhibitors displayed a wide range of specificity.

KEYWORDS

Cartilage proteases proteoglycan degradation; protease isoinhibitors; chondrosarcoma.

INTRODUCTION

It is well established that the proteoglycan of the cartilage matrix is destroyed in articular diseases. It has been suggested that hyperactivity of lysosomal proteases produced by the tissue cells could initiate cartilage degradation. The involvement of several lysosomal cathepsins has been demonstrated (Dingle, 1976). Consequently, however, one would expect extracellular proteolytic activity to be correlated with the normal cartilage components turnover. An attempt was therefore made to test these assumptions.
The resistance of cartilage to tumor invasion has been related to the presence of antiproteolytic activities in this tissue (Eisenstein, 1973).

RESULTS AND DISCUSSION

Proteolytic Activity in Swarm Rat Chondrosarcoma

A rat tumor which originally arose spontaneously, and has been maintained by subcutaneous implantation was used (Choi, 1971). Crude preparations obtained by low ionic strength extraction exhibit proteolytic activity towards azocasein (Fig. 1) This proteolytic activity was found to be associated with a hexuronic-rich fraction.

345

H. Levy, I. Shaked and G. Feinstein

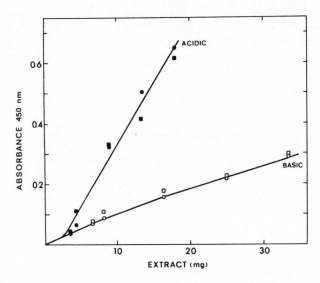

Fig. 1. The azocaseinolytic activity of chondrosarcoma extracts.

Proteoglycan monomers (PGM) were prepared and purified by established methods (Oegema, 1975). The PGM preparation showed a single symmetric Schlieren peak in the analytical ultracentrifuge, as well as a single peak in gel filtration. When a purified PGM preparation was incubated at 37°C for 24 hr two new Schlieren peaks were observed. Self degradation of PGM was abolished by the addition of protease inhibitors such as LBTI. (Fig. 2) The proteoglycan associated proteolytic activity indicated by these findings was further studied using azocasein as substrate. Two pH optima were observed, at 7.0 and 8.5, respectively. (Fig. 3)

The proteolytic activities were inhibited by several inhibitors of serine proteases (e.g. LBTI, see Fig. 4), by low molecular weight microbial inhibitors (e.g. leupeptin, Fig. 5) as well as by organomercury compounds. The data is summarized in Table 1.

The preliminary findings seem to suggest, in our opinion, that the proteolytic activity associated with PGM is most probably not related to cystein protease(s). Attempts to dissociate the proteolytic activity from the PGM molecules were, so far, unsuccessful. This could be possibly explained as due to instability of the dissociated enzyme species as suggested by our finding that degradation of the chondroitin sulfate side chains with chondroitinase abolished the proteolytic activity (Table 2). It thus seems reasonable to assume that the proteolttic activity is associated with the PGM molecule via interactions involving chondroitin sulfate side chains. In further attempts to investigate the nature of the association between the enzyme and the PGM molecules it was shown that divalent cation cations enhanced the proteolytic activity (Fig. 6) while monovalent cations led to its inhibition (Fig. 7) Moreover, the addition of several organic solvents led to the enhancement of the PGM associated proteolytic activity (Fig. 8).

The combined data seems to emphasize the fact that the nature of the association between the enzymic activity and the PGM molecule is rather complex and requires further investigation.

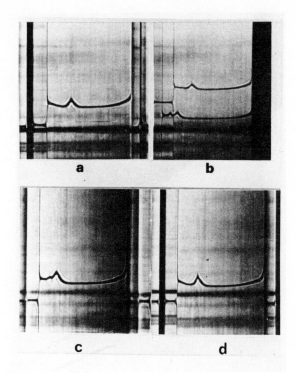

Fig. 2. Schlieren patterns of proteoglycan following different treatment.

a. Proteoglycan monomer, control.

b. Upper peak, proteoglycan (2 mg/ml) incubated at 37°C for 24 hr, pH 7.0 with LBTI (0.8 mg/ml).
Lower peaks - proteoglycan incubated at 37°C, pH 7.0, for 24 hr in the absence of added inhibitor

c. Proteoglycan incubated at 37°C, pH 8.5, for 24 hr, no inhibitor.

d. Proteoglycan incubated at 37°C, pH 8.5, for 24 hr, in the presence of LBTI (0.8 mg/ml).

H. Levy, I. Shaked and G. Feinstein

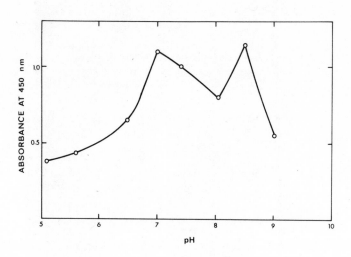

Fig. 3. pH profile of azocasein digestion by proteoglycan.

TABLE 1   Inhibition of proteoglycan associated proteases.
(75 µg proteoglycan were used per assay.)

Inhibitor	Final Concentration µg/assay	Residual activity %	
		pH 7.0	pH 8.5
None		100	100
SBTI	150	50	60
LBTI	300	50	50
Leupeptin	90	57	68
Pepstatin	70	100	100
Chymostatin	70	100	100
Elastatinal	70	100	100
PHMB	240	67	80
Dip-F	800	>100	>100
ε-ACA	1000	100	100
Benzamidin	300	100	100

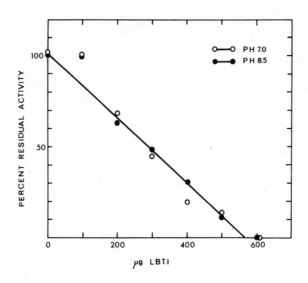

Fig. 4. Inhibition with Lima Bean Trypsin Inhibitor (LBTI).
(75 µg PGS were used per assay)

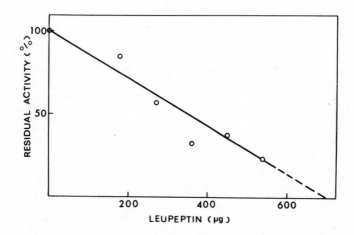

Fig. 5. Inhibitor with Leupeptin (75 µg PGM were used per assay)

H. Levy, I. Shaked and G. Feinstein

TABLE 2 Effect of removal of chondroitin sulfate side chains
on proteoglycan proteolytic activity

	Hexuronic acid μg/mg PG	Proteolytic activity (A_450) pH 7.0	Proteolytic activity (A_450) pH 8.5
Proteoglycan (control)	46.7	0.75	0.75
Chondroitinase treated proteoglycan	26.6	0.00	0.00

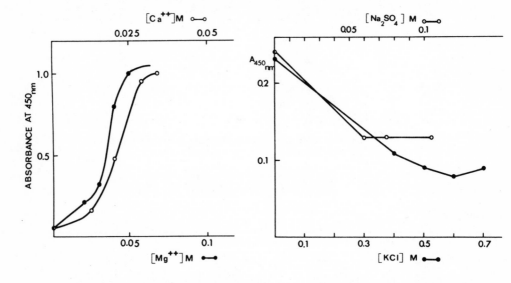

Fig. 6. The effect of Ca^{+2} and Mg^{+2} on the proteolytic activity associated with PGM (75 μg PGM per assay)

Fig. 7. The effect of KCl and Na_2SO_4 on the proteolytic activity associated with PGM. (75 μg PGM were used per assay)

Antiproteolytic Activity in Swarm Chondrosarcoma

Extracts from the chondrosarcoma tissue were passed through a trypsin-Sepharose column (details are published elsewhere, see Levy, 1980). The column was eluted with 0.57 M NaCl:HCl pH 2.0 buffer. Five bands possessing protease inhibitory activity were obtained upon PAG-electrophoresis at pH 8.6 of the affinity column eluate (Fig. 9) The isoelectric points of all five bands were found to lie in the range 3.5 to 4.7. Immunodiffusion experiments showed that these five inhibitory protein species shared common antigenic determinants, strongly suggesting the possibility of their being isoinhibitors. SDS-electrophoresis of the pooled inhibitors yielded two major bands of estimated molecular weights 16,500 and 20,000, respectively.

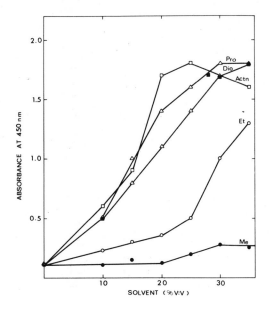

Fig. 8. The effect of organic solvents on the proteolytic
activity associated with PGM.

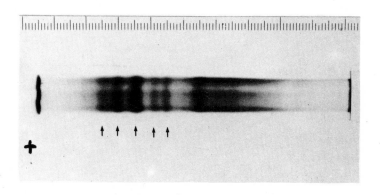

Fig. 9. Polyacrylamide electrophoresis of trypsin -
Sepharose eluate.

CONCLUDING REMARKS

The possibility that a proteolytic activity can be associated or bound to a bio-polymer could be of great biological significance in the normal turnover of such molecules. The phenomenon described in this communication i.e. proteolytic activity associated with proteoglycan monomer, may indicate the existence of such mechanisms.

The isoinhibitors isolated from the Swarm rat chondrosarcoma seem to belong to a new class of protease inhibitors that differ from Aprotonin since they have an acidic isoelectric point. Moreover, they were active against trypsin as well as non trypsin-like enzymes e.g. chymotrypsin, neutrophil elastase and chymotrypsin-like protease, but not against pancreatic elastase, which differs from the neutrophil elastase in its specificity (Levy, 1979). Such inhibitors could play a critical role in protecting cartilage from endogeneous and exogeneous proteolytic attack.

REFERENCES

Choi, H.U., Meyer, K. and Swarm (1971). Mucopolysaccharide and protein - polysaccharide of a transplanatable rat chondrosarcoma. Proc. Natl. Acad. Sci. USA, 68, 877-879.

Dingle, J.T. (1976). In D.W. Ribbons, and K. Brew (Eds.), Proteolysis and Physiological Regulation, Academic Press, pp. 339-355.

Eisenstein, R., Sorgente, N., Soble, L.W., Miller, A., and Kuettner, K.E. (1973). The resistance of certain tissues to invasion. Am. J. Pathol., 73, 765-768.

Levy, H., and Feinstein, G. (1979). The digestion of the oxidized B chain of insulin by human neutrophil proteases: elastase and chymotrypsin-like protease. Biochem. Bioph. Acta, 567, 35-42.

Levy, H., Shaked, I., and Feinstein, G. (1980). Protease isoinhibitors from Swarm rat chondrosarcoma. Conn. Tis. Res., in press.

Oegema Jr., T.R., Hascall, V.C., and Dziewiatkowski, D.D. (1975). Isolation and characterization of proteoglycan from the Swarm rat chondrosarcoma. J. Biol. Chem., 250, 6151-6159.

ACKNOWLEDGEMENT

We are grateful to Dr. Leon Goldstein for his critical review of the manuscript.

THE KALLIKREIN—KININ SYSTEMS*

Reinhard Geiger

*Abteilung für Klinische Chemie und Klinische Biochemie in der Chirurgischen
Klinik der Universität München, Nussbaumstrasse 20, D-8000 München 2, FRG*

ABSTRACT

The kallikrein-kinin system consists of four major components, the
kininogenase or kallikreins, the kininogens, the kinins and the kininases. Present evidence suggests that one has to discriminate between a plasma kallikrein system and a tissue kallikrein-kinin system. Plasma kallikrein and its major substrate, the high molecular
weight kininogen, are chiefly involved in blood coagulation, whereas
tissue kallikreins with their preferential substrate, the low molecular weight kininogen, are assumed to be involved in several basic
regulatory functions of the organism. This paper will center on biochemical and physiological aspects of the plasma kallikrein-kinin and
tissue kallikrein-kinin system.

KEYWORDS

Kininogen, kininogenase, kallikrein, kinin, bradykinin, kallidin, kininase, angiotensin-I converting enzyme, carboxypeptidase N, kallikrein inhibitor

INTRODUCTION

More than 50 years ago, in 1925, a new field in biochemistry was discovered, namely the kallikrein-kinin system. Working on the so-called
postoperative reflex anuria, the Munich surgeon E. K. Frey found that
intravenous injection of dialysed urine caused a reversible reduction
of blood pressure in dogs (Frey, 1926). Convinced that this effect
was due to a hitherto unknown substance he asked the Nobel price win-
ner, Richard Willstätter, for cooperation who recommended his cowork-
er H. Kraut. Frey and Kraut tried to isolate the active principle in
urine and called it Frey or F substance. They observed during their
studies that contamination by blood reduced the blood pressure lowering effect of urine. This observation marked the beginning of the
work on the F substance inactivator, today known under bovine basic
kallikrein-trypsin inhibitor (Kunitz), aprotinin or Trasylol (Fritz
and others, 1974). The inactivated F substance could be reactivated

---

Dedicated to Professor Dr. H. Holzer on the occasion of his 60th birthday.

by acidification, moreover, acidified blood or serum showed F sub-
stance-like activity. These observations led Frey and Kraut to the
hypothesis that F substance might be a circulation hormone and ini-
tiated the search for the producing gland. E. Werle, who joined the
group of Frey and Kraut in 1930, found large amounts of F substance
in bovine, porcine and human pancreas (Kraut, Frey and Werle, 1930).
This was the reason why F substance was named "kallikrein" after the
Greek synonym for pancreas. In 1937 Werle showed that kallikrein was
not a hormone but an enzyme which liberates an effector substance
from an inactive precursor in plasma (Werle, Götze and Keppler,1937).
The effector substance, a hypotensive and muscle contracting peptide,
was named kallidin and the precursor, from which it was liberated,
obtained the name kallidinogen. Ten years later, in 1948, bradykinin
- liberated from the plasma precursor bradykininogen - was discovered
independently by Rocha e Silva, Beraldo and Rosenfeld (Rocha e Silva,
Beraldo and Rosenfeld, 1949) during experiments concerned with the
liberation of hypotensive substances by the action of snake venoms
on plasma proteins. Later on, both kallidinogen and bradykininogen
turned out to be identical, they are known now under the term kininogen. After the initial discoveries decades elapsed before the characterization of kininases (kinin destroying enzymes) by Erdös and
colleagues (Erdös, 1979a) completed the system.

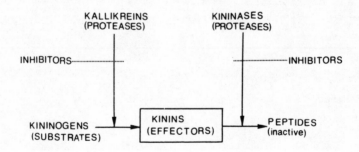

Fig. 1. Components of the kallikrein-kinin systems
and their interactions

The kallikrein-kinin system consists of four major components (Fig.1):
(i) the kininogenases or kallikreins, (ii) the kininogens, (iii) the
kinins, and (iv) the kininases. Present evidence suggests that we
have to discriminate between a plasma kallikrein-kinin system and a
tissue kallikrein-kinin system. Plasma kallikrein and its preferred
substrate, the high molecular weight kininogen, are chiefly involved

in blood coagulation (Movat, 1978), whereas tissue kallikreins with their preferential substrate, the low molecular weight kininogen, are assumed to be involved in several basic regulatory functions of the organism (Fink and others, 1980).

BIOCHEMISTRY OF THE KALLIKREIN-KININ SYSTEM

Kallikreins or Kininogenases

Kallikreins (E.C. 3.4.21.8) are serine proteinases which liberate vasoactive peptides (bradykinin, kallidin) from kininogens by limited proteolysis and have no or only little proteolytic activity on other proteins (Erdös and Wilde, 1970; Erdös, 1979b). The term "kininogenase" comprises in addition proteolytic enzymes such as trypsin, plasmin, acrosin, etc. which primarily cleave their natural substrates but are also capable of releasing kinins from kininogens (Erdös and Wilde, 1970; Erdös, 1979b). Kallikreins and the other kininogenases can be differentiated from each other at least in one respect: kallikreins cause a decrease in blood pressure in vivo after intravenous injection of rather low amounts (0.008-0.1 µg/kg of a tissue kallikrein, 0.6-1.25 µg/kg of plasma kallikrein), whereas similar amounts of other kininogenases have no effect. The blood pressure effect is caused by kinins liberated by the action of the kallikreins on kininogens despite the presence of excess but relatively slowly reacting inhibitors in the case of plasma kallikrein or due to a very low inhibitory potential in the case of the tissue kallikreins. In contrast to the kallikreins, kininogenases such as trypsin or plasmin are normally faced with a strong, fast reacting inhibitory potential in plasma so that only relatively high amounts of these enzymes cause a blood pressure decrease.

Tissue kallikreins, which are also called organ or glandular kallikreins, differ from the circulating plasma kallikreins in several respects (for review see ref. Geiger and Fritz, 1981; Fiedler and others, 1981), Table 1. Therefore, both kinds of kallikreins will get different EC numbers in future (A. J. Barrett, personal communication, 1980).

TABLE 1   Properties of Plasma and tissue kallikreins

	PLASMA KALLIKREIN-KININ SYSTEM	TISSUE KALLIKREIN-KININ SYSTEM
ENZYME	PLASMA KALLIKREIN (MW: 100 000)	TISSUE KALLIKREINS (MW: 25 000 - 35 000)
SUBSTRATE	HMW KININOGEN ⟷ LMW KININOGEN	
KININ LIBERATED	BRADYKININ	KALLIDIN
OCCURENCE & FUNCTION	BLOOD; FACTOR OF THE CLOTTING SYSTEM (TOGETHER WITH HMW KININOGEN)	BODY FLUIDS (BLOOD, URINE, SEMINAL PLASMA), SALIVARY GLANDS, PANCREAS, KIDNEY, SMALL AND LARGE INTESTINE; INVOLVED IN VARIOUS BIOLOGICAL PROCESSES

The most important features of the plasma kallikreins are: (i) the
release of bradykinin from kininogen, (ii) the presence of potent
and relatively rapidly reacting inhibitors in plasma such as the C1-
inactivator and α_2-macroglobulin besides the slowly reacting inhibitors α_1-antitrypsin (Fritz and others, 1972) and antithrombin III,
(iii) the preference for Factor XII (Hageman factor) besides for HMW
kininogen as natural substrates. Characteristic features of glandular kallikreins are: (i) the release of kallidin (i.e. Lys-bradyki-
nin) from the kininogens, (ii) the absence of potent fast reacting
inhibitors in plasma, (iii) the potent blood pressure lowering effect
in vivo by less then 0.1 µg/kg amounts if intravenously injected.
Tissue kallikreins are found in pancreas, pancreatic juice, salivary
glands, saliva, small and large intestine, kidney and urine (Erdös
and Wilde, 1970; Erdös 1979b; Zimmermann, Geiger and Kortmann, 1979),
but also in plasma or serum (Geiger and others, 1980). They occur in
pancreatic tissue and juice as prokallikrein (Fiedler and Gebhard,
1980) but in salivary tissue and juice only in the active form. In
urine, both latent and active forms are found (Corthron and others,
1979; Geiger, Stuckstedte and Fritz, 1980).
Tissue kallikreins are capable of hydrolysing acylated derivatives
of basic amino acids, mainly arginine esters, but they hydrolyse the
corresponding p-nitroanilides, which are good substrates for trypsin
(Geiger and Fritz, 1981), only slowly. However, certain peptide p-
nitroanilides like DValLeuArgNHNp are cleaved at nearly the same rates as BzArgOEt. For kallidin liberation glandular kallikreins have
to cleave kininogens at two positions. In bovine kininogen one peptide bond is cleaved between arginine and serine, the other between
methionine and lysine. A plausible explanation for the rapid cleavage of the Met-Lys-bond in the kininogen molecule in comparison to
the rather slow hydrolysis of methionine-containing model substrates
cannot be given yet (Geiger and Fritz, 1981; Geiger, Stuckstedte,
Fritz, 1980; Fiedler and Leysath, 1979).

The activity of glandular kallikreins is strongly inhibited by aprotinin (Geiger, Stuckstedte and Fritz, 1980) and structurally homologous proteinase inhibitors of the Kunitz (aprotinin) type (Fritz,
Fink and Truscheit, 1979). Soybean trypsin inhibitor, a potent inhibitor of trypsin, plasma kallikrein and other serine proteinases
(Fritz, Fink and Truscheit, 1979), does not inhibit glandular kallikreins at all. Like other serine proteinases glandular kallikreins
are also inhibited by diisopropyl fluorophosphate (Fiedler and
others, 1981; Zimmermann, Geiger and Kortmann, 1979), benzamidine inhibits competitively (Geiger and Fritz, 1981). Chloromethyl ketones
of arginine peptides reacting rapidly and specifically with glandular kallikreins (Table 2) have been synthetized (Kettner and others,
1980). Recently, α_1-protease inhibitor (α_1-antitrypsin) was identified as the only, slowly reacting inhibitor of human and porcine
tissue kallikreins present in plasma (Geiger and others, 1981).
Assay methods differing considerably in specificity and sensitivity
are available to quantitate glandular kallikreins. For estimation of
glandular kallikreins in biological samples biological test systems
(dog blood pressure assay), kinin-liberating assays, photometric
assays using synthetic substrates and radioimmunoassays (Geiger and
Fritz, 1981; Mann and others, 1980) are available. Whereas formerly
human urine was used as kallikrein standard, at present the use of
highly purified porcine pancreatic kallikrein (Bayer AG, West-Ger-
many) is recommended. 1 KU (kallikrein unit), originally defined as
the blood pressure reducing effect caused by 5 ml dialysed human
urine, corresponds to approximately 0.9 µg of this kallikrein prepa-

ration (porcine pancreatic kallikrein) or 157 mU (substrate:BzArgOEt,
pH-stat method (Fiedler and others, 1981).

TABLE 2   Kinetic constants of the inactivation of
          tissue kallikrein by chloromethyl ketones
          (from Kettner and others, 1980; 25°C, pH 7.0)

Chloromethyl ketone	K_I (uMol)
Gly-Val-ArgCH$_2$Cl	370[A]
Ala-Leu-LysCH$_2$Cl	1 040[A]
Ac-Phe-ArgCH$_2$Cl	56.0[B]
Pro-Phe-ArgCH$_2$Cl	45.0[B]
Ac-Pro-Phe-ArgCH$_2$Cl	120[B]
Ser-Pro-Phe-ArgCH$_2$Cl	180[B]
Phe-Ser-Pro-Phe-ArgCH$_2$Cl	21.2[B]

[A] PORCINE PANCREATIC KALLIKREIN
    (FROM FIEDLER AND OTHERS, 1977A)

[B] HUMAN URINARY KALLIKREIN
    (FROM KETTNER AND OTHERS, 1980)

Glandular kallikreins of the pig have been investigated in most detail. The enzymes have been isolated from pancreas, submandibular
glands, and from urine. Submandibular, native pancreatic and urinary
kallikrein consist of a single polypeptide chain and are named, therefore, α-kallikrein. From autolysed pancreas homogenates kallikrein is
obtained primarily in a two-chain ß-form obviously derived from the
α-form by cleavage of an internal peptide bond. Possibly, some further amino acid residues are also removed at the site of cleavage.
An additional minor form, γ-kallikrein, containing two internal
splits has also been found. Each of these tissue kallikreins occurs
in several multiple forms (separable by electrophoretic methods),
probably due to differences in their carbohydrate content. The amino
acid compositions of porcine tissue kallikreins from pancreas, submandibular glands, and urine are practically identical. The whole
amino acid sequence of porcine pancreatic ß-kallikrein has been determined. The N-terminal amino acid sequences of porcine submandibular and urinary kallikreins are identical with that of the pancreatic
enzyme. Apparent molecular weights derived from SDS electrophoresis,
gel filtration and ultracentrifugation experiments have been determined between 28 000 and 40 000 daltons for porcine glandular kallikreins (Fiedler and others, 1981).

Of the tissue kallikreins of man, only the urinary enzyme is well
characterized as a single chain α-kallikrein with a molecular weight
of 34 000 and 41 000 daltons, respectively (molecular weights were
determined by SDS electrophoresis, different molecular weight values
correspond to different carbohydrate contents; Geiger, Stuckstedte
and Fritz, 1980). Data for human submandibular and pancreatic kalli-

krein are scarce. In Fig. 2 structural data of human urinary kallikrein in comparison to porcine pancreatic kallikrein and trypsin are
shown, demonstrating the high degree of homology between porcine tissue kallikreins and human urinary kallikrein.

Kallikrein, human urin		Ile-Val-Gly-Gly-Trp-Glu-Cys-Glu-Gln-His-Ser-Gln-Pro-Trp-Gln-
Kallikrein, porcine pancreas		Ile-Ile-Gly-Gly-Arg-Glu-Cys-Glu-Lys-Asn-Ser-His-Pro-Trp-Gln-
Trypsin, porcine pancreas		Ile-Val-Gly-Gly-Tyr-Thr-Cys-Ala-Ala-Asn-Ser-Ile-Pro-Tyr-Gln-

Kallikrein, human urin		-Ala-Ala-Leu-Tyr-His-Phe-Ser-Thr-Phe-Gln-Cys-Gly-Gly-Ile-Leu-Val-
Kallikrein, porcine pancreas		-Val-Ala-Ile-Tyr-His-Tyr-Ser-Ser-Phe-Gln-Cys-Gly-Gly-Val-Leu-Val-
Trypsin, porcine pancreas		-Val-Ser-Leu-Asn-Ser-Gly-Ser-His-Phe-——-Cys-Gly-Gly-Ser-Leu-Ile-

Fig. 2. N-terminal amino acid sequence of human
        urinary kallikrein, porcine pancreatic
        kallikrein and porcine trypsin (Hermodson
        and others, 1973)

Cleavage rates or synthetic substrates (e.g. AcPheArgOEt) catalysed
by human urinary kallikrein can be significantly increased (1.5 to
2.5 fold) by addition of amphiphiles to the assay system. Human serum albumin induces a similar stimulative effect (twofold increase)
(Geiger, Stuckstedte and Fritz, 1980). These observations indicate
that the conformation of a hydrophobic domain present at the kallikrein molecule has an influence on its catalytic activity. Another
indication in this respect is the high affinity of the urinary kallikrein to phenyl- or octyl-Sepharose. These results favor the suggestion that human urinary kallikrein is a membran-associated enzyme.

Recently, the primary structures of nerve growth factor γ-subunit
(γ-NGF) from mice submandibular glands and of rat tonin were shown
to possess a high degree of homology to the amino acid sequences of
porcine pancreatic and human urinary kallikrein (Seidah and others,
1978; Bradshaw, 1980). Human as well as porcine tissue kallikreins
from pancreas, submandibular und urine are indistinguishable by
immunological methods like radial immunodiffusion or radioimmunoassay. This demonstrates the close similarity of the three kallikreins of each species. In the radioimmunoassay even a weak cross-
reactivity between porcine pancreatic and human urinary kallikrein
has been observed (Fink and others, 1980).

Plasma kallikrein, a glycoprotein with a molecular weight of about
100 000 (Heber, Geiger and Heimburger, 1978) and isoelectric points
between pH 8 and 9, occurs in plasma as proenzyme. During contact activation plasma kallikrein is activated proteolytically by Factor
XIIa (activated Hageman factor) and the 28 000 molecular weight fragment of Factor XIIa, the most potent prokallikrein activator known
so far (for a review see Colman and Bagdasarian, 1976). For quantitation in plasma either photometric assays with synthetic substrates
(e.g. BzProPheArgNHNp, Heber, Geiger and Heimburger, 1978; or DPro-
PheArgNHNp, Laemmle and others, 1979) after activation or immuno-
assays (Colman and Bagdasarian, 1976) are used. Obviously, inhibition of plasma kallikrein proceeds so slowly in the presence of a
sufficient high concentration of the synthetic substrate during activation that quantitation of prokallikrein is possible in this way.
Activated plasma kallikrein is faced two proteinase inhibitors of
physiological relevance, the C1inactivator and α2-macroglobulin.

Inhibition of plasma kallikrein may also be achieved with antithrombin III in the presence of heparin and α_1-antitrypsin, however, these reactions proceed too slowly to be of biological significance (Movat, 1978; Fritz and others, 1972).

Kininogens

Kininogens are acidic plasma glycoproteins from which vasoactive polypeptides are released by the action of proteolytic enzymes such as kallikreins or kininogenases. In plasma of various species two kininogen forms are detectable (Erdös and Wilde, 1970; Erdös, 1979b), a high molecular weight (HMW kininogen) and a low molecular weight one (LMW kininogen). The bovine kininogens are the best characterized kininogens till now (Kato and others, 1977). Both bovine kininogens are single-chain glycoproteins with a carbohydrate content of 19.8 % (LMW form) or 12.6 % (HMW form). The molecules remaining after removal of kinins consist of a heavy (H) and a light (L) chain connected by a disulfid bridge (Fig. 3). The H chains of both LMW and HMW kininogen are closely related as regards their primary structure and immunological cross-reactivity, whereas the L chains are structurally and immunologically different. The special structural feature responsible for the clotting activity of HMW kininogen is the stronly basic, histidine-rich peptide region present between the L chain and the kinin portion. HMW kininogen is bound via this basic region to negatively charged surfaces. As HMW kininogen occurs in plasma in associated form with both plasma prokallikrein and Factor XI, both proenzymes can be bound via HMW kininogen to the negatively charged surface in close vicinity to the Hageman factor. In this way the factors responsible for initiation of the clotting cascade are locally concentrated to ensure, for example, a rapid local clot formation at the site where a vessel has been wounded.

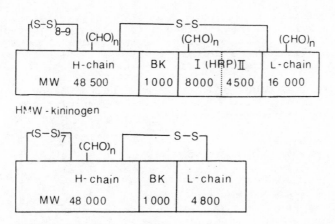

Fig. 3. Schematic representation of the major structural parts of bovine LMW and HMW kininogen (Kato and others, 1977). BK = bradykinin, HRP = histidine-rich peptide (I = fragment 1, II = fragment 2), MW = molecular weight, S-S = disulfide bridge, CHO = carbohydrate chain.

Recently, first structural data of human HMW kininogen were published
(Dittmann and others, 1980). Obviously, both human kininogens are relatively similar to the corresponding bovine ones.

Likewise HMW and LMW kininogen are excellent substrates for both
plasma and tissue kallikreins. However, the affinity of plasma kallikrein to its natural clotting partner, HMW kininogen, is still higher
than to LMW kininogen (Komiya, Kato and Suzuki, 1974) whereas tissue
kallikreins liberate kallidin more rapidly from LMW kininogen than
from the HMW form.

Quantitation of the total kininogen content in plasma is possible by
measuring kallikrein- or trypsin-induced kinin liberation followed
by determination of kinins with biological tests or high performance
liquid chromatography, by radioimmunoassay or enzyme immunoassay
(Geiger and Fritz, 1981). Separate determination of native, i.e. kinin-containing LMW and HMW kininogen is also possible but still a
very time-consuming procedure (Uchida, Oh-Ishi and Katori, 1977).
Discrimination of HMW and LMW kininogen may be performed with specific antibodies directed against the L chains or L chain and histidine-rich peptide. However, kinin-containing and kinin-free kininogens
cannot be differentiated in this way.
Kininogens occur besides in plasma also in human mucus secretions
(seminal plasma, gastric mucus, cervical mucus), urine and inter-
stitial fluid (Werle and Zach, 1970; Geiger, Feifel and Haberland,
1977). Most probably, only LMW kininogen is excreted into mucus
fluids and urine.

Kinins

Kinins are pharmacologically highly active oligopeptides. They increase capillary permeability, induce pain, exert chemotactic activity towards leucocytes, stimulate smooth muscles, dilate arterioles
(thus lowering blood pressure) and constrict venoles. They liberate
catecholamines, guanosine monophosphate and cyclic adenosine monophosphate and stimulate prostaglandin synthesis. Furthermore, they
belong to the most potent mediators of inflammation known so far
(Erdös, 1979b). In Table 3 minimal doses of kinins are compiled necessary to cause a pharmacological effect.

TABLE 3   Minimal doses of kinins inducing the given pharmaco-
          logical effect (Frey, Kraut and Werle, 1968)

	EFFECT	DOSE
GUINEA PIG ILEUM	CONTRACTION	1 NG/ML
RAT UTERUS	CONTRACTION	0.1 - 0.2 NG/ML
RAT DUODENUM	RELAXATION	1 NG/ML
RABBIT DUODENUM	CONTRACTION	1 NG/ML
BLOOD PRESSURE (CAT, RABBIT, RAT)	DECREASE	0.2 - 0.5 µG/KG
BRONCHIAL MUSCULAR SYSTEM (GUINEA PIG)	CONSTRICTION	0.5 - 1.0 µG/KG
CAPILLARY PERMEABILITY (GUINEA PIG)	INCREASE	0.1 NG IN 0.1 ML
LUNG VESSELS (RABBIT)	CONSTRICTION	1 - 5 µG
CORONARY VESSELS (GUINEA PIG)	DILATATION	2 - 3 NG/ML
BRAIN VESSELS (MAN)	DILATATION	1 µG/KG
HEART-RATE, CARDIAC OUTPUT	INCREASE	0.2 - 1 µG/KG
MAN	FLUSH HEART-HAMMERING	0.1 - 0.4 µG/KG
CANTHARIDIAN BULLA	PAIN	0.1 - 1.0 µG/ML

In principle, bradykinin and kallidin cause the same biological effects but differ in the quantitative response in relation to the given system (Frey, Kraut and Werle, 1968). The biological effects of kinins can be potentiated by so called "bradykinin-potentiating peptides" (Sampaio and others, 1977). These peptides act primarily as inhibitors of kininase II, i.e. deferring kinin degradation (cf. the next chapter).

The biological effects of the kinins may be mediated by specific receptors (Erdös, 1979b). Recently, Claeson et al. have described the inhibition of kinin effects by small peptides (Claeson and others, 1979), derivatives of synthetic kallikrein substrates containing a partial sequence of the bradykinin molecule, e.g. DProPheArg-O-heptyl. However, most of the biological effects of the kinins seem to be mediated via prostaglandin synthesis (McGiff and others, 1972). Obviously, bradykinin can activate phospholipase A_2, thus initiating production of either PGE_2 or $PGF_{2\alpha}$ (Fig. 4), the latter two prostaglandins exerting quite opposite biological effects, for example, dilatation of arterioles by PGE_2 and constriction of venoles by $PGF_{2\alpha}$.

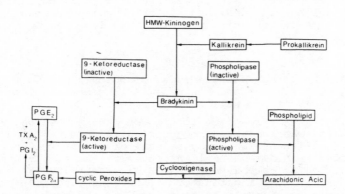

Fig. 4. Stimulation of prostaglandin synthesis by kinins: schematic representation of the presently suggested pathways.

It seems likely that the conversion of PGE_2 into $PGF_{2\alpha}$ is also mediated by kinins depending on their local concentration. As different tissue cells differ significantly in the prostaglandin pattern synthetized, the biological effect caused by kinins may differ from tissue cell to tissue cell. In any case, the biological effect caused by kinins in a special cell type may be the resultant of a direct action of kinins on specific receptors and the stimulation of the production of other local tissue hormones. Regarding this complex reaction possibilities, it is not surprising that the biological effects caused by kinins differ from cell type to cell type and even in the same cell type from species to species. Kinin-related compounds have been found also in venoms of insects and amphibian skin (Erdös, 1979b).

Leukokinins, peptides with kinin-like activity but totally different peptide structure, are released by the action of the acidic pro-

teinase cathepsin D (from leucocytes or neoplastic cells) on a specific plasma protein, the leukokininogen (Erdös, 1979b). The latter
itself is present in plasma as a prosubstrate which has to be converted to the substrate by a trypsin-like proteinase before it can be
cleaved specifically by cathepsin D to release the leukokinins. Remarkably, the trypsin-like proteinase is inhibited by aprotinin.

Kinins can be quantitated by bioassays, for example, by measuring the
contraction of rat uterus or guinea pig ileum, by enzyme or radioimmunoassay and by high performance liquid chromatography (Geiger and
Fritz, 1981). Formerly, the amount of kinins present in human plasma
was reported to be approximately 3 ng/ml (Talamo, Haber and Austen,
1969). However, very recent studies indicate that the kinin concentration in plasma should be much lower (Rabito, Scicli and Carretero,
1980) as not yet identified kinin-like material interfered in both
the biological assays and the radioimmunoassay. The same holds true
for the urinary kinin levels reported so far.

In the organism kinins are inactivated very rapidly by kininases.
After only a single passage through the lung, more than 90 % of a given amount of a kinin in blood is inactivated. In the circulation of
dogs, half life was determined to be 16 s for bradykinin and 19 sec
for kallidin (Ferreira and Vane, 1967). However, kinin degradation in
vitro and in vivo can be significantly deferred by peptides (bradykinin-potentiating peptides) and synthetic compounds inhibiting the
action of kininases, especially of kininase II.

Kininases

Kininases are kinin degrading enzymes of which kininase I (carboxypeptidase N, EC 3.4.12.7) and kininase II (angiotensin-I converting
enzyme, EC 3.4.15.1) are well characterized.

Kininase I, a tetrameric glycoprotein composed of two pairs of identical subunits, was isolated from porcine and human plasma (Erdös,
1979b). It inactivates kinins by removal of the basic C-terminal arginine residue. Kininase I is identical with the anaphylatoxin inactivator. Due to its carboxypeptidase B-like activity it proved to be
a potent physiological inactivator of the anaphylactic peptides C3a
and C5a as well as of proteohormones and of epidermal and nerve
growth factors (Erdös, 1979b).

Kininase II is identical with the angiotensin-I-converting enzyme. It
occurs especially concentrated in tissue of lung and the kidney, where
it is bound to the endothelial cells of the vessels or tubules. It is
also found in other tissues and even in plasma and urine. Tissue bound
kininase II is a membrane-associated glycoprotein and has to be solubilized by detergents before isolation (Erdös, 1979b).

Kininase II has an important dual function: it inactivates kinins and
converts angiotensin I to angiotensin II by removing Phe-Arg, or His-
Leu from the C-terminus of these oligopeptides. The affinity of kininase II to bradykinin ($K_m \sim 8.5 \times 10^{-7}$ mol/l) is much higher than to
angiotensin I ($K_m \sim 7 \times 10^{-5}$ mol/l). Most remarkably, bradykinin has
just the opposite biological effects compared to angiotensin II, for
example, bradykinin dilates arterioles, thus decreasing the blood
pressure whereas angiotensin II constricts arterioles thus increasing
the blood pressure. The dual function of kininase II has an important
consequence for medical therapy: if a kininase II inhibitor is applied in vivo, the kinin level increases, whereas the angiotensin II
concentration decreases. This is the rational of the therapeutic

effect of kininase II inhibitors in treatment of hypertensive patients (cf. below).
Kinin-inactivating potency has been reported also for proteases and
peptidases such as chymotrypsin, carboxypeptidase B and others (Erdös, 1979b).

BIOLOGICAL ASPECTS OF THE KALLIKREIN-KININ SYSTEMS

The plasma kallikrein-kinin system

Activation of the intrinsic pathway of blood coagulation leads simultaneously to the generation of bradykinin. Both coagulation and
kinin generation require participation of the Hageman factor (Factor
XII). Previous studies provided evidence that during contact activation of the clotting cascade Factor XII, plasma prokallikrein, HMW
kininogen and negatively charged surfaces are involved (Erdös, 1979b).
Present knowledge of the reaction cascade is as follows: HMW kininogen, circulating at least in part associated with plasma prokallikrein and Factor XI, binds plasma prokallikrein and Factor XI to a
negatively charged surface (formed, for example, by exposure of collagen fibres of a wounded vessel) in close vicinity to Factor XII
(Movat, 1978). A binding-induced conformational change leads to the
exposure of the active site of Factor XII so that activation of plasma prokallikrein by a limited proteolytic cleavage can occur. Active
plasma kallikrein converts in turn Factor XII to XIIa by limited proteolysis, Factor XIIa activates then proteolytically Factor XI to XIa
(Fig. 5). Factor XII becoming active by conformational change can also convert Factor XI to XIa proteolytically. The preferred pathway,
conversion of Factor XI to XIa by conformational change-activated
Factor XII or by Factor XIIa formed from Factor XII by limited proteolysis, depends on the nature of the thromboplastin and surface
structure available at the wounded tissue or vessel (Fujii, Moriya
and Suzuki, 1979; Movat, 1978; Erdös, 1979b).

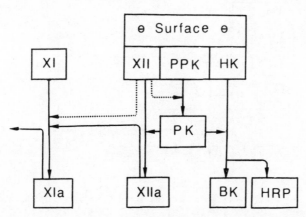

Fig. 5. Scheme of the initial phase of solid phase activation of
        the clotting cascade (intrinsic pathway). PPK=plasma pro-
        kallikrein, PK=plasma kallikrein, HK=HMW kininogen, XII=
        Factor XII, XI=Factor XI, BK=bradykinin, HRP=histidine-
        rich peptide

Besides its significance as a cofactor for solid phase activation of
the intrinsic pathway of blood coagulation, HMW kininogen functions
also as substrate for activated plasma kallikrein. Thus, simultaneously to the activation of the clotting cascade, plasma kallikrein
liberates bradykinin from HMW kininogen and, at least in bovine plasma, also the histidine-rich peptide. Release of the latter basic peptide from human HMW kininogen has not been conclusively shown yet
(Erdös, 1979b; Movat, 1978).

Bradykinin might exert various effects at the site of injury functioning either as mediator of tissue repair or of inflammatory reactions. For example, the released bradykinin could increase the local
blood flow and thus the substrate (e.g. clotting factors, etc.) supply,
it could stimulate prostaglandin synthesis in blood platelets and
other body cells and support in this way tissue repair. But it could
also produce pain, act chemotactically or lead to oedema formation.
Though it is assumed that the bradykinin released during activation
of the clotting cascade functions primarily in tissue repair, its
true functional significance in this process is not yet known.

Another product of the activation reaction at the solid phase is the
histidine-rich peptide cleaved off from bovine HMW kininogen simultaneously with bradykinin (cf. Fig. 5). This basic peptide was shown to
be involved in a negative feed-back mechanism by inhibiting accumulation of Factor XII at the site of injury, as it occupies primarily
the exposed negatively charged surface.

Via Factor XIIa the plasma kallikrein-kinin system effects indirectly
also the fibrinolytic and complement systems. Factor XIIa, activated
by plasma kallikrein, activates the plasminogen proactivator, which
leads to the conversion of plasminogen to plasmin, the key enzyme of
fibrinolysis (Erdös, 1979b; Movat, 1978), Fig. 6. Furtheron, Factor
XIIa or plasmin can activate complement C1 to $\overline{C1}$ and thus the complement cascade via the classical pathway (Austen and Fearon, 1979),Fig.6.

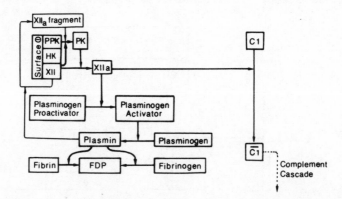

Fig. 6. Schematic representation of the relations existing be-
        tween the following systems: clotting cascade, kalli-
        krein-kinin system, fibrinolysis and complement casca-
        de. PPK=plasma prokallikrein, PK=plasma kallikrein, HK
        =HMW kininogen, C1=complement factor 1, XII=Factor XII,
        FDP=fibrin degradation products.

Of special interest is another possible pathway, namely the participation of plasmin in the positive feed-back activation of plasma prokallikrein and Factor XII. Plasmin converts Factor XIIa proteolytically to a Factor XIIa fragment (formerly also called $P_f$dil) of M_r
28 000 which has two characteristic features: (i) it does not stick
to negatively charged surfaces because of the separation of the solid
phase anchor part of Factor XII or XIIa and (ii) it proved to be the
most potent activator of plasma prokallikrein identified so far. This
discovery is also of clinical significance as it could be shown recently that the presence of this 28 000 M_r XIIa fragment in therapeutically administered plasma fractions may cause severe clinical complications due to a rapid activation of plasma prokallikrein (Erdös,
1979b; Movat, 1978).

The significance of plasma kallikrein for the activation of Factor
XII was inferred from studies of plasma from an individual with Flet-
cher trait. Plasma from such an individual suffering from plasma prokallikrein deficiency clots only after a prolonged lag period. HMW
kininogen deficiency was reported for five cases so far (Fitzgerald,
Williams, Flaujeac, Reid and Fujiwara trait; Erdös, 1979b; Movat,1978).
Remarkably, patients with plasma prokallikrein and/or kininogen deficiency do not show a clinically significant bleeding tendency. However, genetically determined deficiency in the C1 inactivator, the
major inhibitor of plasma kallikrein, may be accompanied by increased
risk for developping an angioneurotic oedema (Laurell and others,1976)
In such a patient, the acute attack seems to be accompanied by both
plasma kallikrein and complement activation. In patients suffering
from C1 inactivator deficiency the formation of complexes between antithrombin III and plasma kallikrein in the presence of heparin could
also be demonstrated.

The tissue kallikrein-kinin system

The physiological role of the tissue kallikrein-kinin system is still
largely unknown. The same holds true for the mechanism of action by
which administered exogenous tissue kallikrein causes the various
effects reported so far, for example: stimulation of intestinal absorption, cell proliferation, and the increase in sperm motility and
sperm count in patients suffering from asthenozoospermia and oligozoospermia, the enhancing effect on muscular glucose uptake observed
in patients with maturity onset diabetes, the normalisation of blood
pressure measured in patients suffering from essential hypertension.

Intestinal absorption: The effects of tissue kallikreins, mainly of
porcine pancreatic kallikrein, on the function of the gastrointesti-
nal tract, especially regarding absorption of various substances,
have been studied by several authors. Meng and Haberland (Meng and
Haberland, 1973) could demonstrate a significant increase in the passage of d-glucose through segments of rat small intestine in the
presence of tissue kallikrein in concentrations between 0.001 and
0.1 KU/ml, i.e. 3×10^{-11} and 3×10^{-9} mol/l. The uptake of 3-O-me-
thyl-glucose measured simultaneously was not influenced by the exogenous kallikrein (Fig. 7).

A stimulative effect of exogenous kallikrein on the absorption of water, electrolytes and hoxoses in rat intestine was observed by Dennhardt and Haberich (Dennhardt and Haberich, 1973) (Fig. 8).

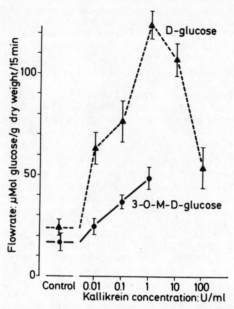

Fig. 7. Kallikrein-stimulated transport of d-glucose through
rat intestine. Glucose was added to the mucosal side
of the incubate and the glucose transport followed
from the mucosal to the serosal side (from Meng and
Haberland, 1973).

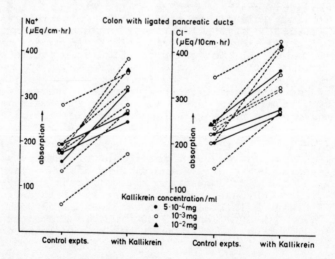

Fig. 8. Influence of various kallikrein concentration on net Na^+
and Cl^- fluxes in rat colon. The absorption of Na^+ and
Cl^- was measured before and after kallikrein was introduced at the mucosal side of the intestine. For further
experimental details, see Dennhardt and Haberich, 1973.

Furtheron, an influence of exogenous kallikrein on valine (Moriwaki
and others, 1973) influx and on intestinal absorption of vitamin
B_{12} in man was reported (Rumpeltes, Koeppe and Pribilla, 1975).

Glucose utilization: Dietze and Wicklmayr recently presented evidence that muscular substrate metabolism is significantly affected by
tissue kallikreins or kinins during muscular work and hypoxia in humans. The increase in muscular glucose uptake observed during isolated forarm exercise was inhibited in the presence of the kallikrein
inhibitor aprotinin (Wicklmayr and others, 1979; Dietze and others,
1978) - probably due to local inhibition of a kallikrein - but raised
again to the original values if bradykinin was administered in addition to aprotinin (Fig. 9). The amino acid uptake into muscle cells
showed a similar response to bradykinin administration (Schifmann
and others, 1980). Furtheron, after treatment of maturity-onset dia-
betics with bradykinin (13.3 ng/min) the pathological glucose tole-
rance was improved (Wicklmayr and others, 1980). And most remarkably,

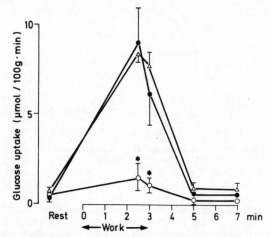

Fig. 9. Muscular glucose uptake during mild rhythmic isometric
        forarm exercises: during saline i.v. ( ), kallikrein-
        trypsin inhibitor i.v. (o) and additional intrabrachial-
        arterial bradykinin ( ) (from Dietze and others, 1978).

intensive-care patients resistant to insulin-stimulated glucose utilization responded to simultaneous infusion of bradykinin (30 min)
again with glucose uptake and a significant reduction of nitrogen excretion indicating also an improved utilization of the parenterally
administered amino acids (Schifmann and others, 1980). In summary,
bradykinin given in physiological doses, which are about 100-1 000
times lower than doses necessary to cause a pharmacological effect,
can improve muscular carbohydrate and protein metabolism by increasing the insulin sensitivity in insulin deficiency and in insulin resistance (Wicklmayr and others, 1980). Furthermore, nearly all the
biological effects induced by insulin can be mimiced also by bradykinin (Dietze and others, 1980). It seems likely, therefore, that kinins exert an insulin-like activity or act synergistically with insulin, and that tissue kallikreins are involved indirectly, by the liberation of kinins, in the various processes mediated by insulin.

<u>Fertilization:</u> An involvement of the tissue kallikrein-kinin system in fertilization processes is evident from studies stimulated originally by Haberland, Stüttgen and Schirren and performed recently in more detail by Schill et al. (Schill, 1977). After daily treatment of infertile patients suffering from asthenozoospermia or oligozoospermia with (3 x 200 KU) porcine pancreatic kallikrein for 7 weeks, an increase in the number of spermatozoa with a maximum 2 to 3 months after onset of therapy was observed (Fig. 10) (Schill, 1977; Schirren, 1977). An improvement of the quantitative and qualitative sperm motility with a maximum at the end of the therapy was also found (Fig. 11). Three months after termination of the therapy the

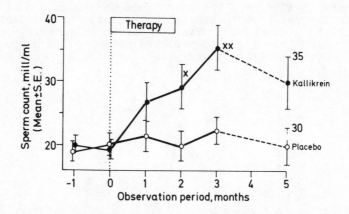

Fig. 10. Sperm count of 90 men with idiopathic oligozoospermia under kallikrein therapy compared to the placebo group over a period of 7 weeks. (600 KU per day were orally administered) From Schill, 1977.

pathological pretreatment values were measurable again. Conception rates raised significantly under kallikrein therapy (Schill, 1979). The basis of these observations might be - besides improved spermatogenesis - the stimulation of sperm motility (in ejaculates containing spermatozoa with reduced motility) by kinins or kallikreins and, consequently, an acceleration of sperm penetration through cervical mucus as shown also in studies in vitro (Schill and others, 1976).

In this connection emphasis should be put also on the kininogenase nature of acrosin (Palm and Fritz, 1975), the trypsin-like sperm acrosomal proteinase (Müller-Esterl, Kupfer and Fritz, 1980). Whereas in the male genital tract the potent acrosin inhibitors present should prevent an acrosin-induced liberation of kinins from kininogens, it seems possible that acrosin escaping from disintegrating spermatozoa in the femal genital tract secretions may liberate kinins and thus enhance sperm motility and also sperm migration by passive transport due to uterus contractions. In fact, Schill et al. could demonstrate that acrosin can enhance sperm migration through cervical mucus, which also contains kininogen, after prior removal of seminal plasma. As kininases are also present in both male and female genital tract secretions, the various components of the kallikrein/acrosin-kininogen-kinin-kininase system may also be involved in fertilization in vivo.

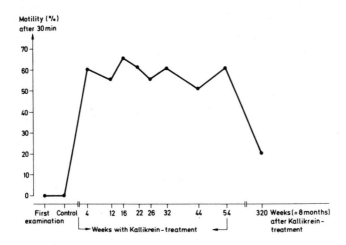

Fig. 11. Influence of kallikrein administration on sperm
         motility in men with idiopathic oligozoospermia.
         Oral dosages of 6 x 100 KU per day were administe-
         red for 54 weeks. From Schirren, 1977.

Kidney function and blood pressure regulation: Of special significance is the involvement of tissue kallikreins or kinins in kidney
function, i.e. in the regulation of the electrolyte/water balance as
well as in renal haemodynamics and thus blood pressure regulation.
In the kidney all the components of the kallikrein-kinin system and
of the renin-angiotensin system are locally concentrated to perform
the haemostatic equilibrium (Fig. 12):

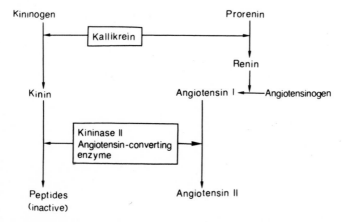

Fig. 12. Relations between the kallikrein-kinin system
         and the renin-angiotensin system

The key enzyme regulating the balance between both systems is kininase II, the angiotensin-I converting enzyme. It can generate on one side angiotensin II, a highly potent vasoconstrictor increasing blood pressure and decreasing blood flow and excretion of Na^+ and water and, on the other side, bradykinin, a highly potent vasodilator exerting just the opposite effects. As prostaglandins and steroid hormones like aldosteron are also involved in kidney function, it is not surprising that the relations between all these systems are very complicated and deserve further careful studies to clarify the significance and the way of interactions of the various pathways in more detail. This concerns also most recent observations claiming that plasma or tissue kallikreins are responsible for the conversion of prorenin to renin in vivo (Derkx and others, 1979; Sealey, Atlas and Laragh, 1978). Nevertheless, the positive therapeutic effect observed during treatment of hypertensive patients with captopril (Laragh and others, 1980), a strong synthetic inhibitor of kininase II, clearly indicates that our present knowledge of the basic mechanisms of these systems is real though many details have still to be figured out.

This view is supported also by the results of clinical trials obtained under administration of tissue kallikrein in patients suffering from essential hypertension. It is now generally accepted that in patients with established essential hypertension urinary kallikrein excretion is lower than in normatensive individuals (Margolius and others, 1971). After treatment of hypertensive patients with porcine pancreatic kallikrein, a reduction of the elevated blood pressure combined with a normalization of the excretion of endogenous urinary kallikrein was observed (Fig. 13) (Overlack and others, 1980). A double-blind trial performed recently confirmed these observations. Although the origin of the urinary kallikrein is not yet known - it may be produced either by the kidney (distal tubule (Nustad, Vaaje and Pierce, 1975)) or simply represent filtered tissue kallikrein/prokallikrein from the circulation - the involvement of tissue kallikreins or the tissue kallikrein-kininogen-kinin-kininase system in kidney function seems to be established by the given observations.

The mode of action of exogenous and endogenous tissue kallikrein: In order to clarify what happens between oral intake of glandular kallikrein (wrapped up in acid-resistant capsules to allow passage through the stomage without denaturation) and the observed effects, investigations concerning intestinal absorption, the occurrence of tissue kallikreins in the circulation and their interaction with plasma proteins have to be performed. In a first attempt Fink et al. could demonstrate that exogenous tissue kallikrein can be absorbed by the rat intestine in small amounts and can be detected as kallikrein-like antigen in blood and lymph (Fink and others, 1980). The presence of endogenous tissue kallikrein-like antigens in human, pig and rat plasma or serum is now well established (Fink and others, 1980). Furthermore, using immunoaffinity chromatography, an active tissue kallikrein being identical in all properties with urinary kallikrein could be recently isolated from human plasma (Geiger and others, 1980). The presence of active tissue kallikrein in the circulation would not be surprising, because inhibition by α_1-antitrypsin, the only tissue kallikrein inhibitor identified in human plasma, proceeds extremely slowly (Geiger and others, 1981), i.e. in the course of hours and not minutes or seconds as is the case with other proteinase/inhibitor systems.

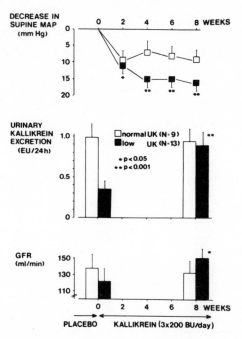

Fig. 13. Effect of oral kallikrein administration on mean arterial blood pressure (MAP), 24 hrs urinary kallikrein
excretion and glomerular filtration rate (GFR) in hypertensive patients with low and normal urinary kallikrein excretion. EU=urinary excretion, UK=urinary kallikrein, BU=biological units or kallikrein units (KU)
(from Overlack and others, 1980).

In conclusion, tissue kallikreins, which have access to the circulation either by absorption through the intestine or via the micro-
circulation of the glands, may maintain a physiological kinin level
in the circulation or in other body fluids by continuous liberation
of kallidin from kininogen ubiquitously present throughout the organism. The therapeutic effect of exogenous kallikrein might be explained by an increase of the kinin level, caused either locally or
systemically, if the endogenous tissue kallikrein-kinin system is im-
paired. However, verification of the concept needs further detailed
studies.

Outlook: Although we have already detailed knowledge of the properties of the various components of the kallikrein-kininogen-kinin-kininase systems, many details of their biological functions are still
obscure. This concerns especially the tissue kallikrein-kinin system
and its interactions with other systems of the organism. Of special
medical interest for future studies is the mode of action of the
tissue kallikrein-kinin system in kidney function and blood pressure
regulation as well as in carbohydrate and protein metabolism.

Concerning basic research activities, the similarity of tissue kalli-kreins to the arginine esterases of growth factors is very exciting. Evidence presented by Bothwell et al. (Bothwell, Wilson and Shooter, 1979) indicates that the proteinase subunits of growth factors have potent kininogenase or kallikrein activity as well as amino acid compositions similar to those of the glandular kallikreins. Recently, the primary structure of nerve growth factor (NGF)γ-subunit from sub-mandibular gland was elucidated (Bradshaw and others, 1980). We were really surprised to see that there is a higher degree of sequence ho-mology between mice γ-NGF and porcine pancreatic kallikrein than be-tween mice γ-NGF and porcine trypsin, indicating that mice NGF γ-sub-unit is structurally more closely related to tissue kallikrein than to trypsin. The relation of tissue kallikreins and growth factor esterases, as regards their function, are a fascinating topic for new research activities.

ACKNOWLEDGEMENT

I'm indebted and grateful to Prof. H. Fritz for reading over the ma-nuscript, for many helpful discussions and the generous support du-ring my work.

ABBREVIATIONS

DValLeuArgNHNp   = D-valyl-L-leucyl-L-arginine-p-nitroanilide
BzArgOEt   = Nα-benzoyl-L-arginine ethyl ester
BzProPheArgNHNp   = Nα-benzoyl-l-prolyl-L-phenylalanyl-L-arginine-
                     p-nitroanilide
DProPheArgNHNp   = D-prolyl-L-phenylalanyl-L-arginine-p-nitro-
                     anilide

REFERENCES

Austen, K. F. and Fearon, D. T. (1979). A Molecular Basis of Activa-tion of the Alternative Pathway of Human Complement. Adv. Exp. Med. Biol. 120B, 3-17
Bothwell, M. A., Wilson, W. H. and Shooter, E. M. (1979). The Rela-tionship between Glandular Kallikrein and growth Factor-pro-cessing proteases of Mouse Submaxillary Gland. J. Biol. Chem. 254, 7287-7294
Bradshaw, R. A., Grant, G. A., Thomas, K. A. and Eisen, A. Z. (1980). Mouse NGF γ Subunit and crab collagenase: two serine proteases of unusual function. Proteins and Related Subjects 28, 119-122, Pergamon Press, Oxford
Claeson, G., Fareed, J., Larsson, C., Kindel, G., Arielly, S., Si-minsson, R., Messmore, H. L. and Balis, J. U. (1979). Inhibition of the contractile Action of bradykinin on isolated smooth musc-le preparations by Derivatives of low molecular weight peptides. Adv. Exp. Med. Biol. 120B, 691-713
Colman, R. W. and Bagdasarian, A. (1976). Human Kallikrein and Pre-kallikrein. Methods Enzymol. 45, 303-314
Corthron, J., Imanari, T., Yoshida, H., Kaizu, T., Pierce, J. V. and Pisano, J. J. (1979). Isolation of Prokallikrein from Human Uri-ne. Adv. Exp. Med. Biol. 120B, 575-579
Dennhardt, R. and Haberich, F. J. (1973). In Kininogenases (Haber-land, G. L., Rohen, J. W., eds.) Effect of Kallikrein on the ab-sorption of Water, Electrolytes, and Hexoses in the Intestine of

rats. pp. 81-88. Schattauer Verlag Stuttgart

Derkx, H. M., Tang-Tjiong, H. L., Man in't Veld, A. J., Schalekamp,
M. P. A. and Schalekamp, M. A. D. H. (1979). Activation of In-
active Pläsma Renin by Tissue Kallikreins. J. Clin. Endrocrinol.
Metabol. 49, 765-769

Dietze, G., Wicklmayr, M., Böttger, J. and Mayer, L. (1978). Inhibition of Insulin Action on Glucose Uptake into Skeletal Muscle
by a Kallikrein-Trpyisn-Inhibitor. Hoppe Seyler's Z. Physiol.
Chem. 359, 1209-1215

Dietze, G., Wicklmayr, M., Böttger, J., Schifmann, R., Geiger, R.,
Fritz, H. and Mehnert, H. (1980). The Kallikrein-Kinin System
and Muscle Metabolism: Biochemical Aspects. Agents and Actions
10, 335-338

Dittmann, B., Lottspeich, F., Henschen, A. and Fritz, H. (1980).
Structural and Functional Aspects of Human High-Molecular Weight
(HMW) Kininogen. Proteins and Related Subjects 28, 213-216

Erdös, E. G. (1979). Kininases. Handb. Exp. Pharmacol. 25, Suppl.
(Erdös, E. G., ed.) pp. 428-488, Springer Verlag, Berlin

Erdös, E. G. and Wilde, A. F. (1970). Bradykinin, Kallidin and Kallikrein. Handb. Exp. Pharmacol. 25, Springer Verlag, Berlin

Erdös, E. G. (1979b). Bradykinin, Kallidin and Kallikrein. Handb.
Exp. Pharmacol. 25, Suppl., Springer Verlag, Berlin

Ferreira, S. H. and Vane, J. (1967). The Disappearance of bradykinin and eledoisin in the circulation and vascular beds of the
cat. Br. J. Pharmacol. 30, 417-424

Fiedler, F., Fink, E., Tschesche, H. and Fritz, H. (1981). Porcine
Glandular Kallikreins. Meth. Enzymol. in press

Fiedler, F. and Gebhard, W. (1980) Isolation and Characterization of
Native Single-Chain Porcine Pancreatic Kallikrein, Another Pos-
sible Precursor of Urinary Kallikrein. Hoppe Seyler's Z. Physiol.
Chem. 361, 1661-1671

Fiedler, F. and Leysath, G. (1979). Substrate specificy of porcine
pancreatic kallikrein. Adv. Exp. Med. Biol. 120A, 261-271

Fink, E., Geiger, R., Witte, J., Biedermann, S., Seifert, J. and
Fritz, H. (1980). In Enzymatic release of vasoactive peptides
(Gross, F., Vogel, G., eds.) 101-115, Raven Press, New York.
Biochemical, Pharmacological and Functional Aspects of Glandu-
lar Kallikrein

Frey, E. K. (1926). Zusammenhänge zwischen Herzarbeit und Nierentä-
tigkeit. Arch. Klin. Chir. 142, 663-669

Frey, E. K., Kraut, H. and Werle, E. (1968). Das Kallikrein-Kinin-
System und seine Inhibitoren. Ferdinand Enke Verlag, Stuttgart

Fritz, H., Wunderer, G., Kummer, K., Heimburger, N. and Werle, E.
(1972). α_1-Antitrypsin und C1 Inaktivator: Progressiv-Inhibito-
ren für Serumkallikrein von Mensch und Schwein. Hoppe Seyler's
Z. Physiol. Chem. 353, 906-910

Fritz, H., Fink, E. and Truscheit, E. (1979). Kallikrein inhibitors.
Fed. Proc. 38, 2753-2759

Fritz, H., Tschesche, H., Greene, L. and Truscheit, E. (1974). Proteinase inhibitors, Springer Verlag, Berlin

Fujii, S., Moriya, H. and Suzuki, T. (1979). KININS-II, Biochemistry,
Pathophysiology and Clinical Aspects. Adv. Exp. Med. Biol. 120A
+ 120B

Geiger, R., Stuckstedte, U., Clausnitzer, B. and Fritz, H. (1981).
Progressive inhibition of human glandular (urinary) kallikrein
by human serum and identification of the progressive antikalli-
krein as α_1-antitrypsin. Hoppe Seyler's Z. Physiol. Chem. 362,
317-325

Geiger, R. and Fritz, H. (1981). Human Urinary Kallikrein. Meth. En-
     zymol., in press
Geiger, R., Stuckstedte, U. and Fritz, H. (1980). Isolation and Cha-
     racterization of Human Urinary Kallikrein. Hoppe-Seyler's Z.
     Physiol. Chem. 361, 1003-1016
Geiger, R., Feifel, and Haberland, G. L. (1977). A Precursor of Ki-
     nins in the Gastric Mucus. Hoppe Seyler's Z. Physiol. Chem. 358,
     931-933
Geiger, R., Clausnitzer, B., Fink, E. and Fritz, H. (1980). Isolation
     of an Enzymatically Active Glandular Kallikrein from Human Plasma
     by Immunoaffinity Chromatography. Hoppe Seyler's Z. Physiol. Chem.
     361, 1795-1803
Heber, H., Geiger, R. and Heimburger, N. (1978). Human Plasma Kalli-
     krein: Purification, Enzyme Characterization and Quantitative De-
     termination in Plasma. Hoppe Seyler's Z. Physiol. Chem. 359, 659-
     669
Hermodson, M. A. Ericsson, L. H., Neurath, H. and Walsh, K. A. (1973).
     Determination of the Amino Acid Sequence of Porcine - Trypsin by
     Sequenator Analysis. Biochem. 12, 3146-3153
Kato, H., Han, N., Iwanaga, S., Hashimoto, N., Sugo, T., Fujii, S.
     and Suzuki, T. (1977). In Kininogenases 4 (Haberland, G. L., Ro-
     hen, J. W. and Suzuki, T., eds.) 63-72, Schattauer Verlag, Sutt-
     gart. Mammalian Plasma Kininogens: Their Structur and Functions
Kettner, C., Mirabelli, C., Pierce, J. V. and Shaw, E. (1980). Acti-
     ve Site Mapping of Human and Rat Urinary Kallikrein by Peptidyl
     Chloromethyl Ketones. Arch. Biochem. Biophys. 202, 420-430
Komiya, M., Kato, H. and Suzuki, T. (1974) Bovine Plasma Kininogens.
     III. Structural Comparison of High Molecular Weight and Low Mo-
     lecular Weight Kininogens. J. Biochem. 76, 833-845
Kraut, H., Frey, E. K. and Werle, E. (1930). Der Nachweis eines Kreis-
     laufhormons in der Pankreasdrüse. Z. Physiol. Chem. 189, 97-106
Laemmle, B., Eichlisberger, R., Marbet, G. A. and Duckert (1979).
     Amidolytic Activity in Normal Human Plasma Assessed with Chromo-
     genic Substrates. Thromb. Res. 16, 245-254
Laragh, J. H., Case, D. B., Atlas, S. A. and Sealey, J. E. (1980).
     Captopril Compared with Other Antirenin System Agents in Hyper-
     tensive Patients: its Triphasic Effects on Blood Pressure and its
     Use to Identify and Treat the Renin Factor. Hypertension 2, 586-
     593
Mann, K., Geiger, R., Göring, W., Lipp, W., Fink, E., Keipert, B.
     and Karl, H. J. (1980). Radioimmunoassay of Human Urinary Kalli-
     krein Determination of Human Urinary Kallikrein, II. J. Clin.
     Chem. Clin. Biochem. 18, 395-401
Margolius, H. S., Geller, R., Pisano, J. J. and Sjoerdsma, A. (1971).
     Altered Urinary Kallikrein Excretion in Human Hypertension.
     Lancet II, 1063-1065
McGiff, J. C., Terragno, N. A., Malik, K. U. and Lonigro, A. J.
     (1972). Release of a prostaglandin E-like Substance from Canine
     kidney by Bradykinin. Cir. Res. 31, 36-43
Meng, K. and Haberland, G. L. (1973). In Kininogenases (Haberland, G.
     L., Rohen, J. W., eds.) pp 75-80, Schattauer Verlag, Stuttgart.
     Influence of Kallikrein on Glucose Transport in the Isolated Rat
     Intestine.
Moriwaki, C., Moriya, H., Yamaguchi, K., Kizuki, K. and Fujimori
     (1973). In Kininogenases (Haberland, G. L., Rohen, J. W., eds.)
     pp 57-66, Schattauer Verlag, Stuttgart, Intestinal Absorption of
     Pancreatic Kallikrein and some Aspects of its Physiological Role
Movat, H. Z. (1978). The Kinin System: its Relation to Blood Coagu-
     lation, Fibrinolysis and the formed Elements of the Blood. Rev.

Physiol. Biochem. Pharmacol. 84, 143-202

Müller-Esterl, W., Kupfer, S. and Fritz, H. (1980). Purification and Properties of boar Acrosin. Hoppe Seyler's Z. Physiol. Chem. 361, 1811-1821

Nustad, K., Vaaje, K. and Pierce, J. V. (1975). Synthesis of kallikreins by rat kidney slices. Br. J. Pharmacol. 53, 229-234

Overlack, A., Stumpe, K. O., Ressel, C., Kolloch, R., Zywzock, W. and Krück, F. (1980). Decreased Urinary Kallikrein activity and elevated blood pressure normalized by orally applied Kallikrein in essential Hypertension. Kli. Wo. 50, 37-42

Palm, S. and Fritz, H. (1975). In Kininogenases 2 (Haberland, G. L., Rohen, J. W., Schirren, C., Huber, P., eds.) pp 17-22, Schattauer Verlag, Stuttgart, Comonents of the Kallikrein-Kinin-System in Human Midcycle Cervical Mucus and Seminal Plasma

Rabito, S. F., Scicli, A. G. and Carretero, O. A. (1980). In Enzymatic Release of Vasoactive Peptides (Gross, F., Vogel, H. G., eds) pp 247-258, Raven Press, New York, Immunreactive Glandular Kallikrein in Plasma

Rocha e Silva, M., Beraldo, W. T. and Rosenfeld, G. (1949). Bradykinin, a Hypotensive and Smooth Muscle stimulating Factor Released from Plasma Globulin by Snake Venoms and by Trypsin. Am. J. Physiol. 156, 261-273

Rumpeltes, H., Koeppe, P. and Pribilla (1975). In Kininogenases 3 (Haberland, G. L., Rohen, J. W., Blümel, G., Huber, P., eds.)pp 63-72, Schattauer Verlag Stuttgart. Influence of a Kallikrein-Preparation (Bay d 7687) on the Intestinal Absorption of Vitamin B12 in Man.

Sampaio, M. U., Reis, M. L., Fink, E., Camargo, A. C. M. and Greene, L. (1977). SP-Sephadex Equilibrium Chromatography of Bradykinin and Related Peptides: Application to Trypsin-Treated Human Plasma. Anal. Biochem. 81, 369-383

Schifmann, R., Wicklmayr, M., Boettger, J. and Dietze, G. (1980). Insulin-like Activity of Bradykinin on Amino acid Balances Across the Human forearm. Hoppe Seyler's Z. Physiol. Chem. 361, 1193-1199

Schill, W. B., Grösser, A., Preissler, G. and Wallner, O. (1976). In Human fertilization (Ludwig, H., Tauber, P. F., eds.) pp 106-113 Georg Thieme Verlag Stuttgart. In Vitro Stimulation of Cervical Mucus Spermatozoa Penetration by Components of the Kallikrein-Kinin-System

Schill, W. B. (1977). In Kininogenases 4 (Haberland, G. L., Rohen, J. W., Suzuki, T., eds.) pp 251-280, Schattauer Verlag Stuttgart. Kallikrein as a Therapeutical Means in the Treatment of Male Infertility.

Schill, W. B. (1979). Treatment of iodiopathic Oligozoospermia by Kallikrein: Results of a double-blind study. Arch. Androl. 2, 163-170

Schirren, C. (1977). In Kininogenases 4 (Haberland, G. L., Rohen, J. W., Suzuki, T., eds.) pp 247-250 Schattauer Verlag Stuttgart. Clinical Experiences with Kallikrein on Subfertile Males.

Sealey, J. E., Atlas, S. A. and Laragh, J. H. (1978). Human urinary kallikrein converts inactive to active renin and is a possible physiological activator of renin. Nature 275, 144-145

Seidah, N. G., Routhier, R., Caron, M., Chretien, M., Demassieux, S., Boucher, R. and Genest, J. (1978). N-Terminal amino acid sequence of rat tonin: homology with serine proteases. Can. J. Biochem. 56, 920-925

Talamo, R. C., Haber, E. and Austen, K. F. (1969). A Radioimmunoassay for bradykinin in plasma and synovial fluid. J. Lab. Clin. Med. 74, 816-827

Uchida, Y., Oh-Ishi, S. and Katori, M. (1977). In Kallikrein 4 (Haberland, G. L., Rohen, J. W. and Suzuki, T., eds.) pp 99-105 Schattauer Verlag, Suttgart. A Determination Method for Plasma Kininogen and its Application

Werle, E., Götze, W. and Keppler, A. (1937). Über die Wirkung des Kallikreins auf den isolierten Darm und über eine neue darmkontraktierende Substanz. Biochem. Z. 289, 217-233

Werle, E. and Zach, P. (1970). The distribution of kininogen on the serum and tissues of rats and other mammals. J. Clin. Chem. Clin. Biochem. 8, 186-189

Wicklmayr, M., Dietze, G., Günther, B., Schifmann, R., Böttger, J., Geiger, R., Fritz, H. and Mehnert, H. (1980). The Kallikrein-Kinin System and Muscle Metabolism: Clinical Aspects. Agents and Actions 10, 339-343

Wicklmayr, M., Dietze, G., Mayer, L., Böttger, T. and Grunst, J. (1979) Evidence for an involvement of kinin liberation in the priming Action of insulin on Glucose uptake into Skeletal Muscle Febs Letters 98, 61-65

Zimmermann, A., Geiger, R. and Kortmann, H. (1979). Similarity between a Kininogenase (Kallikrein) from Human Large Intestine and Human Urinary Kallikrein. Hoppe Seyler's Z. Physiol. Chem. 360, 1767-1773

Laurell, A. B., Mårtensson, U. and Sjöholm, A. (1976). Electroimmunoassay of C1 Inactivator and C4 in Hereditary Angio Neurotic Edema (H.A.N.E.). Clin. Immunol. Immunopathol. 5, 308-313

Species Index

Subject Index